Lecture Notes in Computer Science 2079

Edited by G. Goos, J. Hartmanis, and J. van Leeuwen

Springer

Berlin
Heidelberg
New York
Barcelona
Hong Kong
London
Milan
Paris
Singapore
Tokyo

Edmund Burke   Wilhelm Erben (Eds.)

Practice and Theory of Automated Timetabling III

Third International Conference, PATAT 2000
Konstanz, Germany, August 16-18, 2000
Selected Papers

Springer

Series Editors

Gerhard Goos, Karlsruhe University, Germany
Juris Hartmanis, Cornell University, NY, USA
Jan van Leeuwen, Utrecht University, The Netherlands

Volume Editors

Edmund Burke
University of Nottingham
School of Computer Science and Information Technology
Automated Scheduling, Optimisation and
Planning Research Group
Jubilee Campus, Wollaton Road, NG8 1BB Nottingham, UK
E-mail: ekb@cs.nott.ac.uk

Wilhelm Erben
FH Konstanz, University of Applied Science
Department of Computer Science
Brauneggerstr. 55, 78462 Konstanz, Germany
E-mail: erben@fh-konstanz.de

Cataloging-in-Publication Data applied for

Practice and theory of automated timetabling : third international
conference ; selected papers / PATAT 2000, Konstanz, Germany, August 16 -
18, 2000. Edmund Burke ; Wilhelm Erben (ed.). - Berlin ; Heidelberg ; New
York ; Barcelona ; Hong Kong ; London ; Milan ; Paris ; Singapore ; Tokyo :
Springer, 2001
    (Lecture notes in computer science ; Vol. 2079)
    ISBN 3-540-42421-0

CR Subject Classification (1998): F.2.2, G.1.6, G.2, I.2.8

ISSN 0302-9743
ISBN 3-540-42421-0 Springer-Verlag Berlin Heidelberg New York

Springer-Verlag Berlin Heidelberg New York
a member of BertelsmannSpringer Science+Business Media GmbH

http://www.springer.de

© Springer-Verlag Berlin Heidelberg 2001
Printed in Germany

Typesetting: Camera-ready by author, data conversion by PTP-Berlin, Stefan Sossna
Printed on acid-free paper   SPIN 10839265   06/3142   5 4 3 2 1 0

Preface

This volume is the third in an ongoing series of books that deal with the state of the art in timetabling research. It contains a selection of the papers presented at the 3rd International Conference on the Practice and Theory of Automated Timetabling (PATAT 2000) held in Constance, Germany, on August 16–18th, 2000. The conference, once again, brought together researchers, practitioners, and vendors from all over the world working on all aspects of computer-aided timetable generation. The main aim of the PATAT conference series is to serve as an international and inter-disciplinary forum for new timetabling research results and directions. The conference series particularly aims to foster multi-disciplinary timetabling research. Our field has always attracted scientists from a number of traditional domains including computer science and operational research and we believe that the cross-fertilisation of ideas from different fields and disciplines is a very important factor in the future development of timetabling research. The Constance conference certainly met these aims. As can be seen from the selection of papers in this volume, there was a wide range of interesting approaches and ideas for a variety of timetabling application areas and there were delegates from many different disciplines.

It is clear that while considerable progress is being made in many areas of timetabling research, there are a number of important issues that researchers still have to face. In a contribution to the previous PATAT conference, George M. White said:

> If a single goal could be formulated to describe the aspirations of time-tabling research groups, it would be this: An automatic timetabling system should formulate complete descriptions of which students and which teachers should meet, at what locations, at what times and should accomplish this quickly and cheaply while respecting the traditions of the institutions and pleasing most of the people involved most of the time.[1]

This is discussing university course timetabling, but it also applies to other timetabling application areas. One of the problems with many existing timetabling systems (from across the range of application areas) is that the system is often tailored to the needs of a particular institution or user. However, one individual organisation's requirements are often very different from those of another organisation in the same area. One of the themes of the Constance conference is

---

[1] George M. White and Junhan Zhang: Generating Complete University Timetables by Combining Tabu Search with Constraint Logic. In: Edmund Burke and Michael Carter (eds.): Practice and Theory of Automated Timetabling II, 2nd International Conference, PATAT '97, Toronto, Canada, August 1997, Selected Papers. Lecture Notes in Computer Science, Vol. 1408. Springer-Verlag, Berlin Heidelberg New York 1998

that the attempt to operate at a higher level of generality is becoming an ever more important timetabling research direction.

A major feature of the 3rd PATAT conference was the diversity of papers from a variety of timetabling fields. The PATAT conferences have always welcomed papers from across the timetabling spectrum but previous events have attracted the majority of their contributions from the educational sector. In this volume, other areas such as employee timetabling, sports timetabling, and transportation timetabling are more strongly represented. Indeed, we have devoted a section of the book to employee timetabling. This diversification motivated us towards organising the whole volume around application areas rather than on the solution methods and techniques employed (in contrast to earlier volumes). This is because we think that people are usually more interested in one specific timetabling sector than in a particular type of approach. We also include a section on Practical Considerations and General Issues containing papers that are of interest to all researchers and practitioners.

The Conference Series

The conference in Constance was the third in a series of international conferences on the Practice and Theory of Automated Timetabling (PATAT). This volume contains a selection of papers from that conference. The first conference was held in Edinburgh in August/September 1995. Selected papers from this also appeared in the Springer Lecture Notes in Computer Science series. The full reference is

> Edmund Burke and Peter Ross (eds.): Practice and Theory of Automated Timetabling, 1st International Conference, Edinburgh, UK, August/September 1995, Selected Papers. Lecture Notes in Computer Science, Vol. 1153. Springer-Verlag, Berlin Heidelberg New York 1996.

The second conference in the series was held in Toronto in August 1997. Selected papers from this appeared in

> Edmund Burke and Michael Carter (eds.): Practice and Theory of Automated Timetabling II, 2nd International Conference, PATAT '97, Toronto, Canada, August 1997, Selected Papers. Lecture Notes in Computer Science, Vol. 1408. Springer-Verlag, Berlin Heidelberg New York 1998.

The fourth conference in the series will be held in Ghent, Belgium, in August 2002. Future conferences will be held every two years. For further information about the conference series, contact the steering committee (whose members are listed below) or see http://www.asap.cs.nott.ac.uk/ASAP/ttg/patat-index.html.

Acknowledgements

Like its two predecessors, the conference in Constance (Konstanz) was very successful. Old acquaintances were renewed and new friends were made. Many people worked very hard to make the conference a valuable, enjoyable, and interesting event. We would like to express our sincere thanks to all the members of the organising committee (listed below). Their help and friendliness in efficiently and effectively dealing with all sorts of problems was very much appreciated.

All the papers that appear in this volume have been through a very rigorous and careful review process. All submissions to the conference were fully refereed in a first round. A second round of refereeing, specifically for this volume, was carried out after the conference. Many thanks go to the members of the programme committee (listed below) who refereed the papers during these two rounds.

We are also very grateful to the staff of Springer-Verlag for their support and encouragement. As series editor of the Lecture Notes in Computer Science series, Jan van Leeuwen was particularly helpful throughout the duration of this project, as he was with the previous two volumes. We would also like to thank Piers Maddox for the excellent job he made of copy editing this volume. His hard work is very much appreciated.

Special thanks go to Alison Payne for all the secretarial support she has given us, particularly during the second round of refereeing and during the preparation of this volume. We would also like to thank Diane French for her secretarial support during the first round of refereeing.

Of course, it is the group of authors, presenters and delegates who ultimately determine the success of a conference. Our thanks go to them for the enthusiasm and support they have given to us and indeed, to the PATAT series of conferences. Finally, we would like to thank the steering committee (listed below) for their continuing work in bringing us this and future PATAT conferences. We apologise for any omissions that have been inadvertently made. So many people have helped with this conference and with the series of conferences that it is difficult to remember them all.

May 2001

Edmund Burke
Wilhelm Erben

The 3rd International Conference on the Practice and Theory of Automated Timetabling Programme Committee

Edmund Burke (co-chair) University of Nottingham, UK
Wilhelm Erben (co-chair) FH Konstanz – University of Applied Sciences, Germany
Victor Bardadym   L and H Speech Products NV, Belgium
Patrice Boizumault   Ecole des Mines de Nantes, France
Peter Brucker   University of Osnabrück, Germany
Michael Carter   University of Toronto, Canada
David Corne   University of Reading, UK
Peter Cowling   University of Nottingham, UK
Andrew Cumming   Napier University, UK
Patrick Decausmaecker KaHo St Lieven, Gent, Belgium
Kathryn Dowsland   University of Wales Swansea, UK
Jacques Ferland   Université de Montréal, Canada
Emma Hart   Napier University, UK
Alain Hertz   EPF-Lausanne, Switzerland
Martin Henz   National University of Singapore, Singapore
Jeffrey Kingston   University of Sydney, Australia
Gilbert Laporte   Ecole des Hautes Etudes Commerciales, Canada
Vahid Lotfi   University of Michigan-Flint, USA
Michael Magazine   University of Cincinnati, USA
Amonon Meisels   Ben-Gurion University, Beer-Sheva, Israel
Thiruthlall Nepal   ML Sultan Technikon, Durban, South Africa
James Newall   University of Nottingham, UK
Ben Paechter   Napier University, UK
Sanja Petrovic   University of Nottingham, UK
Peter Ross   Napier University, UK
Andrea Schaerf   Università di Udine, Italy
Jan Schreuder   University of Twente, The Netherlands
Wolfgang Slany   Technische Universität Wien, Austria
Jonathan Thompson   Cardiff University, UK
Michael Trick   Carnegie Mellon University, Pittsburgh, USA
Dominique de Werra   EPF-Lausanne, Switzerland
George White   University of Ottawa, Canada
Michael Wright   Lancaster University, UK
Jay Yellen   Rollins College, USA
Masazumi Yoshikawa   NEC Corporation, Tokyo, Japan

The 3rd International Conference on the Practice and Theory of Automated Timetabling Organising Committee

Fachbereich Informatik, Fachochschule Konstanz, Germany

Wilhelm Erben	Chair
Verena Becker	Secretarial support
Sabine Dusteroft	Technical support
Mike Frohlich	Secretarial support
Armin-Peter Hohl	Registration desk
Hansjorg Isele	Technical support
Karin Wendling	Secretarial support
Jutta Wenger	Registration desk
Jutta Zweschper	Secretarial support

The International Series of Conferences on the Practice and Theory of Automated Timetabling (PATAT) Steering Committee

Edmund Burke (Chair)	University of Nottingham, UK
Ben Paechter (Treasurer)	Napier University, UK
Victor Bardadym	L and H Speech Products NV, Belgium
Michael Carter	University of Toronto, Canada
Patrick Decausmaecker	KaHo St Lieven, Gent, Belgium
David Corne	University of Reading, UK
Wilhelm Erben (co-chair)	FH Konstanz – University of Applied Sciences, Germany
Jeffrey Kingston	University of Sydney, Australia
Gilbert Laporte	Ecole des Hautes Etudes Commerciales, Canada
Amnon Meisels	Ben-Gurion University, Beer-Sheva, Israel
Peter Ross	Napier University, UK
Dominique de Werra	EPFL, Switzerland
George White	University of Ottawa, Canada
Anthony Wren	University of Leeds, UK

Contents

Course and School Timetabling

Examination Timetabling

Employee Timetabling

Other Timetabling and Related Problems

Practical Considerations and General Issues

Other Timetabling Presentations

Course and School Timetabling

A Multiobjective Genetic Algorithm for the Class/Teacher Timetabling Problem

Marco P. Carrasco[1,3] and Margarida V. Pato[2,3]

[1] Escola Superior de Gestão, Hotelaria e Turismo, Algarve University,
Largo Eng. Sárrea Prado n.21, 8500 859 Portimão, Portugal,
pcarras@ualg.pt,
[2] Instituto Superior de Economia e Gestão, Technical University of Lisbon,
Rua do Quelhas n.6, 1200 781 Lisboa, Portugal,
mpato@iseg.utl.pt
[3] Centro de Investigação Operacional, Faculdade de Ciências, University of Lisbon

Abstract. The drawing up of school timetables is a slow, laborious task, performed by people working on the strength of their knowledge of resources and constraints of a specific institution. This paper begins by presenting the timetabling problems that emerge in the context of educational institutions. This is followed by a description of the basic characteristics of the class/teacher timetabling problem. Timetables are considered feasible provided the so-called hard constraints are respected. However, to obtain high-quality timetabling solutions, other conditions should be satisfied in this case – those of soft constraints – which impose satisfaction of a set of desirable conditions for classes and teachers. A multiobjective genetic algorithm was proposed for this timetabling problem, incorporating two distinct objectives. They concern precisely the minimization of the violations of both types of constraints, hard and soft, while respecting the two competing aspects – teachers and classes. A brief description of the characteristics of a genetic multiobjective meta-heuristic is presented, followed by the nondominated sorting genetic algorithm, using a standard fitness-sharing scheme improved with an elitist secondary population. This approach represents each timetabling solution with a matrix-type chromosome and is based on special-purpose genetic operators of crossover and mutation developed to act over a secondary population and a fixed-dimension main population of chromosomes. The paper concludes with a discussion of the favorable results obtained through an application of the algorithm to a real instance taken from a university establishment in Portugal.

1 Introduction

In most educational institutions, the timetabling activity assumes considerable importance. First, as a result of this task, a set of attributions determining the entire daily activity of all the human resources involved, from students and teachers to employees, is drawn up. Secondly, alongside, a pattern of the use of the institution's physical resources is defined. Rational management of such

E. Burke and W. Erben (Eds.): PATAT 2000, LNCS 2079, pp. 3–17, 2001.

resources is of major importance at the pedagogical and administrative level. Usually, the timetabling data changes substantially from year to year and consequently the previous timetable cannot be used as a starting solution. So this complex task of building school schedules is performed from scratch, each year, by a team of technicians. This is based on the knowledge and experience acquired of the institution and leads to a set of acceptable schedules, through a trial and error procedure.

Distinct types of timetabling problems emerge with regard to schools, and depend on the elements to be scheduled (exams or regular courses). This paper focuses on a particular regular course problem: that of generating weekly timetables involving the assignment of a set of lessons (assignments of teachers to classes within the respective subjects) to classrooms and time periods – the class/teacher timetabling problem (CTTP).

In the operation research domain the CTTP has led to numerous approaches (de Werra 1985, 1997). In fact, this timetabling problem substantially differs from institution to institution, not only in view of the set of conditions required for the final solution, but also due to the dimension of the respective instances. Moreover, as a rule such situations generate high-complexity optimization problems. A practical solution is difficult to encounter when instances are highly dimensioned, which accounts for the application of heuristics to the field. Within this topic, the evolutionary approach provided by genetic algorithms pointed to interesting ways of tackling timetabling problems.

In this paper, a multiobjective genetic heuristic for the CTTP is explored. Section 2 presents the problem under study by describing its main characteristics, whereas Section 3 summarizes the basic foundations of the genetic algorithms as well as referring to a special multiobjective genetic algorithm – the nondominated sorting genetic algorithm (NSGA). Adaptation of the NSGA to the CTTP is discussed in Section 4, and Section 5 shows the computational results obtained from application of the proposed model to a real instance. Finally, Section 6 closes the paper with some conclusions and significant considerations.

2   The Class/Teacher Timetabling Problem

Generally, in teaching institutions two main types of timetabling problems arise: one for exams (the examination timetabling problem) and the other for regular courses (consisting of the course timetabling problem and the CTTP).

The examination timetabling problem is similar in most schools and consists of scheduling the exams for a set of courses, over a limited time period, while avoiding the overlapping of exams for each student, as well as seeking the largest spread over the examination period. A survey of this problem may be found in Carter and Laporte (1996) or in Burke et al. (1996). As to regular courses, the Course Timetabling Problem, typical of universities with flexible curricula, sets out to schedule all lectures included within the set of the institution's courses, while minimizing the overlaps of lectures for courses with common students (see, for instance, Downsland (1990); Kiaer and Yellen (1992)).

The other problem, which is precisely the subject of this study, the CTTP, is usually found in schools with less modular curricula. Such is the case in most high schools and some universities, where the majority of the predefined lessons ascribed to each course are compulsory for the respective classes. Unlike the course timetabling problem, where the CTTP is concerned, the overlapping of lessons is not allowed. Here, the goal consists of scheduling the set of lessons while satisfying a range of constraints specific to the institution. As a result, the CTTP itself emerges in slightly diverse forms and dimensions according to the features of the school (curricula, number of teachers and students, number of rooms and characteristics, geographical layout of the buildings, history of scheduling procedures, etc.) and system (high school or university). The specific problem also depends on the rules imposed by the national educational policy. This issue has always been formulated through combinatorial optimization problems that are very difficult in terms of complexity. In fact, even the simplified versions of the CTTP proved to be NP-hard (Even et al. 1976).

The above considerations and the high-dimension instances normally found in real timetabling problems may explain the continuous interest in studying the CTTP. In fact, several resolution techniques have been applied, from graph-theory-based algorithms (de Werra 1996), to binary programming (Tripathy 1984; Birbas et al. 1997) and constraint-based approaches (Yoshikawa et al. 1994). Moreover, meta-heuristics based on simulated annealing (Abramson 1982), neural networks (Gislén et al. 1992), genetic algorithms (Colorni et al. 1991) and tabu search (Costa 1994) have all been applied with success to specific cases, under study by the respective authors.

In general terms, the CTTP described in this paper resembles all other CTTPs and can be defined as the problem of scheduling a set of lessons (here, a lesson is a prior assignment of one or more rigid classes of students to a teacher and a subject) over a set of time periods and suitable rooms, while satisfying a wide range of constraints. These can be divided into two levels of importance: the hard constraints and the soft constraints.

1. The hard constraints represent compulsory restrictions associated with the feasibility of a solution for the CTTP:
   (a) Each teacher and class is assigned to no more than one lesson and one room at a time period.
   (b) Each class curriculum is respected, i.e. all lessons are scheduled.
   (c) The daily bound on the number of lessons for each subject is not exceeded due to pedagogical reasons.
   (d) Rooms are suitable for the lessons assigned. Each lesson may require a specific type of room with enough seats and/or special resources.
   (e) The teaching shift for each class is respected.
2. The soft constraints represent a second level of non-compulsory restrictions, which refers to aspects that both teachers and classes would like to see satisfied. These conditions are desirable and closely related to the final quality of the solution: a high-quality timetabling solution should satisfy the maximum number of these conditions:

(a) Each class and teacher should have a lunch break lasting at least one hour.
(b) The occurrence of gaps between teaching periods should be minimal, for classes and teachers.
(c) Class and teacher preferences should be satisfied, whenever possible.
(d) The number of teacher and class shifts between teaching locations should be minimized.
(e) The number of days occupied with lessons should be minimized.

These are the basic features considered in our methodology to solve the CTTP. Other particular conditions may be incorporated in the algorithm described in detail in Section 4.

3   Multiobjective Genetic Algorithms

During the last decade, evolutionary techniques such as genetic algorithms (GAs) have been extensively used to solve complex optimization problems. The reference model introduced by John Holland (1975) triggered a wide interest in the application of these types of heuristics, which mimic the natural evolutionary characteristics present in the biological species, with the purpose of solving optimization problems efficiently. Several evolutionary approaches rely on modifications made to Holland's initial model, while preserving the essential evolutionary characteristics. Next, there follows a description of the basic aspects of a standard GA for multiobjective optimization.

The conception of a GA necessarily begins with the choice of the chromosome encoding for each individual, which is the structure that represents each solution of the problem to be solved. Traditionally the encoding is the binary one, theoretically supported by the schema theorem formulated by Holland. In practice, albeit applicable to any problem, in some issues the binary encoding is not the most suitable one. It leads to significant computing time expense due to additional readjustment procedures. For this reason, some authors propose more natural representations, i.e. non-binary alphabets (Goldberg 1989). These problem-oriented encodings may improve certain aspects of the GA, such as the search ability for good solutions. They may also enable one to conceive genetic operators which are specifically suited to the problem and require no hard time-consuming readjustments.

Having defined the appropriate representation for the solutions of the optimization problem, an initial population of individuals or respective chromosomes is generated. This is normally achieved by resorting to random procedures. Next, a fitness function must be created to evaluate each chromosome. This function assigns a fitness value to each chromosome according to the quality of the respective solution. In general it reflects the level of optimization of the problem objective for the corresponding solution.

In conventional GAs, evaluation of the individuals reflects only the optimization of a single objective, but the nature of some real-world optimization problems may benefit from an approach guided by more objectives. Habitually, in the interest of the solution, the multiple objectives are in some way combined through an aggregating function that is a linear combination of the objective functions. However, when the different objectives are not comparable and mixable in terms of penalization, or the definition of the penalizing coefficients is difficult because some objectives tend to dominate the others, this approach seems inappropriate.

To tackle multiobjective optimization problems by using GAs, several other methodologies have been applied, as mentioned in detail by the following surveys: Fonseca and Fleming (1995), Veldhuizen and Lamont (2000). The most widely utilized are the so-called Pareto-based techniques employing the concept of Pareto optimality to evaluate each chromosome in relation to the multiple objectives. The algorithm used in the current paper was derived from the NSGA proposed by Srinivas and Deb (1994), which applies the original Pareto-based ranking procedure implemented by Goldberg (1989) combined with a fitness sharing scheme. The main distinction consists in the implementation of a secondary population procedure to improve the solution quality.

The population is then subjected to the genetic operators of selection, crossover and mutation. During the selection phase, pairs of individuals are chosen according to their fitness. These pairs are submitted to a crossover operator, which, through a combination scheme, generates a new pair that may substitute its progenitors in the population. The mutation operator then introduces random, sporadic modifications, that is mutations, in the genetic content of some randomly chosen chromosomes.

All chromosomes of the population for the next generation are evaluated once more and the procedure is repeated up to a maximum number of generations. Here the intention is to gradually improve the quality of the population. Together, these operators support the search mechanism of a GA by seeking the good solutions suggested by the fitness function.

The structure of a generic NSGA is summarized in Figure 1. Note that the final result of a NSGA may not be one single best solution but a set of Pareto front solutions.

4 The NSGA for the CTTP

The CTTP is, as mentioned above, a complex problem which, within the current approach, assumes the configuration of a multiobjective optimization problem with two main objectives. Nevertheless, it should be noted that this approach is suitable for incorporating more than two objectives such as administration, space workload or security objectives. In fact, following a detailed inspection of the nature of the so-called soft constraints, one may see that they compete with each other for teachers and classes. For instance, a tentative imposition of only the soft constraints related to the teachers would certainly produce timetables with high violations for classes. To prevent this and simultaneously enforce satisfaction of the hard constraints, a multiobjective GA is devised according to the lines of Figure 1, with two distinct objectives conceived to express CTTP con-

BEGIN
Step 1 ▷ **Initialization**
   Setup of an initial population of N chromosomes
Step 2 ▷ **Evaluation**
   Calculation of objective values for each chromosome
   Updating of Pareto front and rankings
   Evaluation of all chromosome's fitness
   Undertaking of fitness sharing
Step 3 ▷ **Selection, Crossover and Mutation**
   Application of genetic operators
   Replacement of chromosomes by offspring
⇑ **Loop to Step 2** ▷ Until the maximum number of generations is reached
END (Output ← Pareto front chromosomes)

Fig. 1. Outline of a generic NSGA

straint violations associated mainly with classes and teachers, respectively. The algorithm will attempt to find a set of feasible and high-quality timetabling solutions. This is performed through a constraint violation penalizing scheme that distinctively penalizes the hard and soft constraints. The penalizations were defined by using a trial and error procedure, which indicated that the algorithm is quite sensitive to the selection of these parameters. Moreover, the method deals with a main population as well as an elitist secondary population.

4.1 Chromosome Encoding

In practice, timetabling plans for the resources involved in a CTTP, such as classes, teachers or rooms, are usually represented in matrix form, where at least one of the axes always represents a time scale. This idea led to the option of a natural representation of the CTTP chromosome, encoded through a non-binary alphabet within a matrix. Although one could also use a binary encoding for the CTTP, this choice would require an additional effort to verify and impose the constraints.

   Let the number of available rooms and time periods be R and T, respectively, and each integer element of the $R \times T$ matrix a label that represents a lesson (class(es)/teacher/subject) assigned to the respective room (row) and time period (column). Here, lessons are the basic elements to be scheduled and can be of different durations as well as having distinct requirements.

   As an illustration, consider the curricula of two classes consisting of 12 lessons, presented in Table 1, where each row corresponds to a lesson and specifies the class(es) and teacher involved, the respective subject, duration and a list of required rooms. Figure 2, exemplifies a chromosome matrix with 3 rows (rooms) and 20 columns (time periods) relative to these curricula.

   This specific chromosome encoding allows one to aggregate the corresponding timetables for classes and teachers in a single matrix and all the lessons previously defined in the curricula of classes are scheduled within the matrix

Table 1. Example of curricula

Lesson	Classes	Teacher	Duration	Subject	Room required
M1	c_1 and c_2	t_1	1 period	Maths	r_2
M2	c_1	t_1	2 periods	Maths	r_1 or r_2
M3	c_2	t_1	2 periods	Maths	r_1 or r_2
E1	c_1	t_2	2 periods	Economics	r_3
E2	c_2	t_2	1 period	Economics	r_3
E3	c_1 and c_2	t_3	2 periods	Economics	r_2
H1	c_1	t_2	1 period	History	r_3
H2	c_2	t_2	1 period	History	r_3
H3	c_2	t_2	1 period	History	r_3
S1	c_1	t_1	1 period	Statistics	r_1 or r_3
S2	c_2	t_1	1 period	Statistics	r_1 or r_3
S3	c_1 and c_2	t_1	1 period	Statistics	r_2

chromosome. Thus, each chromosome represents a solution for the CTTP, i.e.

```
                            Time periods
             Monday    Tuesday  Wednesday  Thursday    Friday
           1  2  3  4  5  6  7  8  9 10 11 12 13 14 15 16 17 18 19 20
        r1                S2  M2        S1
Rooms   r2    M1                                M3              S3  E3
        r3          E1       H1 H2                   H3 E2

Teacher 1t   M1             S2  M2        S1 M3              S3
Teacher 2t         E1       H1 H2                   H3 E2
Teacher 3t                                                          E3
Class  c1    M1  E1         H1     M2     S1                S3  E3
Class  c2    M1             S2 H2              M3  H3 E2     S3  E3
```

Fig. 2. Example of chromosome encoding ($R \times S$ matrix) and corresponding timetables

a set of timetables, that automatically satisfies condition 1(b). However, it may include violations for the other hard constraints, thus representing a non-feasible solution.

The conception of this multiobjective GA is to search for high-quality solutions – solutions violating at the minimum the soft constraints – while penalizing severely violations of hard constraints, i.e. non-feasible solutions. There are two main reasons for this option. First, the approach is very flexible and permits swift implementation of further constraints that may appear in other real CTTPs. The second reason concerns the computational effort to maintain a population of feasible solutions, requiring the expenditure of significant computing time when repairing the non-feasible solutions produced by the genetic operators. In this way, the task of forcing feasibility is performed gradually via a constraint penalization scheme.

4.2   Fitness Function

As observed above, in the current development for the CTTP, two competitive objectives emerge from the soft and hard constraints described in Section 2:

- minimization of the penalized sum of violations mainly occurring in the constraints referring to the teachers, teacher-oriented objective;
- minimization of the penalized sum of violations occurring in the constraints mainly related to classes, class-oriented objective.

Next, there follow some details for the calculation of these two objective function values for each solution, i.e. a set of timetables encoded through a matrix chromosome. Assuming one has a chromosome, the satisfaction of those aspects included in the soft constraints (2(a)–(e)) and hard constraints (1(a), 1(c), 1(d) and 1(e)) is checked. The amount of violations is counted (both for teachers and classes) and, by using a constraint violation penalization scheme, the respective values of the two objectives considered are produced. However, the penalization procedure deals with hard and soft constraints differently. As for soft constraints, which are non-obligatory for timetabling feasibility, the penalizations assigned are small compared with those of hard constraints, thus reflecting solution unfeasibility. In addition, the violations of hard constraints are not linked with teachers or classes alone, but with the solution globally. Therefore, the penalizations of the hard constraints are added to both objective values simultaneously.

At first, each solution of the population is classified in terms of dominance by other solutions. As usual, a nondominated solution does not allow the existence of any other better for all the objectives. In such a case, any improvement on one of the objectives must worsen at least one of the remaining ones. Then, through a ranking process, due to Goldberg 1989, the solution fitness is determined according to the relative solution dominance degree. Basically the Pareto front, formed by the nondominated solutions, receives rank 1 and, from the remaining population, the next nondominated solutions receive rank 2 and so on, until the entire population is ranked. Finally, the fitness of each chromosome is obtained by inverting the chromosome rank level.

The fitness assignment method described above may lead to the loss of diversity within the population due to stochastic errors in the sampling selection procedure. In fact, when faced with multiple equivalent Pareto front solutions, the populations tend to converge to a single solution as a result of the genetic drift. To reduce this effect and keep the Pareto front uniformly sampled, the NSGA includes a procedure of fitness sharing. This is produced by dividing the Goldberg fitness $f(x_i)$ by a value that reflects the number of similar chromosomes around x_i, within the same rank level, as follows:

$$f(x_i)_{shared} = \frac{f(x_i)}{\sum_{j=1}^{N} s[d(x_i,x_j)]} \ . \tag{1}$$

Here, N is the total size of main and secondary populations. The function $s[d(x_i,x_j)]$ states, for the distance between two given chromosomes $d(x_i,x_j)$, the intensity of the sharing process:

$$s[d(x_i,x_j)] = \begin{cases} 1 - \left(\frac{d(x_i,x_j)}{\delta_{share}}\right)^\alpha & \text{if } d(x_i,x_j) < \delta_{share} \\ 0 & \text{otherwise,} \end{cases} \tag{2}$$

where δ_{share} defines the frontier of the neighborhood and α acts as a strength-sharing parameter. The measure of distance $d(x_i,x_j)$ is evaluated in the genotypic space by analyzing the similarities of the pair of solutions depending on the use of the same time period, for each lesson:

$$d(x_i,x_j) = \frac{\sum_{l=1}^{L} t(l,x_i,x_j)}{L} \ , \tag{3}$$

where

$$t(l,x_i,x_j) = \begin{cases} 1 \text{ if lesson } l \text{ occupies the same time period in solution } x_i \text{ and } x_j, \\ 0 \text{ otherwise,} \end{cases}$$

$$\tag{4}$$

and L is the total number of lessons to be scheduled.

Other alternative measures of distance can be applied, as proposed by Burke et al. (1998), with distinct implications on the global diversity of the population. Due to the large number of lessons often involved in the CTTP and the resulting computational effort required to compute the similarity within the population, we opted for the above absolute position distance measure.

The fitness sharing procedure is applied to each chromosome rank by rank. Through the assignment of dummy fitness values, the minimum fitness value of the current rank is always higher than the maximum fitness value of the next rank.

4.3   Main and Secondary Populations

The main population of 200 chromosomes is initially built by using a constructive random heuristic. The heuristic procedure relative to each chromosome starts by scheduling those lessons that are more difficult to allocate, according to a previously defined sorted list, while trying to satisfy the maximum of constraints. This degree of difficulty is measured in terms of lesson duration, preferences of teacher and classes involved and room requirement. The longer and more resource-demanding lessons are usually more difficult to schedule. This ends up with a chromosome that is a set of timetables satisfying at least the hard constraint 1(b).

Due to the stochastic nature of a GA, some high-quality solutions can be definitely lost during the algorithm execution as a result of sampling. Hence, the use of an elitist secondary population composed of some of the Pareto front solutions found so far, may increase the performance of the multiobjective GA (Zitzler 1999). The developed NSGA for the CTTP includes a secondary population, as illustrated in Figure 3. At the beginning of generation k, suppose

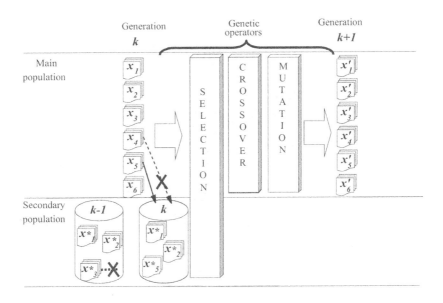

Fig. 3. Example of main and secondary populations updating scheme

that the main population is formed by six chromosomes (x_1 to x_6), and the secondary population includes three chromosomes (x_1^* to x_3^*). After the ranking process within the main population, performed for the fitness calculation, the non-dominated solutions (assume they are x_4 and x_5) are candidates to enter the secondary population. However, they do not always enter, but only if they are sufficiently distinct from the solutions that are stored therein. Suppose that x_4 is very similar to x_2^*, according to the measure $d(x_4, x_2^*)$, then it must be rejected. Next, each secondary population chromosome retained from generation $k - 1$ (x_1^* to x_3^*) and candidate chromosome (x_5) is checked for non-dominance, keeping only the non-dominated ones in the secondary population for generation k. In this example, assume that x_5 dominates x_3^*, thus causing the removal of x_3^*.

4.4 Selection, Crossover and Mutation Operators

Operating over the above main and secondary populations of timetabling chromosomes and using a standard roulette wheel procedure, the selection operator chooses 100 pairs of chromosomes to be combined by the crossover.

As for the crossover, a customized genetic operator is created to effectively combine the contents of a pair of parent chromosomes. Figure 4 will be used to describe this procedure. Having chosen two chromosomes from the above

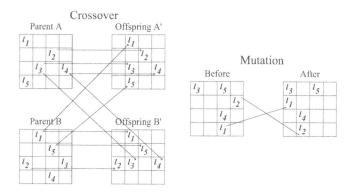

Fig. 4. Example of crossover and mutation operator procedures

populations (parent A and parent B), with the selection operator, two A' and B' offspring are created. The procedure begins by randomly defining a number c of lessons to be received through inheritance by each offspring, within 80 to 95% of the total number of lessons. The offspring A' is obtained from parent A selecting c lessons that belong to the teachers with the best timetables (according to the partial value of constraint violation penalizations for each teacher). Suppose that these lessons are represented in Figure 4 by l_2, l_3 and l_4. The remaining lessons are taken from parent B (l_1 and l_5) and are eventually subjected to a repair procedure to avoid overlapping. In this example the location of l_5 was already occupied by l_2. Consequently, a new location for l_5 must be found. Offspring A' is therefore teacher-guided, because it includes those lesson assignments that benefit the teachers. Offspring B' is obtained through a similar process but for classes. Therefore B' is, in this way, class-guided. Both offspring A' and B' are inserted in the transitory population during 100 crossover operations.

This specific crossover operator, which continuously transfers to the offspring a fraction of the best timetables produced so far, both for teachers and classes, strives to gradually improve the two distinct objectives within the population.

The mutation operator illustrated in Figure 4, can be described as follows. For each chromosome of the transitory population created by selection and crossover, the mutation procedure randomly takes m lessons. This amount of lessons may vary up to 1% of the total number of lessons L. In the example, where m equals two, consider that these lessons are represented by l_1 and l_2. Such lessons are removed from the actual locations and then, from the set of the remaining locations, the procedure finds new locations that cause the minimal increase in total penalization.

At this point, the current generation ends and the transitory population is moved to the main population for the next generation. Note that the GA parameters were set by trial and error and remained appropriate to all the CTTP instances tackled.

5   Computational Results

The above-described algorithm was tested on real timetabling situations, namely the first semester instance of the Escola Superior de Gestão, Hotelaria e Turismo of the Algarve University, for the last academic year. Experiments with other instances also tested from this school showed similar behavior.

This instance is characterized in the following way: 716 lessons to schedule, 29 available rooms on two distinct campuses, 108 teachers, 40 classes and a week of 50 time periods.

The proposed algorithm was applied from the initial main population of 200 timetabling solutions until the maximum number of 1000 generations was reached. It was tested in four different versions. The idea behind this was to test the effect of the fitness sharing procedure and the secondary population on the quality of the final solutions: versions 3, 4 with a secondary population and versions 2, 4 with fitness sharing, corresponding to NSGAs. Version 1 was used for purposes of comparison.

The population of Pareto front timetables obtained at the end of a single run of each GA version is shown in Figure 5. Also represented is the hand-made solution, equally expressed in terms of teacher and class-oriented penalizations. Note that the lowest Pareto front, representing the best solutions for teachers, came from version 3. As shown in this figure, the version 4 algorithm provided the best solutions for classes.

In addition, Tables 2 and 3 present some details on the computational results obtained from each algorithm version.

As for the number and feasibility of Pareto front solutions, version 3 generated the largest set containing 81 timetabling solutions, 76 of which satisfied all hard constraints (columns 5 and 6 of Table 2). When analyzing column 4 of the same table, in reporting the mean similarity degree that expresses how repeatedly each lesson is assigned to the same time period within the Pareto front solutions, one can see that the lowest value was obtained for version 4, with an average use of 34% of the same time periods. Such behavior is due to the fitness sharing that produces a small level of similarity in versions 2 and 4. Also, both versions show a wide and spread Pareto front set, while versions with the secondary population scheme, versions 3 and 4, obtained a more numerous Pareto

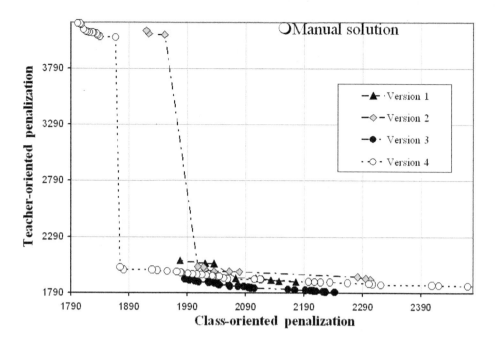

Fig. 5. Pareto front results for the algorithm versions

front. In fact, algorithm versions 3 and 4 achieved the largest sets of feasible solutions (76 and 55 solutions compared with the 12 and 14 solutions obtained in versions 1 and 2, respectively), allowing the decision-maker to choose the more appropriate ones. A detailed analysis of the set of feasible solutions, referred to in column 6 of Table 2, expressed in terms of soft constraints violations, is presented for each algorithm version, in Table 3. Columns 2 to 6 show the average number of soft constraints not satisfied (violated) per feasible solution, while columns 7 and 8 present the average penalizations relative to the set of feasible solutions obtained for classes and teachers. The results clearly point to the fact that versions 3 and 4 with a secondary population are superior to the standard NSGA and to the simplified GA (versions 2 and 1 respectively), both in terms of soft constraint violations and solution penalizations. In all cases the execution time to complete the 1000 generations was similar, about 4 h on a Pentium III 500Mhz PC with 64 MB of memory. It should be added that the execution time is compatible with the situation tackled, whereas the construction of this manual solution involved one week's work by a three person team.

Finally, comparison of the solutions obtained by the algorithm versions with the manual solution (displayed graphically in Figure 5) demonstrates that almost all solutions are superior in the two objectives to the manual solution.

Table 2. Algorithm versions: similarity, feasibility and execution time

Version			Pareto front solutions			Execution
Secondary	Fitness	Mean		No.		time
Number	population	sharing	similarity	No.	Feasible	(min)
			degree			
(1)	(2)	(3)	(4)	(5)	(6)	(7)
1	No	No	66%	13	12	213
2	No	Yes	39%	20	14	225
3	Yes	No	67%	81	76	249
4	Yes	Yes	34%	78	55	261

Table 3. Pareto front feasible solution: mean soft constraint violation

Algorithm	Soft constraints violations					Class	Teacher
version	2(a)	2(b)	2(c)	2(d)	2(e)	penalization	penalization
(1)	(2)	(3)	(4)	(5)	(6)	(7)	(8)
1	0	297	6.2	13.8	405.6	2056.25	2339.25
2	0	277	3.0	20.3	429.1	2074.75	2505.17
3	0	212	1.5	6.0	392.6	1971.88	2103.51
4	0	210	8.1	8.1	399.2	2004.35	2370.21

6   Final Remarks

Overall, the results obtained from the proposed NSGA with the elitist secondary population on the CTTP are promising when applied to a real instance, and lead to the following conclusions. First, the algorithm was able to successfully optimize the two distinct objectives considered. It thus achieved a set of quality timetabling solutions provided, at the end, by the Pareto front solution set. Secondly, this algorithmic approach produced much better timetables than the actual manual timetables, with a significantly reduced effort. On the basis of this algorithm, a software application (named GAHOR) is under development to edit, manage and automatically solve real problems.

In future the authors intend to apply this multiobjective GA to solve much larger CTTP instances, which are already being tackled. This will be attained through the refinement of the algorithm's performance by using improved genetic operators, while taking into account the parallel capability of GAs in order to decrease computational time.

References

Abramson, D.: Constructing School Timetables Using Simulated Annealing: Sequential and Parallel Algorithms. Manage. Sci. **37** (1982) 83–113

Birbas, T., Daskalaki, S., Housos, E.: Timetabling for Greek High Schools. J. Oper. Res. Soc. **48** (1997) 1191–1200

Burke, E.K., Elliman, D.G., Weare, R.: Examination Timetabling in British Universities – A Survey. In: Burke, E.K, Ross, P. (eds.): Practice and Theory of Automated Timetabling. Lecture Notes in Computer Science. Springer-Verlag Berlin Heidelberg New York, **1153** (1996) 76–90

Burke, E.K., Newall, J., Weare, R.: Initialization Strategies and Diversity in Evolutionary Timetabling. Evol. Comput. **6** (1998) 81–103

Carter, M.W., Laporte, G.: Recent Developments in Practical Examination Timetabling. In: Burke E.K, Ross P. (eds.): Practice and Theory of Automated Timetabling. Lecture Notes in Computer Science. Springer-Verlag Berlin Heidelberg New York, **1153** (1996) 3–21

Colorni, A., Dorigo, M., Maniezzo, V.: Genetic Algorithms and Highly Constrained Problems: The Timetabling Case. Proc. 1st Int. Conf. on Parallel Problem Solving from Nature. Lecture Notes in Computer Science, **496** (1991) 55–59

Costa, D.: A Tabu Search Algorithm for Computing an Operational Timetable. Eur. J. Oper. Res. **76** (1994) 98–110

de Werra, D.: An Introduction to Timetabling. Eur. J. Oper. Res. **19** (1985) 151–162

de Werra, D.: Some Combinatorial Models for Course Scheduling. In: Burke, E.K., Ross, P. (eds.): Practice and Theory of Automated Timetabling. Lecture Notes in Computer Science. Springer-Verlag Berlin Heidelberg New York, **1153** (1996) 296–308

de Werra, D.: The Combinatorics of Timetabling. Eur. J. Oper. Res. **96** (1997) 504–513

Dowsland, K.A.: A Timetabling Problem in which Clashes are Inevitable. J. Oper. Res. Soc. **41** (1990) 907–918

Even, S., Itai, A., Shamir, A.: On the Complexity of Timetable and Multicommodity Flow Problems. SIAM J. Comput. **5** (1976) 691–703

Fonseca, C.M., Fleming, P.J.: An Overview of Evolutionary Algorithms in Multiobjective Optimization. Evol. Comput. **3** (1995) 1–16

Gislén, L., Peterson, C., Söderberg, B.: Complex Scheduling with Potts Neural Networks. Neural Comput. **4** (1992) 806–831

Goldberg, D.E.: Genetic Algorithms in Search, Optimization and Machine Learning. Addison-Wesley, Reading, MA (1989)

Holland, J.: Adaptation in Natural and Artificial Systems. Proc. 1st Int. Conf. on Parallel Problem Solving from Nature. University of Michigan Press, Ann Arbor, MI (1975) (Reprinted by MIT Press)

Kiaer, L., Yellen, J.: Weighted Graphs and University Course Timetabling. Comput. Oper. Res. **19** (1992) 59–67

Srinivas, N., Deb, K.: Multiobjective Optimization Using Nondominated Sorting in Genetic Algorithms. Evol. Comput. **2** (1994) 221–248

Tripathy, A.: School Timetabling – a Case in Large Binary Integer Linear Programming. Manage. Sci. **30** (1984) 1473–1498

Veldhuizen, D.A., Lamont, G.B.: Multiobjective Evolutionary Algorithms: Analyzing the State-of-the-Art. Evol. Comput. **8** (2000) 125–147

Yoshikawa, M., Kaneko, K., Nomura, Y., Watanabe, M.: A Constraint-Based Approach to High-School Timetabling Problems: A Case Study. Proc. 12th Conf. on Artif. Intell. (1994) 1111–1116

Zitzler, E.: Evolutionary Algorithms for Multiobjective Optimization: Methods and Applications. Ph.D. Thesis, Swiss Federal Institute of Technology (ETH), Zurich, Switzerland (1999)

Some Complexity Aspects of Secondary School Timetabling Problems

H.M.M. ten Eikelder and R.J. Willemen

Technische Universiteit Eindhoven
Department of Mathematics and Computer Science
P.O. Box 513, 5600 MB Eindhoven
The Netherlands
{H.M.M.t.Eikelder, R.J.Willemen}@tue.nl

Abstract. We consider timetabling problems of secondary schools, in which the students can choose their own curricula. Besides finding a time slot and classroom assignment, every student must be assigned to a subject group for each subject in his curriculum. This problem is NP-hard for several independent reasons. In this paper we investigate the borderline between "easy" and "hard" subproblems. In particular, we show that the addition of blocks of size two, i.e. two lessons to be taught at consecutive time slots, or the addition of a constraint on the subject group size changes a subproblem from polynomially solvable to NP-hard.

1 Introduction

It is well known that most timetabling problems are NP-hard. In fact, realistic timetabling problems may have many sources of complexity. Realistic timetabling problems comprise several decisions. If we focus on one decision and some related constraints, this problem often already turns out to be NP-hard. In this way various different subproblems of the realistic timetabling problem can be proven to be NP-hard. Moreover, the removal of a constraint may change a subproblem from NP-hard to "easy", i.e. polynomially solvable. We illustrate this phenomenon using a simplified version of a timetabling problem for the upper levels of Dutch secondary schools.

In this paper we study two special cases. In Section 4 we consider a situation where finding the time slot assignment is easy if no blocks, i.e. lessons to be taught at consecutive time slots, are present. We show that the addition of blocks of size two makes this problem NP-hard. In Section 5 we discuss the subject group assignment. We describe a situation where the construction of a subject group assignment for each student is an easy problem if there is no maximal subject group size constraint. It turns out that addition of a maximal subject group size constraint makes this problem also NP-hard.

2 Dutch Secondary School Timetabling Problems

We first give an informal description of this problem. In the upper level the students can choose, within certain limitations, their own *curriculum* from a

E. Burke and W. Erben (Eds.): PATAT 2000, LNCS 2079, pp. 18–27, 2001.

list of *subjects*. For each subject a given number of *lessons* per week must be followed.

Each lesson in a subject is taught to a subject group. A *subject group* is a group of students that follow all lessons in one subject together. If many students have chosen a subject in their curriculum, several subject groups are created, and students must be assigned to one of these subject groups. The number of subject groups for each subject is fixed in advance by the school management. If for a subject there are two or more subject groups, the lessons for the different subject groups are not necessarily given at the same time. Teaching these lessons at different times gives more possibilities for students to follow this subject.

The *teacher* for each subject group is also determined by the school management. This implies that all the lessons corresponding to a subject group are given by the same teacher. Teachers may work part-time. Hence, they may not be available each time slot.

In general, there exist various types of *classrooms*, and a subject can only be taught in certain types of classrooms. For instance, physical education can only be taught in a gymnasium and chemistry in a chemistry classroom. But other subjects may allow more types of classrooms. For instance, physics can be given in a physics classroom or in a chemistry classroom and a language course can be taught in all types except a gymnasium.

Finally, we are given a number of *time slots* for the whole week. Each time slot is a period of unit time (usually 50 minutes) on a specific day in the week. Time slots are separated by breaks, like lunch. Each lesson lasts one time slot.

The question is to find a *subject group assignment*, i.e. a mapping that gives for each student and each subject from his curriculum the subject group in which the student must follow the lessons for this subject, a *time slot assignment*, i.e. a mapping that assigns lessons to time slots, and a *classroom assignment*, i.e. a mapping that assigns each lesson to one classroom.

The constraints are the usual timetabling constraints, i.e.

- no teacher or student has two lessons at the same time (*lesson conflict constraints*),
- each lesson is taught at a time slot during which the teacher is available (*person availability constraints*),
- suitable classrooms are assigned (*room type constraints*).

In practice many more constraints may occur. For instance, if for a subject two or more subject groups are used, then only subject group assignments with more or less equal numbers of students assigned to each subject group are acceptable. This can be enforced by introducing a *maximal subject group size constraint*.

Another complication may be the occurrence of *blocks*, i.e. two or more lessons for a subject group that must be given consecutively in the same classroom at time slots which are on the same day and not separated by a break. We shall only consider blocks consisting of two lessons.

To prove the complexity results we use two other NP-complete problems. The first one is 3SAT (see Garey and Johnson [8]). In a 3SAT instance a Boolean formula consisting of a number of *clauses* is given, where each clause is the disjunction of three *literals*. A literal is a Boolean variable (e.g., x_1) or the negation of a Boolean variable (e.g., $\bar{x}_1$). The question is whether the Boolean formula is satisfiable, i.e. whether there exist values for the Boolean variables such that all clauses evaluate to *true*. The second NP-complete problem is NOT-ALL-EQUAL-3SAT (NAE3SAT, [8]). This is a variant of 3SAT in which the question is whether there exist values for the variables such that all clauses evaluate to *true* and, furthermore, in each clause not all literals have the same value. Equivalently, do there exist values for the variables such that in each clause only one or two literals are *true*?

3 Literature

Timetabling problems at Dutch secondary schools have been studied by several people in the past. The earliest reference is a paper by Berghuis et al. [1]. In this paper necessary conditions for the existence of a feasible time slot assignment and a heuristic to find such an assignment are given. Their timetabling problem can be considered as the CLASS–TEACHER TIMETABLING problem with extra constraints. The introduction of a student's freedom of choosing his own curriculum in 1963 led to the study of the so-called CLUSTERING problem by Simons [15] and Van Kesteren [16], amongst others. Instead of assigning lessons to time slots directly, lessons are clustered first and then the subject group assignment is carried out. The assignment of lessons and clusters to time slots is considered after finding a solution to the CLUSTERING problem. De Gans [4] and Schreuder and Van der Velde [14] report on the timetabling problems that remain and solution approaches for these problems.

The decision version of problem types corresponding to realistic formulations of Dutch secondary school timetabling problems is clearly NP-complete. Most problem types will have the NP-complete UNIVERSITY COURSE TIMETABLING problem (see, for example, [5] and [12]) as a special case. Also, by introducing groups of students with the same curriculum ("classes") the CLASS–TEACHER TIMETABLING problem with teacher availabilities becomes a special case. This problem has been shown to be NP-complete by Even et al. [7]. There are a number of papers that report on complexity results, either positive (polynomially solvable subproblems) or negative (NP-completeness). The bibliography by Schmidt and Ströhlein ([13]) presents results published up to 1979. De Werra [5], [6] describes the CLASS–TEACHER TIMETABLING problem and UNIVERSITY COURSE TIMETABLING problem and some related problems in terms of graphs (e.g., bipartite multigraph and network flow models). An overview of easy and NP-hard classroom assignment problems is given by Carter and Tovey [2]. For example, they show that assigning lessons to classrooms in the case of room types and block constraints is already NP-complete if we restrict ourselves to only two time slots. Like ourselves, Cooper and Kingston [3] prove NP-completeness of school timetabling problems for a number of independent reasons. Their results comprise teacher assignment and time slot assignment problems with several constraint types. Amongst others, they prove NP-completeness of time slot assignment for blocks of size three. Recently, Giaro et al. [9] have given polynomially solvable subproblems of the NP-complete CLASS–TEACHER TIMETABLING problem without idle time for teachers and students.

4   Blocks and the Time Slot Assignment

Suppose all students have the same curriculum and there is only one classroom, suitable for all lessons. If for each subject there is only one subject group, the subject group assignment and the classroom assignment are trivial. The only remaining problem is the time slot assignment, i.e. we have to assign all lessons to suitable time slots. Let L be the set of lessons, T the set of time slots and let $S \subseteq L \times T$ give the possible time slots for each lesson. Then, in the absence of blocks, a time slot assignment is simply a matching M in the bipartite graph (L, T, S) with each lesson in L matched. In other words, M is a subset of S such that each lesson occurs exactly once and different lessons occur with different time slots. Such a matching can be found in polynomial time [10]. Hence, in this situation finding the time slot assignment is easy.

In the presence of blocks the lessons that form a block must be assigned to time slots that are consecutive, occur on the same day and are not separated by a break. Let $B \subseteq L \times L$ give the pairs of lessons that form a block and let $BT \subseteq T \times T$ be the pairs of time slots suitable for teaching a block. Elements of BT will be called *block time slots*.

Formally this means that the matching M must satisfy the additional condition that, for all $(l_1, t_1), (l_2, t_2) \in M$, we have

$$(l_1, l_2) \in B \Rightarrow (t_1, t_2) \in BT.$$

We shall call the decision version of this extended matching problem BIPARTITE MATCHING WITH RELATIONS (BMR). It turns out that BMR is NP-complete.

Theorem 1. *The problem BMR is NP-complete.*

Proof: Checking whether a given matching satisfies all conditions can easily be done in polynomial time. Hence BMR $\in$ NP. Next we show that the NP-complete problem 3SAT is reducible to BMR. Consider an instance of 3SAT that consists of m clauses with variables $x_1, \ldots, x_n$. Without loss of generality, we assume that each variable occurs in two clauses at least. We now construct an instance of BMR in the following way. With each clause we associate three new blocks, each corresponding to one of the variables in the clause. For clause p we have the blocks l_{pi}, l'_{pi} for $i = 1, 2, 3$. The block l_{pi}, l'_{pi} can only be taught at block time slots T_{pi}, T'_{pi} or at block time slots F_{pi}, F'_{pi}. These possibilities correspond to the values *true* and *false*, respectively, for the corresponding variable. Furthermore, for clause p there is one "clause lesson" cl_p. This is a single lesson (i.e. not a block), that can be taught at three time slots determined in the following way. If the variable corresponding to block l_{pi}, l'_{pi} occurs in the clause without negation,

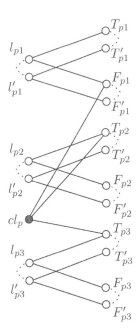

Fig. 1. Construction for the clause $C_p = x_1 \vee \bar{x}_2 \vee \bar{x}_3$; lessons and time slots that are connected by a dotted line form a block and a block time slot, respectively.

the clause lesson cl_p can be given at time slot F_{pi}. If the variable occurs in the clause in the negated form, the clause lesson can be given at time slot T_{pi}. The construction is illustrated by Figure 1, where the construction for the clause $x_1 \vee \bar{x}_2 \vee \bar{x}_3$ is shown.

Note that the necessity of a suitable time slot for the clause lesson implies that the blocks must be assigned to time slots in such a way that the clause holds. More precisely, there exists a time slot assignment (matching) for the situation above if and only if the corresponding clause holds.

There is only one problem with this construction. A variable will in general occur in many clauses, i.e. a variable will correspond to many blocks. To obtain a consistent value for such a variable, either all blocks corresponding to that variable must be matched with block time slots T_{**}, T'_{**}, or all blocks corresponding to that variable must be matched with block time slots F_{**}, F'_{**}. So far this condition does not necessarily hold. To enforce this condition we use the second time slots (T'_{**}, F'_{**}) of the various block time slots. Consider all blocks that correspond to one variable in some arbitrary order, see Figure 2. We now add between two successive blocks an additional "consistency lesson", which can be given at the F'_{**} time slot of the first block, or at the T'_{**} time slot of the second block. We also add such a consistency lesson between the F'_{**} time slot of the last block and the T'_{**} time slot of the first block. The additional lessons are the dark lessons in Figure 2. Clearly a matching with each lesson matched

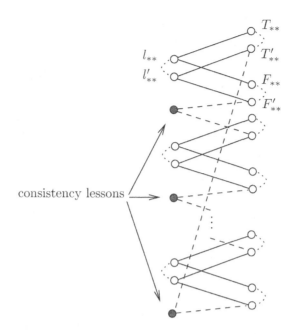

Fig. 2. Enforcing consistency for blocks corresponding to the same variable by introducing consistency lessons.

is now only possible if the blocks are all matched with the T_{**}, T'_{**} time slots or all matched with the F_{**}, F'_{**} time slots. In other words, the various occurrences of a variable now have consistent values. This construction is repeated for each variable x_i.

Now a matching that saturates all lessons and blocks exists if and only if there are values for the variables x_i such that all clauses hold. Hence we have reduced the 3SAT problem to our BMR problem. It is trivial to show that the transformation is polynomial. This completes the proof of the NP-completeness of BMR.   □

We have thus shown that the introduction of blocks makes the time slot assignment NP-hard. Note that the clause lessons can only be given at first time slots of a block time slot, while the consistency lessons can only be given at second time slots of a block time slot. Hence the mechanisms used for making the clauses satisfiable and for obtaining consistent values for the various occurrences of a variable do not influence each other. In fact blocks make it possible to couple these two mechanisms, which is essential for this reduction.

Note that our blocks consist only of two lessons. For blocks of three lessons NP-completeness results have already been given by Cooper and Kingston [3].

5   Maximal Subject Group Size and the Subject Group Assignment

Finally, we show that the addition of the maximal subject group size constraint may make the subject group assignment NP-hard. Consider the situation where each subject has in general several subject groups, with one lesson per week for each subject group. Further we assume that teacher availabilities are so limited that the time slot assignment of the lessons can only be done in one way. If there are enough classrooms available, the classroom assignment is trivial and only the subject group assignment remains. First assume that there is no maximal subject group size constraint. Let V_i denote the set of subjects of student i, T the set of time slots, and let $S_i \subseteq V_i \times T$ indicate at which time slots the lesson for each subject is given. The subject group assignment for student i is now a matching M_i in the bipartite graph (V_i, T, S_i) with each subject in V_i matched. The matching problems of the various students are independent and can be solved in polynomial time.

If the maximal subject group size constraint must hold, the matching problems for the various students are no longer independent, since the total number of matchings of a subject with a time slot is bounded. If the function *Maxgroup* gives the maximal number of students that can follow a lesson in subject v at the same time, then the problem is to find a matching M_i for each student i with the additional constraint that for all subjects v and time slots t

$$\sum_{i|(v,t)\in M_i} 1 \leq \mathit{Maxgroup}(v).$$

We shall call the decision variant of this problem the MULTIPLE BIPARTITE MATCHING WITH MAXIMA (MBMM) problem. This problem also turns out to be NP-complete.

Theorem 2. *The problem* MBMM *is NP-complete.*

Proof: Checking whether given matchings M_i satisfy all conditions can easily be done in polynomial time. Hence MBMM $\in$ NP. Next we show that the NP-complete problem NAE3SAT can be reduced to MBMM.

Consider an instance of NAE3SAT that consists of m clauses with variables $x_1, \ldots, x_n$. We now construct an MBMM instance with

- time slots $t_0, t_1, \ldots, t_m$, so $T = \{t_0, t_1, \ldots, t_m\}$,
- subjects c_i for $1 \leq i \leq n$ and subjects a_{pi}, b_{pi} for $1 \leq i \leq n, 1 \leq p \leq m$,
- all subjects c_i are taught twice, at time slot t_0 and at time slot t_m,
- all subjects a_{pi} and b_{pi} are taught twice, at time slot t_p and at time slot t_{p-1},
- students A_i with subjects c_i and a_{pi} for $1 \leq p \leq m$, so $V_{A_i} = \{c_i, a_{1i}, \ldots, a_{mi}\}$,
- students B_i with subjects c_i and b_{pi} for $1 \leq p \leq m$, so $V_{B_i} = \{c_i, b_{1i}, \ldots, b_{mi}\}$.

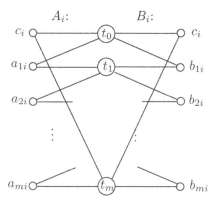

Fig. 3. Matching problems for the students A_i and B_i.

The two matching problems for students A_i and B_i are shown in Figure 3. Since each subject must be matched, we conclude that either student A_i can follow all his a_{pi} subjects ($1 \leq p \leq m$) at time slots t_p and his c_i subject at time slot t_0 or he can follow all his a_{pi} subjects at time slots t_{p-1} and his c_i subject at time slot t_m. Also either student B_i can follow all his b_{pi} subjects ($1 \leq p \leq m$) at time slots t_p and his c_i subject at time slot t_0 or he can follow all his b_{pi} subjects at time slots t_{p-1} and his c_i subject at time slot t_m.

In other words, there exist two possible matchings for student A_i and two possible matchings for student B_i. Now suppose that $Maxgroup(c_i) = 1$. Consequently not both students A_i and B_i can follow the same lesson in subject c_i. Only two possibilities for students A_i and B_i remain, which correspond to the value of the variable x_i as given in the following table:

A_i follows a_{pi} at	t_p	t_{p-1}
A_i follows c_i at	t_0	t_m
B_i follows b_{pi} at	t_{p-1}	t_p
B_i follows c_i at	t_m	t_0
x_i	true	false

So far we have obtained a situation where the possibilities for students A_i and B_i represent the value of the variable x_i. It remains to reduce the possibilities of all students further, such that we have a yes-instance of NAE3SAT if and only if the corresponding instance MBMM is a yes-instance. Recall that in the NAE3SAT problem the question is whether there exist values for the Boolean variables such that each clause holds true and not all literals in the clause are equal.

Consider an arbitrary clause, for instance $C_p = x_i \vee \bar{x}_j \vee x_k$. The question is whether there exist variables such that the following two conditions hold:

– The clause must hold, so we do not want that $x_i = false$ and $x_j = true$ and $x_k = false$. In terms of the students A_* and B_*, we do not want that B_i

follows subject b_{pi} at t_p and that A_j follows subject a_{pj} at t_p and that B_k follows subject b_{pk} at t_p.

– Not all literals in the clause may have the same value. Since at least one literal must be *true*, this simply implies that not all literals may be *true*. Hence we do not want that $x_i = true$ and $x_j = false$ and $x_k = true$. In terms of the students A_* and B_*, we do not want that B_i follows subject b_{pi} at t_{p-1} and that A_j follows subject a_{pj} at t_{p-1} and that B_k follows subject b_{pk} at t_{p-1}.

Note that the three subjects b_{pi}, a_{pj} and b_{pk} are all only taught at time slots t_p and t_{p-1}. Hence we can replace the b_{pi}, a_{pj} and b_{pk} subjects by one new subject. By requiring that for this new subject the maximal group size is at most two, not all three students B_i, A_j and B_k can follow this subject at the same time slot. Hence both above conditions must hold. We follow this procedure for all clauses. For each clause we identify three subjects, chosen as follows. If clause C_p contains variable x_i we select subject b_{pi}, if the clause contains $\bar{x}_i$, we select a_{pi}. By identifying these three subjects and giving the identified subject a maximal group size of two, we enforce that not all literals in the clause are *false* and also that not all literals in the clause are *true*. Hence we have reduced the NAE3SAT problem to our MBMM problem. This completes the proof of the NP-completeness of MBMM.   □

We thus have shown that the addition of the maximal subject group size constraint can change the subject group assignment into a difficult problem. In the proof above we used the maximal group sizes one and two. By introducing more students with the same curricula as A_i and B_i we can reduce NAE3SAT to MBMM instances with more realistic maximal group sizes.

Itai, Rodeh, and Tanimoto [11] have studied the RESTRICTED COMPLETE MATCHING (RCM) problem, which is somewhat related to our MBMM problem. In the RCM problem we are given a bipartite graph, k subsets of edges $E_1, \ldots, E_k$ and k positive integers $r_1, \ldots, r_k$. The problem is to determine whether there exists a complete matching M for the graph, that also satisfies the restrictions $|M \cap E_i| \leq r_i$ for all i. In fact it is not complicated to reduce our MBMM problem to the RCM problem, thus providing an additional NP-completeness proof for the RCM problem.

References

1. Berghuis, J., van der Heiden, A.J., Bakker, R.: The Preparation of School Timetables by Electronic Computer. BIT **4** (1964) 106–114
2. Carter, M.W., Tovey, C.A.: When is the Classroom Assignment Problem Hard? Oper. Res. **40** (Suppl. 1) (1992) S28–S39
3. Cooper, T.B.,Kingston, J.H.: The Complexity of Timetable Construction Problems. In: Burke E.K., Ross P. (eds.): Practice and Theory of Automated Timetabling, 1st Int. Conf., Selected Papers. Springer-Verlag, Berlin Heidelberg New York (1996) 283–295

4. de Gans, O.B.: A Computer Timetabling System for Secondary Schools in the Netherlands. Eur. J. Oper. Res. **7** (1981) 175–182
5. de Werra, D.: An Introduction to Timetabling. Eur. J. Oper. Res. **19** (1985) 151–162
6. de Werra, D.: The Combinatorics of Timetabling. Eur. J. Oper. Res. **96** (1997) 504–513
7. Even, S., Itai, A., Shamir, A.: On the Complexity of Timetable and Multicommodity Flow Problems. SIAM J. Computing **5** (1976) 691–703
8. Garey, M.R., Johnson, D.S.: Computers and Intractability: a Guide to the Theory of NP-completeness. Freeman, San Francisco (1979)
9. Giaro, K., Kubale, M., Szyfelbein D.: Consecutive Graph Coloring for School Timetabling. In: Burke E.K., Erben W. (eds.): Proc. 3rd Int. Conf. on the Practice and Theory of Automated Timetabling (2000) 212–221
10. Hopcroft, J.E., Karp, R.M.: An $n^{\frac{5}{2}}$ Algorithm for Maximum Matchings in Bipartite Graphs. SIAM J. Computing **2** (1973) 225–231
11. Itai, A., Rodeh, M., Tanimoto, S.L.: Some Matching Problems for Bipartite Graphs. Technical Report TR93. IBM Israel Scientific Center, Haifa, Israel (1977)
12. Schaerf, A.: A Survey of Automated Timetabling. Technical Report CS-R9567. Centre for Mathematics and Computer Science, Amsterdam, The Netherlands (1995)
13. Schmidt, G., Ströhlein, T.: Timetable Construction: an Annotated Bibliography. Computer Journal **23** (1980) 307–316
14. Schreuder, J.A.M., van der Velde, J.A.: Timetables in Dutch High Schools. In: Brans J.P. (ed.): Operational Research '84. Elsevier Science Publishers B.V. (North-Holland) (1984) 601–612
15. Simons, J.L.: ABC: Een Programma dat Automatisch Blokken Construeert bij de Vakdifferentiatie Binnen het Algemeen Voortgezet Onderwijs. Technical Report 74107U. NLR, Amsterdam, The Netherlands (1974)
16. van Kesteren, B.: The Clustering Problem in Dutch High Schools: Changing Metrics in Search Space. Internal Report. Leiden Institute for Advanced Computer Science, Leiden University, Leiden, The Netherlands (1999) 99–106

A Generic Object-Oriented Constraint-Based Model for University Course Timetabling

Kyriakos Zervoudakis and Panagiotis Stamatopoulos

Department of Informatics and Telecommunications,
University of Athens,
Panepistimiopolis, 157 84 Athens, Greece
{quasi, takis}@di.uoa.gr

Abstract. The construction of course timetables for academic institutions is a very difficult problem with a lot of constraints that have to be respected and a huge search space to be explored, even if the size of the problem input is not significantly large, due to the exponential number of the possible feasible timetables. On the other hand, the problem itself does not have a widely approved definition, since different variations of it are faced by different departments. However, there exists a set of entities and constraints among them which are common to every possible instantiation of the timetabling problem. In this paper, we present a model of this common core in terms of ILOG SOLVER, a constraint programming object-oriented C++ library, and we show the way this model may be extended to cover the needs of a specific academic unit.

1 Introduction

The construction of university timetables falls under the class of scheduling problems. Its difficulty [8], along with the fact that an educational institution has to tackle it once or twice a year, has drawn the attention of researchers from various fields, even as long ago as the 1960s [15]. Since then, the problem has been tackled with various approaches including graph coloring [23], network flows [6] and operations research methods [1]. During the last years, various artificial intelligence techniques have also been used against the problem, like tabu search [3], [30], simulated annealing [9], [27], genetic algorithms [24], [10] and constraint programming [16], [22], [19], [11], [26].

It is surprising that, even now, after all these years, not all educational institutions use automated tools to construct timetables. A survey regarding this issue in British universities [5] showed that only 21% of the universities used a computer for constructing examination timetables. 37% used a computer only for assisting the process and 42% did not use computers at all. Although these data concern examination timetabling, it seems that the numbers are similar for course timetabling case as well.

One might wonder about the reasons for this situation. In [2], it is mentioned that many institutions have automated solutions, but they are so tailored to each institution's case that they cannot be used by others. It is true that the

E. Burke and W. Erben (Eds.): PATAT 2000, LNCS 2079, pp. 28–47, 2001.
© Springer-Verlag Berlin Heidelberg 2001

differences between institutions are many, especially as far as quality criteria are concerned. That makes the use of one tool that was implemented for one university inappropriate for another. What is more, it is possible for a tool that was developed for one institution to be inapplicable for that same institution if some changes in the curricula occur.

In [2] the basic rules that govern timetables are also mentioned. In spite of the differences between institutions, some basic categories of rules can be recognized. The quality determination criteria can also be easily categorized and many similarities can be identified. For example, in all cases the same hard rules hold, such as that one teacher cannot give more than one lecture simultaneously or that one student cannot attend more than one lecture at a time. Most quality criteria have to do with the distribution of certain groups of lectures in time, like uniform distribution of the lectures in each day of the week, dense scheduling of these lectures for every day, upper bounds in the time that a teacher or a student can be occupied in teaching activities and so on.

Constraint programming is a problem solving methodology that allows the user to describe the data of a problem and the constraints that govern them without explicitly handling these constraints. This methodology fits perfectly to the timetabling problem, as someone might define the involved entities and express, in a declarative way, what is a legal and good timetable. On the other hand, it would be very good, from the software engineering point of view, if the modeling were to follow an object-oriented approach, as the obtained result would then be as general as we would like it to be at the same time as having unbounded opportunities to be specialized as much as we would like, in order to capture any specific timetabling situation. Fortunately, the combination of constraint programming and object-oriented design is provided by an existing commercial C++ library, the ILOG SOLVER [21], [20]. In this paper, a generic model for university course timetabling, which is based on ILOG SOLVER, is presented. The application of this model on the specific course timetabling problem faced by the Department of Informatics and Telecommunications of the University of Athens is also described.

2   University Course Timetabling

Most university timetabling problems are based on the basic student–course model [7]. Let $U = \{u_1, u_2, \ldots, u_{|U|}\}$ be the set of students, $S = \{s_1, s_2, \ldots, s_{|S|}\}$ be the set of courses taught, d_i be the number of teaching periods for course s_i, $S_j = \{s_{j_1}, s_{j_2}, \ldots, s_{j_{|S_j|}}\}$ be the set of courses student j wishes to attend, $T = \{t_1, t_2, \ldots, t_{|T|}\}$ be the set of teachers, $t : S \mapsto T$ a function mapping each course to the teacher that teaches it and $x_{i,k}, 1 \leq k \leq d_i, 1 \leq x_{i,k} \leq p$ the time that the kth teaching period of course s_i is given, with p the number of possible teaching periods of the institution (the maximum length of the timetable). A solution to the problem is any assignment to the variables $x_{i,k}$ such that the following constraints are respected:

– A teacher cannot give more than one course at a time

$$\sum_{i,k}[x_{i,k} = m][t_n = t(s_i)] \leq 1, \quad \forall m, n; \tag{1}$$

– A student cannot attend more than one course at a time

$$\sum_{i,k}[x_{i,k} = m][s_i \in S_j] \leq 1, \quad \forall j, m. \tag{2}$$

The above model gives complete freedom to the student to select the courses he/she takes and is close to the way most universities operate. If, in addition, a set of classrooms $C = \{c_1, c_2, \ldots, c_{|C|}\}$ is added, as well as a constraint

$$\sum_{i,k}[x_{i,k} = m][cl_{i,k} = c_n] \leq 1, \quad \forall m, n \tag{3}$$

where $cl_{i,k}$ is the classroom where the kth teaching period of course s_i is conducted, then the model becomes even more realistic. Within the framework of this basic model, a number of more specialized, but common in practice, constraints can be expressed. For example, the unavailabilities of a teacher can be taken into account by modifying constraint (1):

$$\sum_{i,k}[x_{i,k} = m][t_n = t(s_i)] \leq avail(t_n, m), \quad \forall m, n \tag{4}$$

where $avail(t_n, m)$ equals 1 if teacher t_n is available in period m, or 0 if not. Next, a short introduction to the ILOG SOLVER library is given and, then, an object-oriented constraint programming model of the timetabling problem, based on the student-course model, is presented.

3   Ilog Solver

ILOG SOLVER [21], [20] is a constraint programming object-oriented C++ library. Some information on SOLVER is necessary in order for the model that follows to be understood. The type `IlcInt` corresponds to the C++ `long` type. Integer constraint variables are represented by the class `IlcIntVar`. A command like `IlcIntVar x(m,0,10)` creates a constrained variable x with the integers from 0 to 10 included in its domain. m is an object of the class `IlcManager` to which all constrained variables and constraints between them are connected. The method `IlcIntVar::removeValue(IlcInt a)` removes value a from a variable's domain and the method `IlcIntVar::setValue(IlcInt a)` assigns value a to a variable. The class `IlcIntVarArray` implements an array of constrained integer variables. The call `IlcIntVarArray t(m,10)` creates an array of 10 constrained integer variables which can be accessed, as usually, with the overloaded operator `[]`. Constraints are represented by the class `IlcConstraint` and can be created with the overloaded C++ relational operators. If x and y are objects of the class

IlcIntVar, the call IlcConstraint c(x>y) creates a constraint c that ensures all values in x's domain are greater than the minimum of y's domain. In order for a constraint to be posted, the method IlcManager::add() has to be used. The above constraint, for example, can be posted with the call m.add(c). Posting a constraint ensures that it will be considered by SOLVER's internal constraint propagation engine and that after any modification of a constrained variable, the constraint network will be modified, in order to be brought to a certain consistency degree. SOLVER uses a version of the AC-5 algorithm [28] to bring the constraint graph to an arc-consistent state after any such modification. Finally, SOLVER supports goal programming and gives the user the ability to define any search algorithm of choice, like depth-first search (DFS), limited-discrepancy search (LDS) and so on. We omit the details of the implementation in SOLVER terms of these algorithms, since they are not necessary for the scope of this paper.

4   Object-Oriented Modeling

The model for the university course timetabling problem we present in this section aims to be as generic as possible. However, some simplifying assumptions are commonplace in many universities. For example, the mathematical model presented in Section 2 considers only the number of teaching periods for each subject with no further constraints. For example, a four period subject can be split in two two-hour lectures, one three-hour lecture and one one-hour lecture and so on. It is usual for many universities for subjects to be partitioned in lectures in a preprocessing step and this model follows this assumption. Time is measured in multiples of a predetermined unit of time, further partitioning of which is of no practical importance. This unit can be an hour, half an hour or whatever. In the following, this unit will be referred to as a "teaching period". Also, there is a maximum number of such units within which the timetable has to be constructed. A timetable is constructed for a week with D days and H time units every day. The model, however, can be easily extended to cases where timetables span within more than one week.

The similarity in format of the equations in the student–course model is obvious. Teachers and classrooms are necessary resources for a lecture to take place and each resource cannot support more than one activity at a time. Thus, it is reasonable for the notion of resource to play a central role in the modeling of the problem. It can also be seen that although a student is not actually a resource for a lecture, the equations that define the constraints on students are of the same form, since lectures that are taken by one student cannot be conducted at the same time. In the original student–course model, every student is completely free to take the course he/she chooses. This is true for many universities where modularity is important. In other universities, a student can select from a set of predetermined course offers. In any case, the net effect is that either because of student preferences or because of predetermined management decisions, some sets of lectures are formed and it is demanded that no two lectures of the same set be conducted at the same time. Thus, these sets are modeled like other resources and will be called "lecture groups" from now on.

One more important issue is the criteria according to which the quality of a timetable is measured, since, in practice, this is an optimization problem. However, it would be very difficult to gather all such criteria and present a way of implementing them with the model which follows. We will try to display by example that the implementation of the most usual criteria is not only possible but, indeed, easy within this framework.

4.1 Subjects

```
class c_Subject {
protected:
  IlcIntVarArray StartVariables; };
```

The class c_Subject represents a subject. StartVariables is an array of constrained variables, each one representing the time in which a lecture of this subject is scheduled. This class is used for logistic purposes, but also for calculating preferences regarding the distance between lectures of the same subject.

4.2 Lectures

```
class c_Lecture {
protected:
  IlcInt Duration;
  IlcIntVar Start;
  IlcIntVar Classroom; };
```

The class c_Lecture represents a lecture. It contains the two decision variables for each lecture: the starting time of the lecture and the classroom in which it is conducted.

4.3 Unary Resources

```
class c_UnaryResource {
protected:
  IlcIntVarArray TimeTable;
  IlcIntVarArray LectureTimeUnits;
public:
  void add(IlcIntVar start, IlcInt duration); };
```

As mentioned before, the notion of a unary resource is of central importance in the modeling. The class c_UnaryResource represents a unary resource: that is, a resource that can only support one activity at a time. In order to declare that a lecture uses a certain resource (any activity can of course require more than one resource) the method add has to be called with parameter variables Start and Duration of the corresponding lecture:

```
void c_UnaryResource::add(IlcIntVar start, IlcInt duration) {
  for (IlcInt i=0; i<duration; i++)
    LectureTimeUnits[count++]=start+i; }
```

As can be seen above, for each time unit of a lecture's duration, one variable in array `LectureTimeUnits` is created. For example, if a resource supports three lectures out of which the first has duration 2 time units and starting time `[0..2 4]`, the second duration 1 and starting time `[8..9]` and the third duration 3 and starting time `[12]`, then the array `LectureTimeUnits` will have six elements (the sum of the lectures' durations) which will be `[0..2 4] [1..3 5] [8..9] [12] [13] [14]`. The array `TimeTable` has `D*H` elements which correspond to the `D*H` time units of a timetable. Each one of them takes values within the interval `[-1..d]`, where `d` is the number of elements in the array `LectureTimeUnits`. When all lectures have been added, then the SOLVER function `inverse` is called. This function's declaration is

```
IlcConstraint IlcInverse(IlcIntVarArray f, IlcIntVarArray invf);
```

according to which if f's length is n and $invf$'s length is m then

- If $f[i] \in [0, m-1]$ then $invf[f[i]] == i$,
- If $invf[j] \in [0, n-1]$ then $f[invf[j]] == j$.

The above constraint guarantees that during each time unit the resource supports at most one activity.

4.4   Multiple Resources

```
class c_MultipleResource : public c_UnaryResource {
protected:
  IlcInt Multiplicity;
public:
  void add(IlcIntVar start, IlcIntVar classroom, IlcInt duration); };
```

In most cases, a unary resource can represent teachers and groups of lectures that cannot be given simultaneously, since an activity will either require such a resource or not. But this is not enough in the case of resources such as classrooms, since a lecture can be given in any one of a certain set of classrooms. This requirement is a disjunctive demand of a resource. In our case, all such resources, let us say classrooms, are modeled as one multiple resource with multiplicity equal to the number of the individual resources. The produced class was created with minor modifications on the existing unary resource class:

- The array `TimeTable` has `mult*D*H` elements instead of `D*H`, where `mult` is the multiplicity of the resource. Each `D*H`-tuple of this array corresponds to one of the `mult` individual resources.
- In order for a lecture to be added to a resource, one more parameter is needed along with the starting variable of the lecture and its duration, namely a constrained variable representing the resource to which the lecture will be eventually assigned. In the case of classrooms, this variable represents the classroom in which the lecture will take place.

– The `add` method is modified as follows:

```
void c_MultipleResource::add(IlcIntVar start,
                             IlcIntVar classroom, IlcInt duration) {
for (IlcInt i=0; i<duration; i++) {
  LectureTimeUnits[count++]=start+i+classroom*D*H; } }
```

As can be seen, the elements of the array `LectureTimeUnits` are like the ones in unary resources plus the term `classroom*D*H` which makes each of these elements point to the D*H-tuple of the multiple resource corresponding to the value of the variable `classroom`.

5   Application

We believe that the proposed model is general enough to cover most cases. Its representative potential will be exhibited through one real problem, that of constructing a course timetable for the Department of Informatics and Telecommunications of the University of Athens (DIT/UoA).

5.1   Curricula

The set of given courses and the teaching periods for each course are given. In case the course cannot (or is not desired to) be given in one lecture, then it is partitioned into two or more lectures of predefined duration. In this case, lectures of the same course cannot be given in the same day. Each lecture is taught by one or more teachers. If a course is given by another department, then it is scheduled by that department. It can be required for a lecture to be given in a specific classroom or in one of a subset of the available ones. In this case, a degree of preference between classrooms can be given. The undergraduate curriculum is organized over four years. Each course can be either obligatory for a certain year or belong to a certain group of lectures which are called directions. It is demanded that lectures of the same year which are either obligatory or belong to the same group not be given simultaneously. Also, there is the possibility for teachers to express certain personal constraints on the maximum number of teaching hours in a day and the maximum number of days in which they are occupied in teaching activities.

The above are hard constraints that have to be respected fully in order for a timetable to be feasible. Apart from these, there are also several criteria according to which the quality of a timetable is measured. It is desired that obligatory lectures or lectures of the same direction and year have as few gaps as possible between them during each day. It is also desired for such lectures to be as uniformly distributed during the timetabling period as possible, and it is desired for lectures of the same course to have a reasonable distance in days between them.

5.2   Formulation

Let $L = \{l_1, l_2, \ldots, l_{|L|}\}$ be the set of lectures, $T = \{t_1, t_2, \ldots, t_{|T|}\}$ the set of teachers, $C = \{c_1, c_2, \ldots, c_{|C|}\}$ the set of classrooms and $S = \{s_1, s_2, \ldots, s_{|S|}\}$ the set of courses. There exists a function $sub : L \mapsto S$ which maps every lecture to the corresponding course. Let D be the number of teaching days in a week and H the number of periods in every day. Let d_i be the duration of lecture l_i. We define $x_{i,1} = x_i$, $x_{i,j} = x_{i,j-1} + 1, j = 2, \ldots, d_i$. Let $t(l_i) \subseteq T$ the teachers of lecture l_i and $G = \{g_1, g_2, \ldots, g_{|G|}\}$, $g_i \subseteq L$, the set of lecture groups. The problem is to find the time and classroom for each lecture, so a solution is any mapping $sol : L \mapsto \{0, 1, \ldots, D * H - 1\} \times C, sol(l_i) \mapsto (x_i, cl_i)$ such that the following constraints are respected:

- Any two lectures with one or more common teachers cannot be given simultaneously:

$$\sum_{i,j} [x_{i,j} = k][t_m \in t(l_i)] \leq 1, \quad 0 \leq k \leq D * H - 1, \quad 1 \leq m \leq |T|.$$

- Any two lectures cannot be given simultaneously in the same classroom:

$$\sum_{i,j} [x_{i,j} = k][cl(l_i) = c_m] \leq 1, \quad 0 \leq k \leq D * H - 1, \quad 1 \leq m \leq |C|.$$

- Any two lectures of the same group cannot be given simultaneously:

$$\sum_{i,j} [x_{i,j} = k][l_i \in g_m] \leq 1, \quad 0 \leq k \leq D * H - 1, \quad 1 \leq m \leq |G|.$$

- Two lectures of the same course cannot be given in the same day:

$$sub(l_i) = sub(l_j) \Rightarrow x_i/H \neq x_j/H, \quad 1 \leq i, \quad j \leq |L|.$$

- The quality of a timetable is calculated according to three criteria which are linearly combined with certain weights to produce the objective function which is to be minimized:

  • Uniform distribution of the teaching hours for every lecture group during the week. For each lecture group, this is the difference between the maximum and minimum number of teaching hours in a day during the teaching week:

$$\sum_{g \in G} \left(\max_{0 \leq d \leq D-1} \left\{ \sum_{l_i \in g} d_i [x_i/H = d] \right\} - \min_{0 \leq d \leq D-1} \left\{ \sum_{l_i \in g} d_i [x_i/H = d] \right\} \right).$$

  • Gaps between lectures of the same group during each day:

$$\sum_{g \in G} \sum_{0 \leq d \leq D-1} \max(\{x_i + d_i + 1 : x_i/H = d, l_i \in g\} \cup \{0\})$$
$$- \min(\{x_i : x_i/H = d, l_i \in g\} \cup \{0\}) - \sum_{l_i \in g} d_i$$

- Distances between lectures of the same course. As mentioned before, it is desired for these lectures to have a reasonable distance between them for educational purposes. This is calculated through a user-defined function $pen : [1..D-1] \mapsto \{0, 1, \ldots\}$ as

$$\sum_{s \in S} pen(\max\{x_i/H : sub(l_i) = s\} - \min\{x_i/H : sub(l_i) = s\}).$$

The low-level similarities between this and the student–course model are obvious. The way that the specialities of this case are implemented on top of the core object model is outlined in the next section.

5.3 Subclasses

The basic unary resource class is subclassed to provide classes for the teacher and lecture groups representation. Some extra features are added to provide new characteristics for these classes. In the case of teachers, for example, the number of days with teaching activities and the maximum number of continuous teaching hours in every day are needed. Thus, two new features, which are calculated appropriately, are added to this class:

```
class c_Teacher : public c_UnaryResource {
  IlcIntVar OccupiedDays;
  IlcIntVar MaxContinuous; };
```

In the case of lecture groups, two extra features are needed, namely the measure of uniform distribution of teaching hours through the days of the week and the sum of holes between lectures for all the days of the week:

```
class c_LectureGroup : public c_UnaryResource {
  protected:
  IlcIntVar Difference;
  IlcIntVarArray Holes; };
```

Penalties are associated with distances between lectures of the same subject. These penalties are provided by the user:

```
IlcIntArray Penalties;
if (nbLectures==1)
  Penalty=IlcIntVar(m,0,0);
else {
  IlcIntVar s1=IlcMax(StartVariables)/H;
  IlcIntVar s2=IlcMin(StartVariables)/H;
  Difference=s1-s2;
  IlcIfThen(Difference==k, Penalty==Penalties[k-1]); }
```

Classrooms are just a multiple resource and no extra members are needed for this class.

6   Search

One of the reasons that makes constraint programming attractive is the fact that logical implications stemming from the variables, their possible values and the constraints that are posted upon them are carried out in an implicit way without the programmer's manual intervention. Values that are possible for certain variables are ruled out if they violate any of the constraints. In theory, these implications can be carried out until all inconsistent combinations of values for the variables are ruled out. However, this is a task of exponential complexity, so, in practice, implications are calculated only to a limited extent and the task of finding solutions is accomplished through search. Usually, the search space of a problem is viewed as a tree, where each decision point represents a variable and the edges towards its children are its possible values.

In this case, the variables are the starting times for each lecture. Although there are variables for classrooms too, in our approach, the main decision is the one concerning time. After that decision is made, the lecture is placed on the most appropriate classroom, but that is not regarded as a decision or, in other words, no backtracking occurs for assigning a value to a classroom variable.

Two factors influence the efficiency of the search. The first one is the heuristic rules which involve the selection of a variable to be instantiated next (variable-ordering heuristics) and the selection of a value for that variable (value-ordering heuristics). Heuristics form the actual search tree by labeling each node with a variable and each edge with a value. Usually, the choices of the value-ordering heuristic are depicted by ordering the possible values from left to right, with the leftmost edge being the one selected first by the heuristic. The second factor is the search method which controls the order in which the nodes of the search tree are going to be examined.

6.1   Search Methods

Some of the most popular and efficient search methods were implemented, namely DFS, iterative broadening (IB) [14], LDS [18] and depth-bounded discrepancy search (DDS) [29]. These search methods evolved from the need to exploit the heuristic rules used in the search in the best possible way.

DFS implements the obvious way to explore a search space by examining all the leaves of the search tree from left to right and backtracking if needed. Although constraint propagation prunes hopeless parts of the search space, there is always the danger for such an approach to get trapped in a shallow part of the search tree. One of the reasons for this is that DFS considers each of the decisions of assigning a value to a variable of equal importance. That means that the first choice of a value for a variable made by the value-ordering heuristic is thought of as having the same probability to lead to a solution as the second or any of the latter decisions. IB narrows the search space by searching only the subtree which includes a certain number of the leftmost decisions of the value-ordering heuristic, practically exploring a subtree of limited width. This follows from the assumption that the first choices of the value-ordering heuristic are most probable to lead to a solution. In case this assumption is false, then the width to which the search was limited is incremented and the whole process starts from the beginning.

LDS gives even more importance to the heuristic by assuming that it makes none or just a few errors and using the total number of discrepancies or the deviations from the heuristic's decisions as a guide for the search. In a first iteration, the number of such discrepancies is assumed to be zero, thus only the leftmost decision of the heuristic is considered for each node. If the assumption proves to be false, then the number of discrepancies is incremented assuming that the heuristic might make at most one error and the process is repeated.

DDS is also a discrepancy-based search method. LDS revisits nodes that were examined in earlier iterations and DDS uses an algorithm that examines each node only once.

Both discrepancy-based methods are heavily depending on the heuristic's accuracy. However, it is not always the case that a heuristic making no more than, let us say, 5% erroneous decisions can be implemented. Actually, it is usual for heuristics to get confused deep in the search tree. Usually, this is tackled by adding a lookahead parameter of a certain depth in the method, allowing subtrees of that depth at the bottom of the search tree to be explored fully without taking the discrepancies that occur there into account. The trust of such methods in the heuristic's accuracy is also expressed by the assumption that heuristics usually make errors high in the search tree where decisions are not so informed. In our approach, the opposite was also implemented in LDS's case: that is a version that assumes that the heuristic fails deeply in the search tree.

In our implementation, it was also possible to loosen the confidence on the heuristic. For example, IB increases the width bound by one in every iteration and the same happens with the number of allowed discrepancies in LDS. We implemented versions that allowed bounds to be variable, thus allowing the search to explore wider areas of the search space in every iteration. This seems to contradict the very essence of these methods and that might be true in feasibility problems, but there is a good reason for this; these methods were designed for feasibility problems. In other words, they were designed assuming that the heuristic makes choices towards finding a solution and not necessarily a good one. On the other hand, timetabling problems are usually optimization problems, so the heuristics are targeted towards the quality of the solutions. In the experiments that follow, it will be seen that there is a trade-off between the ease in finding a solution and its quality. Since we are interested in quality, it is reasonable to examine how these search methods can be extended to handle heuristics towards better solution and how that affects their ability to find one.

6.2 Heuristics

The celebrated first-fail [17] and Brelaz [4] variable ordering heuristics are employed in our implementation. Also, some other general-purpose heuristics, the kappa family of heuristics for constraint satisfaction problems, proposed in [13], are also included. The first-fail selection criterion is supposed to follow the rule "in order to succeed, try first where you are most likely to fail". Although it is argued that selecting the variable with the least values in its domain leads to harder subproblems, this heuristic is efficient in many cases. The Brelaz heuristic extends it by tie-breaking on the number of neighboring unbound constraint variables in the constraint graph. The kappa family of heuristics is based on a measure of difficulty of a constraint satisfaction problem (CSP), namely the parameter kappa. The kappa heuristic selects the variable that is supposed to lead to an easier subproblem when instantiated. Since it is costly to compute, two approximations are proposed, the $E(N)$ and the rho heuristics, the former being closer to the original kappa measure. Since these parameters are extensively described in [13], we will describe them only briefly for the sake of completeness. The kappa parameter expresses the constrainedness of a problem, or else the difficulty of finding a solution for it, and is defined as follows; if V is the set of variables, C the set of constraints on those variables, C_i the set of constraints on variable i and m_v the cardinality of variable's v domain, then the parameter kappa of the problem is equal to

$$\kappa = \frac{-\sum_{c \in C} \log(1 - p_c)}{\sum_{v \in V} \log(m_v)}$$

where p_c is the tightness of a constraint involving two variables, that is the percentage of the combinations of values of the two variables that are ruled out by the constraint. Since this parameter is hard to compute, two approximations are provided; $E(N)$, which approximates the expected number of solutions for a problem

$$E(N) = \prod_{v \in V} m_v \times \prod_{c \in C} (1 - p_c)$$

and ρ, which approximates the solution density

$$\rho = \prod_{c \in C} (1 - p_c).$$

These parameters can be used as heuristic information in the following manner: since a problem with bigger kappa is tougher than one with a lower one, we should branch on the variable which would lead to an easier problem if instantiated. The same idea holds for $E(N)$ and ρ as well. The only practical problem in calculating these parameters in a real problem is that they were expressed for binary CSPs. However, we can use the assumption that the dominating constraint in timetabling problems is that no two activities demanding some common resource can happen simultaneously, which in general holds. Thus, we can easily calculate the parameter p_c for any pair of lectures demanding any common resource. Of course, we will have to ignore disjunctive resources – classrooms in our case – in our calculations, but that is not of great importance, since timetabling problems can also be solved in two phases, taking classrooms into account in the second one.

Although the literature is rich in general-purpose variable ordering heuristics, the same does not hold for value ordering ones. That is to be expected in optimization problems, since such heuristics have to do with the quality evaluation criteria and thus depend on the problem instance. In our case, heuristics making decisions according to the objective function of the DIT/UoA were implemented among which was one choosing the value that caused the minimum increase in the lower bound of the objective. Also, heuristics aiming at finding a feasible solution were implemented. Following the idea behind the kappa heuristics and the effectiveness of first-fail, a heuristic that chose the value for the current variable that maximized the product of the values of the remaining variables, as proposed in [12], was also implemented.

7   Experimental Results

Experiments were carried out on a Sun SPARCserver 1000 under SunOS 5.6 and with 256 MB of main memory. The problem to be solved involved 68 lectures that had to be scheduled in five days of nine teaching periods, each within four classrooms. The total duration of the lectures is 187 hours.

The heuristics used for variable selection were first-fail, Brelaz and the kappa family of heuristics Rho, $E(N)$ and Kappa, denoted FF, BR, R, E and K respectively. Many different value selection heuristics were used in preliminary experiments. The most successful proved to be one that chose the value which would lead to the least increase in the lower bound of the minimization objective. Since the objective is represented by a constrained variable and the problem is a minimization one, an increase in the lower bound of the objective value means a certain decrease in the quality of the partial solution constructed until that point.

In Figure 1, a quality (smaller values represent higher qualities) versus time (measured in seconds) plot for DFS with some variable selection heuristics is shown. A major drawback of DFS can be observed there: there is not much improvement after the first solution is found. It can also be seen that the classic heuristics FF and BR are faster in finding a first solution than E and R (no solution was found with K). Drawing conclusions from just one data set is risky, but a first intuition is that the simple heuristics FF and BR can be more efficient in this specific problem than other more sophisticated and generic heuristics because of the following reason: the dominant type of constraint in course timetabling is a disjunctive constraint between lectures which has, more or less, the same effect in every pair of mutually exclusive lectures. Thus, what remains is the number of values in every variable's domain.

The LDS application on timetabling was not problem free. The first plot in Figure 2 shows the progress of search with LDS. It can be observed that the first solution is found rather late (an order of magnitude later than DFS) and the quality is not that good. The last solution found is quite satisfactory, but is found too late. This is a problem that should be expected; such methods were designed to address feasibility problems assuming that special heuristics would

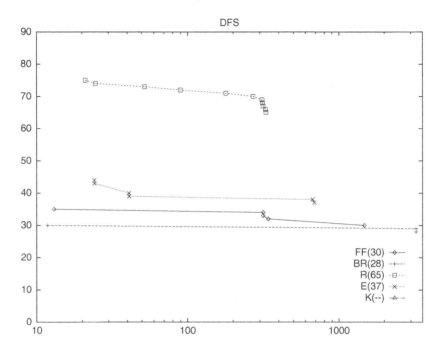

Fig. 1. DFS with varying variable selection heuristics

assist the search for a solution. Truly, the second plot shows the progress of search with LDS and a value selection heuristic targeting on feasibility. It is obvious that such a heuristic leads very quickly to a solution. On the other hand, the solution's quality is not satisfactory at all and that should be expected, since the value selection heuristic focuses on feasibility. In order to overcome these problems, the original method was modified. The third plot shows the progress of search with a modified version of LDS; the method assumes that the heuristic fails lower in the search tree. With that addition alone the search is much faster. The last plot shows one more addition: in every LDS iteration the discrepancy limit is not increased by a step of one, but by a step of three discrepancies. That means that less trust is justified upon the heuristic and its choices and the method is allowed to explore wider areas of the search space. There it can be observed that although LDS is somewhat slower than DFS in the beginning, it soon manages to find solutions of equal quality and, eventually, even better ones having the ability to escape from local minima.

Local search is a search method with great ability to escape from local minima and very popular in optimization problems. In [25], a local search variant for constraint programming is proposed, named large neighborhood search (LNS). The main idea is to iteratively keep a part of a solution stable while leaving the remaining variables unbound and thus performing a full search in a narrowed search space. Results are shown in Figure 3. It can be seen that this method with a good starting solution leads to the best-quality solution of the experiments

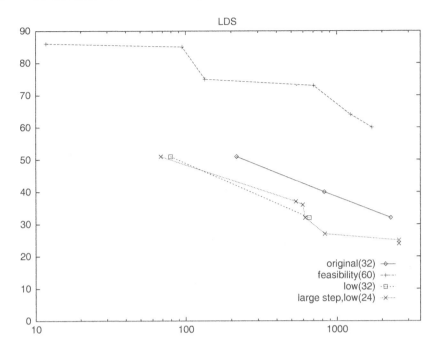

Fig. 2. LDS and variations with FF

presented. However, it should be noted that LNS alone is not able to provide as good solutions as a direct tree search. As Figure 3 shows, LNS needs more than 1000 s to reach a solution of quality slightly worse than the one found by a direct tree search guided by good heuristics in something more than 10 s. That is easily explained: LNS can lead only to minor local improvements. On the other hand, global decisions made by heuristics taking into account the problem as a whole can lead to larger improvements at the beginning of the search. Of course, direct tree search will eventually get stuck in a local optimum and that is the point where post-processing methods like LNS can be effectively used.

Another factor of interest is scalability. In this direction, we decided to use artificial data sets created on the real ones we used. Data sets up to six times bigger than the original data set used were created in the following way: the teachers and classrooms were kept the same while the number of lectures given and the total number of available time slots were multiplied by the corresponding factor. For example, to create a data set twice as large as the original, we created a timetabling problem with 10 instead of 5 days with 136 instead of 68 lectures while keeping all the other elements untouched. Then, of course, it would not be of interest to compare the objective value since the problems would be different. One parameter that could be measured and show how the approach scales with regard to the problem's size is the time needed to reach a first solution. Figure 4 shows how the time needed for reaching a first solution scales for the original problem mentioned above and for problems up to six times larger in size. All runs used plain DFS. The figure shows that time rises no faster than n^3 but faster than n^2. That is to be expected in constraint programming. The complexity of maintaining bound consistency is ed, where e is the number of constraints and d is the cardinality of a variable's domain. Increasing the problem size linearly involves increasing d linearly, but also increasing the number of variables linearly which leads to a quadratic increase of their interrelations or, in other words, constraints between them. This increase in running time is not prohibitive, since algorithms that scale polynomially are generally acceptable. We should also note that since the problem of constructing a timetable for a university occurs twice a year in most cases, there are no real-time constraints on the running time of such an application. Also, the running time can further be reduced by using more sophisticated propagation algorithms for the model's constraints, like network flow algorithms, for example.

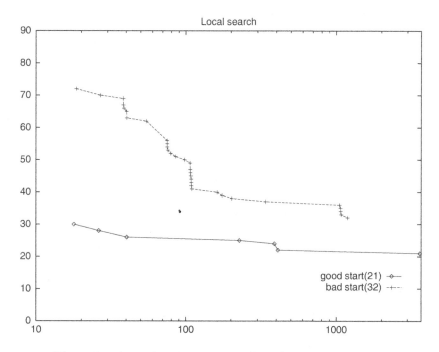

Fig. 3. Local search with varying quality of starting solutions

8   Conclusions

In this paper, we showed that it is possible to define, by exploiting the facilities offered by object-oriented design, a generic model for the university course timetabling problem, which might form the basis for dealing with the timetabling

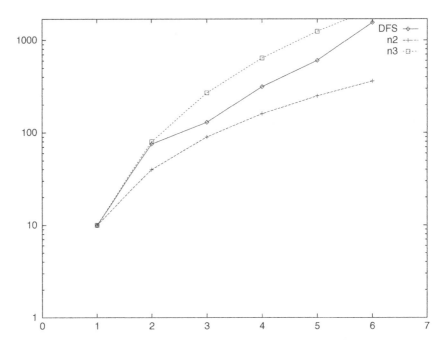

Fig. 4. Scaling of time for the first solution w.r.t. problem size

problems faced by most academic departments. We demonstrated the extensibility of the core model by applying it to the case of the DIT/UoA. The fact that the core model was extended might lead to the conclusion that it is too general, since too much extra needs to be implemented in order to cover any specific case. But, in order to produce the class c_Teacher from c_UnaryResource, 13 lines of code were needed. In the case of c_LectureGroup, 20 lines were needed. The aim of the above model was not to be able to cover any specific case, but to provide the framework on which any case could be covered. From a software engineering point of view, all these little additions for calculating holes between lectures or uniform distribution of lectures during a week or whatever could be stored in libraries and a user could pick from a large collection of such building blocks to construct classes that would represent the entities to model the problem to be solved, minimizing the amount of extra effort.

What is more, the whole idea is based on the constraint programming framework, since the tool employed is the ILOG SOLVER C++ library. Thus, the user is allowed to express hard and soft constraints with the same ease. For example, if one is concerned with the teaching hours of a teacher in a certain day, it is equally easy to post a hard constraint on the maximum number of these hours or add a penalty in the objective function, if that amount exceeds a certain limit, or associate preferences with the values of that variable.

A variety of search methods and variable ordering heuristics were implemented to be used for the search for near-optimum solutions. All search methods which were implemented behaved as they were expected to in theory. DFS was able to find feasible solutions fast, but could not improve them significantly in the long run. LDS, with some minor modifications aiming at the resolution of small conflicts at the bottom of the search tree, showed better long-term behavior, but was sometimes slow in finding a first solution. We expect LDS performance to scale better as the problem size increases. DDS would not be our method of choice, due to its lack of accepting modifications like LDS. IB did not improve DFS's performance significantly.

Concerning variable ordering heuristics, the celebrated first-fail and Brelaz heuristics performed better than any other variable selection heuristic. The negative result of our experiments was the bad performance of the kappa family of heuristics, which in some cases did not even lead to a feasible solution within a reasonable time limit. One explanation might be that the kappa measure is an approximate measure of a CSP's difficulty, which takes into account several factors, but mainly the number of values in each variable's domain and the constraint tightness between pairs of variables, that is the percentage of the total combinations between values of these two variables that are actually legal. In the timetabling problem, however, the tightness between any two variables sharing the same resources is constant. In other words, if some lectures are sharing the same resources the tougher one to schedule would be the one having the fewer possible slots regardless of the number of lectures, explaining why first-fail and Brelaz apply well on our problem.

As for value ordering heuristics, no general heuristic was applied. Heuristics targeted towards the DIT/UoA were used, that is heuristics that take into account holes between lectures, the distribution of lectures during the days of the week and the distances in days between lectures of the same subject. These heuristics worked fine by guiding the search towards feasible solutions of satisfying quality. At first glance, such heuristics do not have to do with feasibility and so the fact that the search finally reached feasible solutions might be considered coincidental. However, one can notice that the objective contains a factor expressing the number of holes between lectures, so that the search is guided towards more compact schedules. If the objective did not contain such a term, then a heuristic taking into account both the objective and the feasibility factor should be used. The feasibility factor could be measured by counting holes between lectures or, even better, with a heuristic calculating the product of the cardinality of all remaining variables' domains, as mentioned in the previous section concerning heuristics.

The problem of using tree search in optimization is that the search, either naive or sophisticated, will eventually get stuck in a local optimum. Using LNS helped in every case with our experiments. In no case, however, could LNS outperform direct tree search alone, so the best way to use it, in our opinion, is as a post-processing method after a normal tree search has been performed and a reasonable time cutoff limit reached.

References

1. Badri, M.: A Two-Stage Multiobjective Scheduling Model for [Faculty-Course-Time] Assignments. Eur. J. Oper. Res. **94** (1996) 16–28
2. Barbadym, V.A.: Computer-Aided School and University Timetabling: The New Wave. In: Burke, E., Ross, P. (eds.): Proc. 1st Int. Conf. on the Practice and Theory of Automated Timetabling. Lecture Notes in Computer Science, Vol. 1153. Springer-Verlag, Berlin Heidelberg New York (1995) 22–45
3. Boufflet, J.P., Negre, S.: Three Methods Used to Solve an Examination Timetable Problem. In: Burke, E., Ross, P. (eds.): Proc. 1st Int. Conf. on the Practice and Theory of Automated Timetabling. Lecture Notes in Computer Science, Vol. 1153. Springer-Verlag, Berlin Heidelberg New York (1995) 327–344
4. Brelaz, D.: New Methods to Color the Vertices of a Graph. JACM **22** (1979) 251–256,
5. Burke, E., Jackson, K., Kingston, J., Weare, R.: Automated University Timetabling: The State of the Art. Comput. J. **40** (1997) 565–571
6. Chahal, N., de Werra, D.: An Interactive System for Constructing Timetables on a PC. Eur. J. Oper. Res. **40** (1989) 32–37
7. de Werra, D.: An Introduction to Timetabling. Eur. J. Oper. Res. **19** (1985) 151–162
8. de Werra, D.: The Combinatorics of Timetabling. Eur. J. Oper. Res. **96** (1997) 504–513
9. Elmohamed, S., Coddington, P., Fox, G.: A Comparison of Annealing Techniques for Academic Course Scheduling. In: Burke, E., Ross, P. (eds.): Proc. 2nd Int. Conf. on the Practice and Theory of Automated Timetabling. Lecture Notes in Computer Science, Vol. 1408. Springer-Verlag, Berlin Heidelberg New York (1997) 92–112
10. Erben, W., Keppler, J.: A Genetic Algorithm Solving a Weekly Course-Timetabling Problem. In: Burke, E., Ross, P. (eds.): Proc. 1st Int. Conf. on the Practice and Theory of Automated Timetabling. Lecture Notes in Computer Science, Vol. 1153. Springer-Verlag, Berlin Heidelberg New York (1995) 198–211
11. Frangouli, H., Harmandas, V., Stamatopoulos, P.: UTSE: Construction of Optimum Timetables for University Courses – A CLP Based Approach. In: Proc. 3rd Int. Conf. on the Practical Applications of Prolog (1995) 225–243
12. Geelen, P. A.: Dual Viewpoint Heuristics for Binary Constraint Satisfaction Problems. In: Proc. 10th Eur. Conf. on Artificial Intelligence (1992) 31–35
13. Gent, I.P., MacIntyre, E., Prosser, P., Smith, B.M., Walsh, T.: An Empirical Study of Dynamic Variable Ordering Heuristics for the Constraint Satisfaction Problem. In: Proc. 2nd Int. Conf. on the Principles and Practice of Constraint Programming (1996) 179–193
14. Ginsberg, M.L., Harvey, W.D.: Iterative Broadening. Artif. Intell. **55** (1992) 367–383
15. Gotlieb, C.: The Construction of Class–Teacher Timetables. In: Proc. IFIP Congress (1962) 73–77
16. Gueret, C., Jussien, N., Boizumault, P., Prins, C.: Building University Timetables Using Constraint Logic Programming. In: Burke, E., Ross, P. (eds.): Proc. 1st Int. Conf. on the Practice and Theory of Automated Timetabling. Lecture Notes in Computer Science, Vol. 1153. Springer-Verlag, Berlin Heidelberg New York (1995) 130–145

17. Haralick, R.M., Elliott, G.L.: Increasing Tree Search Efficiency for Constraint Satisfaction Problems. Artif. Intell. **14** (1980) 263–313,
18. Harvey, W.D., Ginsberg, M.L.: Limited Discrepancy Search. In: Proc. 14th Int. Joint Conf. on Artificial Intelligence (1995) 607–613
19. Henz, M., Wurtz, J.: Using Oz for College Timetabling. In: Burke, E., Ross, P. (eds.): Proc. 1st Int. Conf. on the Practice and Theory of Automated Timetabling. Lecture Notes in Computer Science, Vol. 1153. Springer-Verlag, Berlin Heidelberg New York (1995) 162–177
20. ILOG S.A.: ILOG Solver 4.4: Reference Manual (1999)
21. ILOG S.A.: ILOG Solver 4.4: User's Manual (1999)
22. Lajos, G.: Complete University Modular Timetabling Using Constraint Logic Programming. In: Burke, E., Ross, P. (eds.): Proc. 1st Int. Conf. on the Practice and Theory of Automated Timetabling. Lecture Notes in Computer Science, Vol. 1153. Springer-Verlag, Berlin Heidelberg New York (1995) 146–161
23. Mehta, N.: The Application of a Graph Coloring Method to an Examination Scheduling Problem. Interfaces **11** (1981) 57–64
24. Rich, D.: A Smart Genetic Algorithm for University Timetabling. In: Burke, E., Ross, P. (eds.): Proc. 1st Int. Conf. on the Practice and Theory of Automated Timetabling. Lecture Notes in Computer Science, Vol. 1153. Springer-Verlag, Berlin Heidelberg New York (1995) 181–197
25. P. Shaw. Using Constraint Programming and Local Search Methods to Solve Vehicle Routing Problems. In: Proc. 4th Int. Conf. on the Principles and Practice of Constraint Programming (1998) 417–431
26. Stamatopoulos, P., Viglas, E., Karaboyas, S.: Nearly Optimum Timetable Construction Through CLP and Intelligent Search. Int. J. Artif. Intell. Tools **7** (1998) 415–442
27. Thompson, J., Dowsland, K.: General Cooling Schedules for a Simulated Annealing Based Timetabling System. In: Burke, E., Ross, P. (eds.): Proc. 1st Int. Conf. on the Practice and Theory of Automated Timetabling. Lecture Notes in Computer Science, Vol. 1153. Springer-Verlag, Berlin Heidelberg New York (1995) 345–363
28. van Hentenryck, P., Deville, Y., Teng, C.M.: A Generic Arc-Consistency Algorithm and its Specializations. Artif. Intell. **57** (1992) 291–321
29. Walsh, T.: Depth-Bounded Discrepancy Search. In: Proc. 15th Int. Joint Conf. on Artificial Intelligence (1997) 1388–1393
30. White, G., Zhang, J.: Generating Complete University Timetables by Combining Tabu Search with Constraint Logic. In: Burke, E., Ross, P. (eds.): Proc. 2nd Int. Conf. on the Practice and Theory of Automated Timetabling. Lecture Notes in Computer Science, Vol. 1408. Springer-Verlag, Berlin Heidelberg New York (1997) 187–198

A Co-evolving Timeslot/Room Assignment Genetic Algorithm Technique for University Timetabling

Hiroaki Ueda, Daisuke Ouchi, Kenichi Takahashi, and Tetsuhiro Miyahara

Faculty of Information Sciences, Hiroshima City University
Hiroshima 731-3194 Japan
{ueda, ouchi, takahasi, miyahara}@rea.its.hiroshima-cu.ac.jp

Abstract. We present a two-phase genetic algorithm (TGA) to solve timetabling problems for universities.[1] Here, we use two kinds of populations. The first population is related to class scheduling, and the second one is related to room allocation. These populations are evolved independently, and a cost value of each individual is calculated. Then several individuals with the lowest costs are selected from each population, and these individuals are combined in order to calculate the fitness values. To evaluate the performance of TGA, we apply TGA to several timetabling problems and compare results obtained by TGA with those obtained by the simple GA (SGA). For the timetabling problem based on the curriculum of the Faculty of Information Sciences at Hiroshima City University, TGA finds a solution that satisfies all constraints, but SGA cannot find a feasible solution. From the results for problems generated by an automatic timetabling problem generator, we show that TGA obtains a better solution than the simple GA when the room utilization ratio is high.

1 Introduction

It is well known that the timetabling problem is difficult to solve, since the problem has a huge search space and is highly constrained. Thus, various optimization methods such as simulated annealing techniques, tabu search techniques, and genetic algorithm techniques have been applied to find feasible timetables [1], [2], [3], [4], [5], [6], [7]. In this paper, we present genetic algorithms to obtain a solution to the timetabling problem for universities.

Here, we define a timetabling problem as finding both a class schedule and room allocation which do not conflict with constraints given by users. According to the definition, the main tasks to solve a timetabling problem are class scheduling and room allocation. Since these tasks are closely related, it is hard to complete these tasks separately. However, it seems inefficient to complete these tasks simultaneously. As one method to solve this difficulty, we present a

---

[1] This research is partly supported by Research Grant from Hiroshima City University.

E. Burke and W. Erben (Eds.): PATAT 2000, LNCS 2079, pp. 48–63, 2001.

technique that uses GA in two phases. We call this method TGA. TGA consists of two evolution phases and uses two types of genotypes. In the evolution phases, the GA-based method performs evolutionary operations for two types of populations. Next, two types of populations are combined to calculate the fitness values, and the values are assigned to the populations. Through computer simulations using the curriculum of Hiroshima City University, we show that TGA finds a feasible solution. Moreover, we apply TGA and the simple GA to timetabling problems which are generated by an automatic timetabling problem generator, and we compare the performance of TGA with that of the simple GA.

The rest of this paper is organized as follows. In the next section, we define the timetabling problem which we focus on in our study. In Section 3, we present TGA and some experimental results are shown in Section 4.

2   Timetabling Problem

Here, we focus on the timetabling problem for Japanese universities and define the timetabling problem as finding both a class schedule and room allocation. Thus, not only assumptions for class scheduling but also assumptions for room allocation exist. The following is a list of assumptions based on a typical timetabling problem for Japanese universities.

(A1) A class consists of a number of students in the same year grade in a department.
(A2) A subject is taught once a week.
(A3) Multiple-class subjects are allowed.
(A4) Multiple-period subjects are allowed.
(A5) Subjects taught by multiple-teacher are allowed.
(A6) There are several kinds and sizes of rooms.

Table 1 shows an example of a timetable. Multiple-class subjects such as Subject 1 must be allocated to the same timeslot. Subjects 2, 4 and 6 are examples of multiple-period subjects, and they use consecutive periods. Subjects 4 and 6 are examples of subjects taught by multiple-teacher. In order to obtain a solution that does not conflict with these assumptions, some constraints must be considered. We call these constraints basic constraints. A list of basic constraints that we use is the following.

C1) No teachers appear more than once in a period.
C2) Multiple-class subjects must be allocated to the same periods.
C3) Multiple-period subjects must use consecutive periods.
C4) A room whose size and kind are suitable for a subject should be allocated to the subject.
C5) No rooms appear more than once in a period.

Since we consider not only class scheduling but also room allocation, we deal with three kinds of basic constraints: constraints for class scheduling, room

Table 1. An example of a university timetable

Class	Period 1	Period 2	Period 3	Period 4	Period 5
A1		**Subject-1** **Staff-1** **Medium-1**	Subject-4 All associate professor of Dept. A Computer_room-1		
B-1		Subject2 Staff-2 Computer_room-1	Subject-5 Staff-4 Small-4		
C-1		**Subject-1** **Staff-1** **Medium-1**	Subject-6 Staff-5 and Staff-6 Small-2		
D-1	Subject-3 Staff-3 Small-1	Subject-7 Staff-7 Small-4	Subject-8 Staff-3 Small-1		

allocation and room clashing. In the above list, C1, C2 and C3 are constraints for class scheduling. C4 and C5 are constraints for room allocation and room clashing, respectively.

As well as basic constraints, we consider additional constraints. These constraints are considered as requests of professors. A list of additional constraints made in this study is the following.

C6) Some teachers should not appear at some periods.
C7) Some rooms cannot be used at particular periods.
C8) For some classes, no subjects appear at particular periods.
C9) Some subjects should be taught immediately after a particular subject is taught.

C6 is a constraint for the professor meeting or the department meeting. In Japanese universities, there are some subjects that are taught by part-time teachers, and periods for these subjects tend to be fixed. In order to reserve rooms and periods for these subjects, we use C7 and C8. Some subject pairs should be treated as a multiple-period subject. C9 is a constraint for these subjects. Similar to basic constraints, C6, C8 and C9 are constraints for class scheduling, and C7 is a constraint for room allocation. To find a timetable (a class schedule and room allocation) that satisfies all of the constraints mentioned above is the subject of our timetabling problem.

3   Genetic Algorithms

3.1   The Simple GA

To compare the performance of TGA with that of the most fundamental GA, we apply the simple GA (SGA) to the timetabling problem. The genotype is

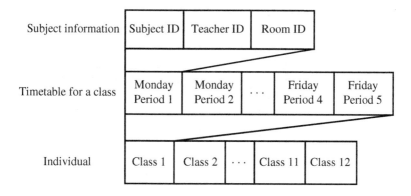

Fig. 1. The genotype for SGA

illustrated in Figure 1. In SGA, genetic operations such as selection, crossover and mutation are applied to individuals repeatedly. These operations are almost the same as those for TGA. Thus, details about these operations are described in the next section.

3.2   The TGA

In TGA, we use two types of genotypes. Figure 2 shows the genotypes. One is the genotype for class scheduling, and the other is the genotype for room allocation. The genotype for class scheduling consists of a timetable for each class, and each timeslot in a timetable consists of subject information. Subject information consists of subject ID and teacher ID. The genotype for room allocation consists of room ID.

Figure 3 shows the flow diagram of the TGA. In the first phase, genetic operations are applied to the population for class scheduling, and the cost values related to constraints for class scheduling are calculated. In the second phase, genetic operators are applied to the population for room allocation. Finally, we combine two types of individuals, and the fitness values of these individuals are calculated. These steps are repeated until the stopping criterion is satisfied. Next, we describe details of each genetic operation.

Class scheduling phase. Here we describe genetic operations for class scheduling.
1) Selection and Reproduction
We employ roulette selection and the elite keeping strategy. An individual with a low cost has a high probability that the individual is selected in the selection step. The elite is a pair of two types of individuals. Here we denote an individual for class scheduling as C_i, and an individual for room allocation as R_j. An individual pair is denoted as $< C_i, R_j >$. When the sum of the costs for $< C_i, R_j >$ is the minimum in the fitness calculation, the pair is treated as the elite. That is,

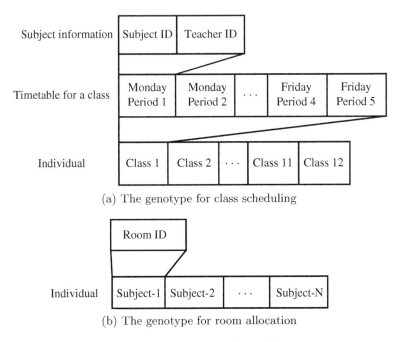

(a) The genotype for class scheduling

(b) The genotype for room allocation

Fig. 2. The genotypes for TGA

genetic operations such as crossover and mutation are not applied to C_i and R_j when $< C_i, R_j >$ is the elite.

2) Crossover

We use one-point crossover and crossover points are restricted to borderlines between timetables for any classes. Figure 4 shows an example of the crossover operation. In the figure, a crossover point is the borderline between Class 2 and Class 3.

After crossover, all subjects must appear exactly once due to this restriction, and it seems that a timetable for each class cannot be changed by the crossover operator. However, this problem is solved by employing a repair operation. By use of the crossover operator, many multiple-class subjects tend to be violated. The repair operator modifies timetables in order to repair the violated multiple-class subjects. Thus, the crossover operator also contributes to changing a schedule of each class.

3) Mutation

In mutation, subject information of a particular timeslot is replaced with the subject information of a randomly selected timeslot. If the subject of the timeslot is a multiple-period subject, then subject information in multiple timeslots is exchanged. Figure 5 shows an example of mutation.

4) Repair

After crossover and mutation, several multiple-class subjects may be violated.

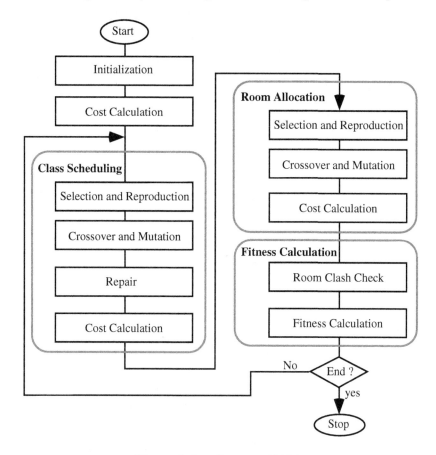

Fig. 3. A flow diagram of TGA

The repair operator tries to repair these subjects by moving the violated subject to a different timeslot if possible. When we cannot repair the violated subject, a penalty is given to the individual to whom the violated subject belongs. We will describe the penalty values when we outline the cost calculation step.

The operator does not repair the violated subjects when some multiple-class subjects or multiple-period subjects are newly violated by repairing the violated subjects. In our method, more complex subjects are repaired earlier than less complex subjects. Here, we define the complexity of a subject as $pC + c$, where p is the number of periods needed by the subject, C is the number of classes in the timetabling problem, and c is the number of classes that the subject is taught. According to the definition, multiple-period subjects that need many periods are assumed to be the most complex subjects, and multiple-class subjects that are taught to many students (classes) are assumed to be more complex subjects. In Table 2, Subject 2 is more complex than Subject 1, since the complexities of Subjects 2 and 1 are 10 and 7, respectively. To explain the behavior of the repair

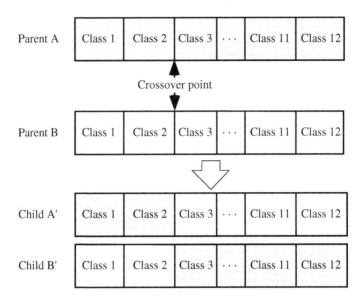

Fig. 4. Crossover for the genotype related to class scheduling

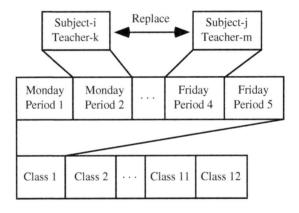

Fig. 5. Mutation for the genotype related to class scheduling

operation, we use this example. In Table 2, multiple-class Subjects 1 and 2 are violated. Since Subject 2 is more complex than Subject 1, repair of Subject 2 is attempted first. The repair operator tries to move Subject 2 for B-1 to periods 3 and 4, or Subject 2 for D-1 is moved to periods 1 and 2. When Subject 2 for B-1 is moved to periods 3 and 4, Subject 7 is newly violated. Thus, Subject 2 for D-1 is moved to periods 1 and 2. In the same manner, Subject 1 for A-1 is moved to period 3. Table 3 shows the timetable after the repair operation, where no multiple-class subjects are violated.

Since the repair operator is a simple operator, its computational costs are not high. However, many violated subjects can be repaired by the operation.

Table 2. An example of a timetable before the repair operation

Class	Period 1	Period 2	Period 3	Period 4	Period 5
A1		**Subject-1** **Staff-1**	Subject-5 Staff-5		Subject-6 Staff-5
B-1	**Subject-2** **Staff-2**		Subject-7 Staff-6 and Staff-7		
C-1			Subject-3 Staff-3	**Subject-1** **Staff-1**	
D-1	**Subject-1** **Staff-1**	Subject-4 Staff-4	**Subject-2** **Staff-2**		Subject-8 Staff-3

Table 3. An example of a timetable after the repair operation

Class	Period 1	Period 2	Period 3	Period 4	Period 5
A1			Subject-5 Staff-5	**Subject-1** **Staff-1**	Subject-6 Staff-5
B-1	**Subject-2** **Staff-2**		Subject-7 Staff-6 and Staff-7		
C-1			Subject-3 Staff-3	**Subject-1** **Staff-1**	
D-1	**Subject-2** **Staff-2**		**Subject-1** **Staff-1**	Subject-4 Staff-4	Subject-8 Staff-3

5) Cost calculation
In the class scheduling phase, only constraints for class scheduling are considered for the cost calculation. In Equation (1) $Scheduling\_cost(C_i)$ is defined to be a sum of penalty values:

$$Scheduling\_cost(C_i) = \sum_{\substack{k \in all\ constraints \\ for\ class\ scheduling}} Penalty\_value(k). \qquad (1)$$

When multiple-period subjects and multiple-class subjects are violated, the penalty values are evaluated according to the extent of its violation. Now, we assume that Subject-T is taught for n classes and the subject needs m consecutive periods. Moreover, we assume the subject uses $p(p \leq n)$ kinds of timeslots in a timetable. The penalty value for violation of the subject depends on m and p, and the value is defined as $(p-1)m$. In Table 2, m and p for Subject 1 are 3 and 1, respectively. Thus, the penalty value for violation of the subject is evaluated as $(3-1)1 = 2$. In the same manner, the penalty value for Subject 2 in the same table is evaluated as $(2-1)2=2$, since Subject 2 uses two kinds of timeslots: i.e. {Period 1, Period 2} and {Period 3, Period 4}.

Room allocation phase. In the room allocation phase, we use similar genetic operations as for the class scheduling phase except for mutation. Thus, we only describe the mutation operator here.

In mutation, we reassign a room to a mutated subject. Figure 6 shows a mutation for Subject-T. When room allocation for a subject is mutated, a room whose size is different to the size of the currently assigned room may be reassigned to the subject, but rooms with different kinds are not assigned to it. For example, we assume that Subject-T in Figure 6 needs a lecture room whose size is medium. When the room allocation for Subject-T is mutated, we may assign a lecture room whose size is not medium: that is, a big or small room may be assigned to the subject, but any computer rooms are not assigned to it.

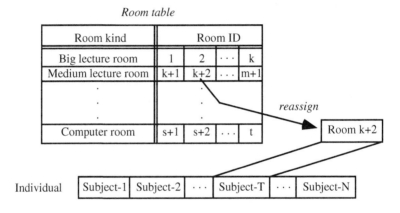

Fig. 6. Mutation for the genotype related to room allocation

Fitness calculation phase. In the fitness calculation step, the fitness values are calculated after the room clash check is done. Figure 7 shows the outline of the room clash check. In the room clash check, n individuals with the lowest costs are selected from two kinds of populations, and selected individuals are combined. Any individuals selected from the population for class scheduling are combined with all individuals selected from the population for room allocation. That is, n^2 individual pairs are made. For each pair, we check whether room clashes occur or not, and a room clashing cost is calculated. Here, we denote the room clashing cost for individual pair $< C_i, R_j >$ as $Room\_Clashing\_Cost(C_i, R_j)$, and we assume $Room\_Clashing\_Cost(C_i, R_j)$ is in proportion to the number of room clashes. By using $Room\_Clashing\_Cost(C_i, R_j)$, we update the cost values of all individuals for both populations. Equations (2) and (3) show the functions which update the cost values of individuals related to class scheduling, where $Class\_Cost(C_i)$ represents the updated cost value of C_i. When C_i is selected at the room clash check, we use Equation (2) for calculation of $Class\_Cost(C_i)$. That is, $Class\_Cost(C_i)$ equals the minimum value of the cost of room clash plus $Scheduling\_Cost(C_i)$. When C_i is not selected at the room clash check, $Class\_Cost(C_i)$ is calculated by Equation (3). In (3), the average value of the minimum costs of room clash is a penalty value for not being selected for the room clash check. When we select all individuals from the population for class scheduling, only (2) is used. However, we select some of individuals in our experiments in order to save computational costs.

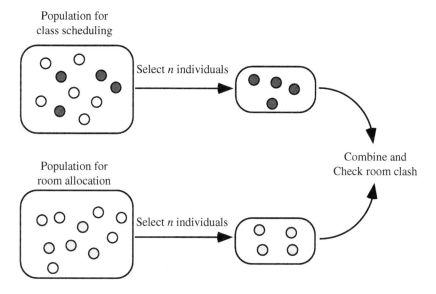

Fig. 7. Overview of the room clash check

As an example of updating the cost values, we consider that there are five individuals C_1–C_5, and C_1–C_3 are selected for the room clash check. We also assume that there are five individuals R_1–R_5, and that R_1–R_3 are selected for the room clash check. In this case, nine individual pairs $< C_i, R_j > (i, j \leq 3)$ are made, and we calculate the room clashing cost for each pair. Then, $Class\_Cost(C_i)$ for $i \leq 3$ is calculated by Equation (2), and $Class\_Cost(C_i)$ for $i \geq 4$ is calculated by Equation (3).

Using the updated cost $Class\_Cost(C_i)$, the fitness value $Class\_fitness(C_i)$ is calculated by Equation (4). The fitness values of individuals for room allocation are calculated in the same manner. The selection and the reproduction steps are performed by using these fitness values.

Although TGA may not be a co-evolution method in the strict sense, it is a co-evolution method in the sense that the fitness values for one population are closely related to those of the other population. For example, $Class\_Cost(C_i)$ may be less than $Class\_Cost(C_k)$ when $Scheduling\_Cost(C_i)$ is greater than $Scheduling\_Cost(C_k)$:

$$Class\_Cost(C_i) = \min_j(Room\_Clashing\_Cost(C_i, R_j))$$
$$+ Scheduling\_Cost(C_i). \qquad (2)$$

$$Class\_Cost(C_i) = \text{average}_k(\min_j(Room\_Clashing\_Cost(C_k, R_j)))$$
$$+ Scheduling\_Cost(C_i). \qquad (3)$$

$$Class\_fitness(C_i) = \max_k(Class\_Cost(C_k)) - Class\_Cost(C_i). \qquad (4)$$

4   Experimental Results

The methods mentioned above were implemented in the C language on a Sun Ultra10 (333MHz). Table 4 shows the timetabling problem that we used. This problem derives from the curriculum of the Faculty of Information Sciences at Hiroshima City University in 1997(HCU_97). There are many multiple-class subjects, some multiple-period subjects, and many room types in the problem. Table 5 lists the additional constraints. Due to these constraints, the problem becomes more complex.

Table 4. Parameters of the timetabling problem

	no. of subjects	93
	no. of teachers	97
	no. of classes	12
Subjects and teachers	no. of school days on a week	5
	no. of periods in a date	5
	no. of multiple-class subjects	21
	no. of multiple-period subjects	4
	no. of small classrooms	19
	no. of medium classrooms	12
	no. of large classrooms	18
Rooms	no. of very large classrooms	1
	no. of language laboratory rooms	3
	no. of laboratory rooms	7
	no. of computer rooms	3

Table 5. The number of additional constraints

Constraint type	No. of constraints
C6	20
C7	3
C8	76
C9	28

Table 6. Penalty values for room allocation

		The size of the allocated room			
		Small	Medium	Large	Very large
	Small	0.0	0.2	0.5	0.5
The adequate size	Medium	1.0	0.0	0.2	0.5
for the subject	Large	1.0	1.0	0.0	0.2
	Very large	1.0	1.0	1.0	0.0

Next, we detail the parameter values in our experiments. Crossover and mutation probabilities are 100% and 1%, respectively. The number of selected individuals at the room clash check is 10. The process is terminated when 500 generations are evolved or all constraints are satisfied. These parameter values are decided by performing preliminary experiments, and we obtain good results when we use these values. Apart from the constraints for room allocation, the penalty value is 1.0 per unsatisfied constraint. The penalty values for room allocation are shown in Table 6. When the size of the allocated room is smaller than the adequate size for a subject, the penalty value is 1.0. When the size of the allocated room is slightly larger than the adequate size for a subject, the penalty value is 0.2. We use 0.5 as the penalty value when the size of the allocated room is much larger than the adequate size for a subject.

Figure 8 shows the changes of the cost values for the elite. The curves indicate average values over ten trials. Obviously, TGA finds a better solution than SGA. Even in early generations, TGA finds the elite with smaller cost. In TGA, we combine two kinds of individuals with the lowest costs, thus TGA finds a better elite in the early evolution phase. Table 7 shows details of the experimental results. The worst (best) case in the table shows the cost of the elite at the final generation in the worst (best) case. The average cost in the table shows the average cost of the elites for ten trials. CPU time is the average CPU time for ten trials. Both for the worst and best cases, we obtain good results by TGA. In the best case, all constraints are satisfied.

Table 8 shows a part of the timetable obtained by TGA. This table shows the schedule and room allocation on Friday. In the table, Sxx represents Subject ID, and SRxx, MRxx, LRxx, LLxx and CR represent room IDs. Boldface subjects such as **S78** are multiple-class subjects, and subjects in italics such as *S99* are multiple-period multiple-class subjects. There are no violations of constraints nor room clashes. Table 9 shows the schedule on Thursday. In the table, Txx

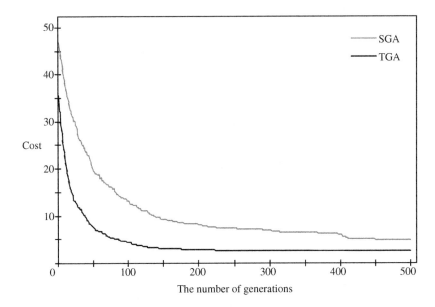

Fig. 8. Changes of the cost values of the elite

Table 7. Costs of the elite for ten trials

	TGA	SGA
The worst case	5.0	7.0
The best case	0.0	1.0
The average cost	2.5	4.8
CPU (s)	23.5	37.5

represents teacher ID, and boldface subjects are taught by professors. In our experiments, we consider the constraints that any professors must not have any classes on Thursday afternoon because of a professor meeting. Thus, there are no subjects taught by professors in the afternoon in Table 9. Subjects in italics such as *S33* and *S34* are treated as multiple-period subjects.

Finally, we applied TGA and SGA to several timetabling problems. These problems were generated by the automatic timetabling problem generator which is implemented in C. This tool generates timetabling problems randomly by using some parameters such as the number of classes, the average number of subjects per class, the room utilization ratio, and so on. Table 10 shows some parameter values that we used. For all problems in this table, the number of classes, the number of school days in a week, and the number of periods in a day are assumed to be 12, 5 and 5, respectively. The room utilization ratio is the probability that a room is used in a period. Thus, the room utilization ratio multiplied by the number of periods in a week gives the average frequency of the use of a room per week. Non-lecture rooms in Table 10 indicate special-purpose rooms such as

Table 8. An example of the timetable obtained by TGA (Friday)

Class	Period 1	Period 2	Period 3	Period 4	Period 5
A1	S4, MR12	S3, LR8			
B1	S18, LR7	S15, SR4		S20, LL2	**S22, MR11**
C1		S27, LL2			
D1					**S22, MR11**
A2			*S99, CR*	*S99, CR*	*S99, CR*
B2		S43, LR6	*S99, CR*	*S99, CR*	*S99, CR*
C2		S51, LR2	*S99, CR*	*S99, CR*	*S99, CR*
D2	S58, LL2		*S99, CR*	*S99, CR*	*S99, CR*
A3	S73, LR2	**S70, LR3**			
B3	**S78, LR1**	**S70, LR3**			
C3	**S78, LR1**				
D3			S89, LR5	S86, LR5	

Table 9. An example of the timetable obtained by TGA (Thursday)

Class	Period 1	Period 2	Period 3	Period 4	Period 5
A1					
B1	**S21, T9**	**S21, T9**			
C1					S24, T104
D1			*S33, T70*	*S34, T70*	S36, T106
A2	S44, T54		S38, T53	S45, T64	
B2	S48, T57		T38, T53	S45, T64	
C2				S45, T64	
D2	S63, T72	S97, T72		S45, T64	S62, T71
A3					S74, T56
B3		**S75, T7**			
C3	S81, T75			S84, T66	
D3	**S91, T27**		S90, T67		S74, T56

computer rooms. Since the difference between TGA and SGA rests in whether or not the genotype is split, we focus on the results when the room utilization ratio is changed. Thus, the only difference between these problems is the parameter values with respect to the room utilization ratios. Other parameter values are chosen in order to make the complexity of the problems equal to that of HCU_97. While the average number of multiple-class subjects is different from the number of multiple-class subjects in Table 4, the total number of timeslots needed by multiple-class subjects in the problems is almost equal to that in HCU_97.

Table 11 shows results for these problems. In the table, CC indicates the class cost of the elite, and RC equals the room cost plus the room clashing cost for the elite. That is, CC plus RC is equal to the total cost of the elite. In P1 and P2, TGA and SGA find the solution that satisfies all constraints. However,

Table 10. Parameters for generated timetabling problems

Parameter type	Problems			
	P1	P2	P3	P4
The average number of subjects per class	17.5	17.5	17.5	17.5
The average number of subjects per teacher	4.0	4.0	4.0	4.0
The average number of multiple-period subjects	3.5	3.5	3.5	3.5
The average number of multiple-class subjects	3.5	3.5	3.5	3.5
The room utilization ratio for lecture rooms	0.1	0.1	0.5	0.5
The room utilization ratio for non-lecture rooms	0.1	0.5	0.1	0.5

Table 11. Results for timetabling problems generated by ATPG

Problems	TGA			SGA		
	CC	RC	CPU (s)	CC	RC	CPU (s)
P1	0.0	0.0	3.6	0.0	0.0	9.1
P2	0.0	0.0	5.1	0.0	0.0	14.8
P3	0.0	0.0	38.4	1.0	3.0	48.8
P4	4.0	4.0	48.0	1.0	10.5	48.0

TGA finds better solutions than SGA when the average room utilization ratio is high.

5   Conclusion

We have described a genetic algorithm with two phases to solve a university timetabling problem. The key points of the TGA are the use of two types of genotypes for the evolution phases and combination of these genotypes for the fitness calculation. In the experiment for the timetabling problem in Hiroshima City University, TGA finds a feasible solution. For problems generated by the automatic timetabling problem generator, TGA does not find a feasible solution when the average room utilization ratio is high. However, we show that TGA find better solutions than the simple GA.

Improvement of the repair operator with the use of more intelligent methods such as a local search technique, and comparisons between TGA and other methods remain as important goals for future work.

References

1. Mitchell, M.: An Introduction to Genetic Algorithms. MIT Press, Cambridge, MA (1998)
2. Carter, M.W., Laporte, G.: Recent Developments in Practical Course Timetabling. In: Burke, E., Carter, M. (eds.): The Practice and Theory of Automated Timetabling: Selected Papers. Lecture Notes in Computer Science, Vol. 1408. Springer-Verlag, Berlin Heidelberg New York (1998) 3–19

3. Fukushima, M., Tanaka, S.: Adaptive Genetic Algorithms for Solving School Timetabling Problems. IEICE Trans. D-I, J82-D-I(6), (1998) 883–885 (in Japanese)
4. Rich, Daivid C.: A Smart Genetic Algorithm for University Timetabling. In: Burke, E., Ross, P. (eds.): The Practice and Theory of Automated Timetabling: Selected Papers. Lecture Notes in Computer Science, Vol. 1153. Springer-Verlag, Berlin Heidelberg New York (1998) 181–197
5. Erben, W., Keppler, J.: A Genetic Algorithm Solving a Weekly Course-Timetabling Problem. In: Burke, E., Ross, P. (eds.): The Practice and Theory of Automated Timetabling: Selected Papers. Lecture Notes in Computer Science, Vol. 1153. Springer-Verlag, Berlin Heidelberg New York (1998) 198–211
6. Burke, E., Newall, J-P.: Multi-Stage Evolutionary Algorithm for Timetable Problem. IEEE Trans. Evol. Comput. (1999)
7. Peachter, B., Rankin, R.C., Cumming, A.: Improving a Lecture Timetabling System for University-Wide Use. In: Burke, E., Carter, M. (eds.): The Practice and Theory of Automated Timetabling: Selected Papers. Lecture Notes in Computer Science, Vol. 1408. Springer-Verlag, Berlin Heidelberg New York (1998) 156–165

A Comprehensive Course Timetabling and Student Scheduling System at the University of Waterloo

Michael W. Carter,

Mechanical and Industrial Engineering,
University of Toronto,
5 Kings College Road,
Toronto, Ontario, Canada
M5S 3G8
carter@mie.utoronto.ca
http://www.mie.utoronto.ca/staff/profiles/carter.html

Abstract. This paper describes a comprehensive course timetabling and student scheduling system that was developed for the University of Waterloo between 1979 and 1985. The system is based on a "demand-driven" philosophy where students first chose their courses, and the system tries to find the best timetable to maximize the number of satisfied requests. The problem is first decomposed into small manageable sub-problems. Each sub-problem is solved in sequence using a greedy heuristic to assign times to sections, and a Lagrangian relaxation algorithm to assign classrooms. Timetable representatives from each department have interactive access to make final modifications. Finally, each student is individually assigned to the combination of course sections that maximizes timetable satisfaction and balances section sizes. The system has been used successfully for 15 years.

1 Introduction

For the purposes of this paper, course timetabling and student scheduling will be defined as the sequence of problems where each meeting of each course is assigned to a classroom and a time, and students are assigned to sections of each course such that they have no conflicts. At Waterloo, we did not automate the assignment of faculty to course sections; that process is normally done manually in advance.

Although the system was implemented 15 years ago, much of the discussion is still very relevant. In a recent survey of practical course timetabling, Carter and Laporte [6] observed that there were very few papers published that described actual implementations of course timetabling. Aubin and Ferland [1] describe a method that used quadratic assignment formulation for both course timetabling and student sectioning. The program was implemented at a large Montreal High School and two university departments (although they did not provide details on the implementation). Paechter et al. [9] and Rankin [10] describe a genetic course timetabling algorithm that is being used for the Computer Science Department at Napier University. Sampson et al. [11] describe a heuristic that was implemented in the Graduate School

E. Burke and W. Erben (Eds.): PATAT 2000, LNCS 2079, pp. 64-82, 2001.
© Springer-Verlag Berlin Heidelberg 2001

of Business at the University of Virginia. The existing university examples were all restricted to a single department or school. There are no published examples of an automated, integrated course timetabling system across the institution. We believe that researchers in practical timetabling can learn a lot from this case.

The University of Waterloo is a comprehensive university with undergraduate and graduate programs in Applied Health Sciences, Arts, Engineering, Environmental Studies, Mathematics and Science. The University offers North America's largest co-operative education program that operates year-round, as well as the more traditional regular program of two consecutive four-month school terms. [12]

In 1985, there were 17 000 undergraduate students and has now grown to close to 20 000. In the Fall term of 1985 (the term we used for our parallel testing and benchmarking) there were 1400 courses offered with about 3000 course sections. (i.e. some courses have multiple sections.)

The system required an estimated 40 person years (1979~1987) to develop. It was designed to run on IBM mainframe equipment, and the database and the programs were all developed in-house. The new system required about 100 new programs to be written and 400 old ones changed. It was implemented between Aug 1985~May 1987. One of the drawbacks of the system is that it has been designed to solve the specific instance of the Waterloo problem; it is not portable, although many of the optimization routines were designed to solve a more general mathematical problem.

In spite of the fact that the system is over 15 years old, we believe that there is still no system that solves the large-scale course timetabling problem with this level of mathematical sophistication.

2 Overview of the System

In North American schools, it is common for students to choose courses from a wide variety of electives. This is particularly true for the humanities and social sciences, but there is often considerable flexibility in upper year science and engineering programs. Waterloo uses a "demand driven" timetabling system: students pre-register for courses from a list of "course offerings" posted in March. The University then constructs a timetable that attempts to minimize potential student conflicts (see Carter and Laporte [6] for a discussion of demand driven timetabling versus master timetabling where the course times are fixed first, and then students pre-register.)

The key milestones in the timetabling/registration cycle (for Fall term) are as follows:

- March: student pre-registration: students submit a list of course requests.
- Early June: initial timetabling: each course is assigned to a day/time slot; initial room assignment: each meeting is assigned to a classroom.
- June~July: on-line timetabling: timetable reps make manual changes.
- August: initial student scheduling: students are assigned to course sections.
- September: on-line scheduling: students revise schedules in a "drop/add" period.

The primary inputs to the process are:

- Courses: list of expected course offerings (with instructors pre-assigned). The departments may also assign preferred meeting times. Historically at Waterloo, 75% of the course offerings are the same as the previous year.
- Student requests: student requests are listed in order of importance with "required" courses first. (In the event that a conflict cannot be resolved, the courses at the end of the list will be dropped.) First year students do not pre-register (until August); so we construct surrogate first year students based on course selection patterns from the previous year. In 1985, just over 50% of the students pre-registered by early June (in time for timetabling).
- Course requests are combined into a "conflict matrix" or "joint demand matrix": the number of students who have requested each pair of courses. Required course conflicts count twice as much as electives.
- Instructors: availability and teaching conflicts.
- Space: we must ensure that there are sufficient rooms of the appropriate size at all times. (room profile)
- Slot system: The University uses a system of standard slot times that are designed to "fit" together nicely. People are not required to use standard slots, but the system introduces a small penalty for courses that do not conform. The list illustrates a few examples of standard slots:

  o   8:30 MWF (3 x 1 hour)
  o   8:30-11:30 T (1 x 3 hours)
  o   8:30-10 TR (2 x 1.5 hours)
  o   8:30 TR (2 x 1 hour)

3   Initial Section Assignments: Homogeneous Sectioning

When courses are offered in multiple sections as they are at Waterloo, it creates a timetabling paradox. Students request a *course*, but timetabling assigns days and times to course *sections*. We cannot assign times to *sections* until we know which students are in each section. But we cannot assign students to *sections* until we know when the sections are timetabled!

Our approach to this problem is called homogenous sectioning. For every course with multiple sections, we group (cluster) the students with *similar* course requests. We construct balanced groups equal to the number of sections. We then assign these groups to sections to minimize the *expected* number of conflicts (based on the preferred times for each section). This is a temporary pre-assignment used only for automated timetabling. After timetabling is complete, we assign students to sections individually to minimize conflicts.

Grouping students based on similar course selections was chosen because it can dramatically *reduce* the number of edges in the conflict matrix. Figure 1 illustrates a simple example. Suppose nine students sign up for Psych 101. Assume three of the students are second year Math, three are first year Psych majors and three are third year Mechanical Engineers. Suppose each student has all of his/her five courses in common with the other students in their group. If we put the Math students in section

1, the Psych majors in section 2 and the Engineers in section 3, then each section will only conflict with four other courses for a total of 30 conflict edges as shown in Figure 1. However, if we mix the students and put one Math student, one Psych major and one Engineer in each section, then all of the Psych 101 sections will conflict with all of the other courses, and this will create a total of 54 conflict edges. When the number of conflict edges is reduced, the quality of the final solution is greatly improved, and it becomes much easier to find conflict-free combinations.

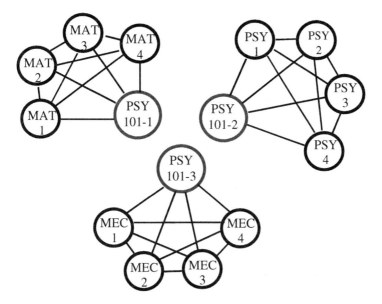

Fig. 1. When the students are assigned to homogeneous sections, the total number of conflict edges is 30. If the students in each program were distributed over the three PSY 101 sections, there would be 54 conflict edges

The algorithm begins by constructing a graph where each vertex represents a student. We draw an edge between each pair of vertices and assign an edge weight that measures the distance between them. The grouping is based on graph clustering where the *distance* between two students varies between 0 and 1 depending on the number of course/section requests that they have in common. Distance 0 means all courses/sections are the same, and 1 means that the two students have nothing in common. The algorithm iteratively combines vertices that are closest to each other without overloading the corresponding section. Whenever two vertices are combined, the new distance to all other vertices is defined as the distance between the two "students" in each pair of vertices of maximum difference. We continue until the number of vertices remaining equals the required number of sections. (As a section "fills up", it is removed from the problem.)

Consider the example in which eight students have requested a course that will be offered in two sections. These students have also requested other courses as shown in Table 1.

In Table 2, the *distance* between each pair of students has been computed. In particular, since students 1 and 2 have identical course selections, they can be combined. The next shortest distance between two groups is 1/7 between pairs (3, 8) and (6, 7). Table 3 illustrates the reduced problem with the new distances between sub-groups.

Table 1. Homogeneous grouping example

Student	Other courses			
1	A	B	C	D
2	A	B	C	D
3	A	D	E	F
4	A	B	F	G
5	A	F	G	H
6	C	D	F	G
7	C	F	G	
8	D	E	F	

Table 2. Distance between pairs of students

	1	2	3	4	5	6	7	8
1	-	0	$4/8$	$4/8$	$6/8$	$4/8$	$5/7$	$5/7$
2	0	-	$4/8$	$4/8$	$6/8$	$4/8$	$5/7$	$5/7$
3	$4/8$	$4/8$	-	$4/8$	$4/8$	$4/8$	$5/7$	$1/7$
4	$4/8$	$4/8$	$4/8$	-	$2/8$	$4/8$	$3/7$	$5/7$
5	$6/8$	$6/8$	$4/8$	$2/8$	-	$4/8$	$3/7$	$5/7$
6	$4/8$	$4/8$	$4/8$	$4/8$	$4/8$	-	$1/7$	$3/7$
7	$5/7$	$5/7$	$5/7$	$3/7$	$3/7$	$1/7$	-	$4/6$
8	$5/7$	$5/7$	$1/7$	$5/7$	$5/7$	$3/7$	$4/6$	-

Table 3. Distances after partial grouping

	1, 2	3, 8	4	5	6, 7
1, 2	-	$5/7$	$4/8$	$6/8$	$5/7$
3, 8	$5/7$	-	$5/7$	$5/7$	$5/7$
4	$4/8$	$5/7$	-	$2/8$	$4/8$
5	$6/8$	$5/7$	$2/8$	-	$4/8$
6, 7	$5/7$	$5/7$	$4/8$	$4/8$	-

Observe that the distances in rows (3, 8) and (6, 7) are the maximums of the distance in rows 3 and 8 or rows 6 and 7 in Table 2 respectively. Continuing the procedure assigns students (1, 2, 3 and 8) to one group, and students (4, 5, 6 and 7) to the other.

In order to test the impact of homogeneous sectioning in practice, we compared three methods of sectioning students: the previous method (sections were assigned to their preferred times with an algorithm for student sectioning similar to the procedure described in Section 8), assigning students to sections randomly, and assigning students to homogeneous sections (see Table 4).

Table 4. Conflict edge reduction with homogeneous sectioning

	Random	Previous	Homogeneous
Conflict Edges	131 212	125 143	103 072
% Increase	21.4%	17.6%	-

Homogeneous sectioning reduced the total number of conflict edges by 17.6% over the previous method, and by 21.4% over a random assignment.

Unfortunately, homogeneous sections are not particularly attractive for academic reasons. In Psych 101 for example, it would be nice to have a mix of students in each section. At Waterloo, we provided the ability for course coordinators to specify the precise allocation of students to sections ahead of time if they are serious about constructing academically balanced sections. In fact, in the common first year Engineering program, they have a special algorithm that distributes students in sections equitably based on gender, program choice, geographic and biographic information.

Of course, the potential reduction in conflict edges is only a side-benefit of homogeneous grouping. The primary consideration is that we need to assign each student to a specific section so that we can define conflicts between each pair of course sections, and use that for timetabling.

Suppose a specific course has four sections. Homogeneous sectioning creates four roughly equal groups of students, but it does not tell us which group should be assigned to which section. The second phase of homogeneous grouping assigns the student groups to sections. Clearly, if there are no constraints on the sections (all equal), then we can assign sections arbitrarily. Normally, each section will have an instructor pre-assigned, and the instructor will have time slot preferences. We constructed an assignment problem where the weight on each edge is equal to the probability that each student in the set would be conflict free if this set were assigned to the given section. For example, for each section, suppose each professor specifies a preferred period and two alternates. We arbitrarily assumed that the section should get it's preferred time 80% of the time, and each alternate 10% of the time. (Similar assumptions were made for all possible preference schemes.) We can then compute the probability that each student would be conflict free with each section.

For example, suppose a particular student has requested a single section course MATH 110, and a multi-section course PSYCH 101. What is the expected number of conflicts if the student is assigned to section 1 of PSYCH 101? The preferred and alternate times are shown in Table 5.

Table 5. Example of expected conflicts

	Preferred time	Alternate times	
PSYCH 101-1	10:30 MWF	9:30 MWF	11:30 MWF
MATH 110	11:30~2:30 M	8:30~11:30M	11:30~2:30 T

If the student is assigned to section 1 of PSYCH 101, then there is a probability of 0.8 that he will be in the preferred time, which will conflict with the first alternate of Math 110 (prob. 0.1). If PSYCH 101 is assigned to the first alternate (prob. 0.1) it will conflict with the first alternate of MATH (prob. 0.1), and the second alternate of PSYCH (prob. 0.1) conflicts with the preferred time of MATH (prob. 0.8). Therefore, the expected number of conflicts for this student, if he is assigned to section 1, is

$$(0.8 \times 0.1) + (0.1 \times 0.1) + (0.1 \times 0.8) = 0.17$$

This is repeated for all students in each group for all of their courses. What is the total number of expected conflicts if the first homogeneous group is assigned to section 1 of PSYCH 101? The same calculation is repeated for each of the homogeneous groups and each section. We then construct an assignment problem: homogeneous groups are assigned to sections to minimize the total number of expected conflicts.

4 Timetable Decomposition

Since we were solving an integrated timetable across the whole University, the size of the problem was a particularly serious issue. With 3 000 independent course sections, the conflict matrix alone has 9 million entries and with an edge density of about 1% for a total of over 90 000 non-zero entries. This was considered too large for optimization. We developed a method to decompose the original problem into a set of smaller problems that could then be solved independently, and later combined optimally (see the working paper by Carter [4] for details). A variation on this procedure has been successfully tested and implemented by Burke and Newell [2].

As a first step in reducing the number of edges, we decided that we should be able to *ignore* any edges where the number of students affected is low (for the purposes of decomposition). Therefore, if a second year Math student wants to take a fourth year German course, we will try to accommodate the request in the final section allocation, but we should not design the sub-problems with this constraint included. We could also remove any conflict edges between a pair of sections that would never overlap in time based on pre-assigned times. For example, a course that must be in the morning cannot conflict with an evening course; and a class pre-assigned to Monday will not conflict with one set on Wednesday. Combining these two concepts, we computed the *expected number of student conflicts* between each pair of sections. It is defined as the number of students requesting both sections multiplied by the (estimated) probability that the two sections will overlap. The probability that the sections overlap uses the principle described in the previous section that the preferred time has a high probability of being chosen, and the alternate periods have a much lower chance. If the expected number of student conflicts is less than one, we delete the edge. This pre-processing reduced the total number of edges significantly (by a factor of 35 in one example). Clearly, this rule could be adjusted.

Decomposition is based on the intuitive assumption that there exists a "natural" partition of the problem into subsets of sections with high interaction. In particular, we would not expect much interaction between advanced language courses and advanced Engineering courses. However, students in both of these programs could be taking Psych 101.

Mathematically, the decomposition algorithm is based on the following concept. Suppose that we can divide the timetabling problem into two pieces called problem A and problem B. Define the edges connecting problems A and B as the "cut set". We will call the decomposition *reasonable* if the two problems can be solved independently. In particular, we want to solve problem A (ignoring B); and then fix those assignments, and solve problem B. If the number of edges in the cut set is small enough, this decomposition should have little or no impact on the solution.

Theoretically, this was modeled as a graph colouring problem assuming that there are no constraints on the available periods for each section. Suppose that you solve the two problems separately and that the solution is represented as a colouring. Is it then possible to find a permutation of the colours in one problem so that there are no conflicts with course section (colour) assignments in the other problem?

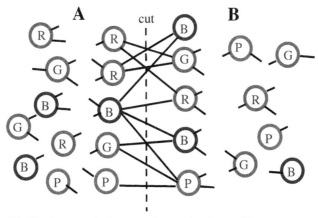

Fig. 2. An example decomposing a colouring problem into two parts

Consider the hypothetical problem illustrated in Figure 2. Suppose there is a large problem that has been divided into two sub-problems, A and B (where only a few of the edges are shown). These two problems are solved independently (using colours Red, Blue, Green and Purple). Observe that there are conflicts between pairs of vertices on either side of the "cut" edges between the two sub-problems. (A conflict occurs if two nodes with the same colour are connected by an edge.) Can we permute the colours in problem B (to an equivalent solution) such that there are no conflicts on the cut set?

Consider only the vertices and edges in the cut set. First, we merge nodes on the same side with the same colour (the two Red nodes in A, and the two Blue nodes in B). The merged nodes inherit the edges from both predecessors. The resulting problem is shown in Figure 3(a). Now, we find the complement set of edges in the cut set (illustrated in Figure 3(b)). An edge appears in the complement if there is no edge in the original graph. If a pair of nodes is connected in the complement, then they are allowed to have the same colour.

We now try to find a *matching* in the complement graph: i.e., find a subset of edges in the complement such that every node in A is connected to a unique node in B. The matching is illustrated in Figure 3(b) as arrows. It is fairly easy to see that, if a matching exists, then the nodes in B can be permuted to take the same colour as the corresponding matched nodes in A: i.e., if *all* nodes in B are permuted such that green becomes blue, red becomes green, blue becomes purple, and purple becomes red, then the solution in B is equivalent, and there will be no conflicts with sub-problem A.

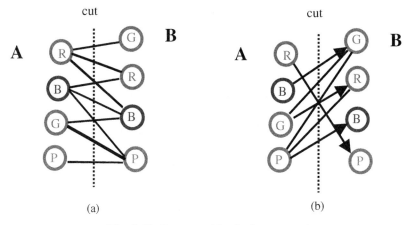

(a)   (b)

Fig. 3. Finding a matching in the cut set edges

We were able to show that, for randomly generated graphs, this is "almost always" true if the density of cut set edges between the two sub-problems is below a defined value (given in [4] as a series of formulae). For example, if there are 20 colours available, and each of the two problems has 40 vertices, then the problems can be solved independently (with probability 99.99%) if the number of edges in the cut set between the two is less than 588 or about 37% density.

In an early version of the decomposition algorithm, we attempted to find all minimum cut sets (using the Gomory~Hu method [8]), but this was not useful on practical problems. (The Gomory~Hu tree was a star-graph where each cut set cut off one vertex!) Instead, we designed an algorithm that searches for groups of course~sections that *should* be timetabled together based on a high internal density of edges. Specifically, the algorithm looks for a large *clique* in the graph (a subset of course-sections where every pair are in conflict). These must certainly be timetabled together.

The program uses an algorithm developed by Carter and Gendreau [5] to find a large clique in a graph. We then search for any single node that has more edges connected to the subset than to the rest of the graph, and add it to the subset: i.e., this rule reduces the number of edges connecting the subset to the outside. We repeat this process until no new vertices can be added to the subset. These vertices are "removed" from the graph, and we look for another clique in the remaining subgraph.

The procedure is illustrated in Figure 4. First, the clique routine finds the four nodes denoted by "C". Every pair of nodes in a clique is connected by an edge. Observe that the node labelled "1" has three edges *into* the clique and only two edges *out* to the rest of the graph. We therefore add node "1" to the subset. Node "2" has three edges into the subset (including "1" now) and two out, so we add "2" to the subset. All remaining nodes have more edges *out* than *in*, so we stop, remove the subset, and repeat the procedure on the remaining graph.

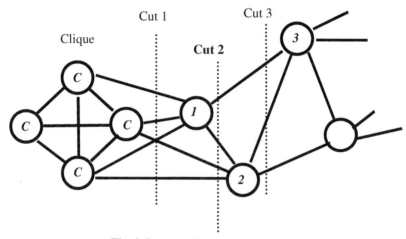

Fig. 4. Decomposition algorithm example

In practice, it was interesting to note that the procedure did not always identify "obvious" sub-problems within a department. For example, one group consisted of 13 third year Health and Kinesiology sections, plus section 04 of Chemistry 123, section 12 of Health 140, and section 01 of Sociology 101. Decomposition had the affect of identifying a large number of small clusters (36) that could be solved independently plus one large central problem! This was not really what we were hoping for; however, the large problem only had 2 500 edges compared to the initial 90 000 plus. We measured the size of the problem in terms of the number of zero~one variables x_{ij} = 1 if section i is assigned to time slot j. Many sections only had a few alternate time slots. On average, the 36 sub-problems had 89 0~1 variables (ranging from 34 to 247). This is still challenging for quadratic 0~1 problems, but not completely hopeless.

5 Automated Course Timetabling

Practical course timetabling is 10% graph theory, and 90% politics! When we first began designing the system, we were warned: "You cannot dictate to professors when they will teach courses!" Consequently, we were told that course timetabling could not work. Or, more precisely, we cannot assume that the timetable can use a clear slate. There are many possible reasons for this. Part-time faculty may be available only on certain days or times. University professors often have other commitments and industrial research projects. Teaching is only a part of their job, and probably less than half of their time is devoted to teaching.

Our solution to this problem was to allow people to constrain the timetable as much (or as little) as they wanted. For each course section: you can specify a "preferred day and time" as indicated in the following examples:

- Must be 10:30 MWF
- Prefer 10:30 MWF, but would accept 9:30 or 11:30 MWF
- Any time in the morning
- Cannot be Friday afternoon
- Any night or weekend slot
- No restrictions.

Our objective was to minimize the total number of student conflicts. (Required course conflicts count double.) The primary constraints were instructor availability, time preferences/restrictions (as described above) and space limitations (room profile). In this phase, we did not attempt to assign specific rooms to each meeting. Instead, for each time period of the week, we constructed a *Room Profile*, as illustrated in Table 6). The profile keeps track of the total number of rooms available by size. We then added constraints to ensure that there would be sufficient space of the appropriate type at all times. For example, if we only have six lecture rooms that hold more than 250 students, we introduce a constraint that no more than six classes over 250 can occur at any one time. Similarly, we constructed separate room profiles for special purpose space. For example, if there are only four physics labs, we constrain the number of concurrent physics lab meetings. Specific rooms are assigned in the room assignment algorithm described in the following section.

The course timetabling algorithm solves the sub-problems created by the decomposition procedure one at a time. We solve the largest (most difficult) sub-problem first, and then these assignments are fixed and we proceed to the next sub-problem; the assignment tries to allocate the new sections taking any adjacent conflicting sections as pre-assignments. Originally, we were hoping that we could apply a true zero~one integer programming algorithm to solve the sub-problems optimally. Unfortunately, the core problem was still too large (in 1985), so in the end we used a fairly simple "greedy" procedure:

1. Assign the most restrictive course section to the best possible time slot.
2. After all course sections have been assigned, apply a 2-opt procedure to look for any potential improvements.

Table 6. Room profile (e.g., Wednesday 10:30~11:00)

Room size	No. available	No. required	Cumulative No. available	Cumulative No. required
1 000	1	0	1	0
500	4	3	5	3
250	1	2	6	5
120	7	5	13	10
100	5	8	18	18
90	10	7	28	25
...	...	...	...	...
...	...	...	...	...
...	...	...	...	...

The most restrictive course section is defined as the section that has the least number of feasible periods. (In particular, there were a large number of pre-assigned sections at Waterloo; so these sections had only one feasible period.) In fact, all course sections with fixed time periods were assigned before any of the sub-problems were solved. Since there are no choices, we might as well use the information when scheduling other sections.

For each feasible period, we calculated the number of student conflicts with other sections that had already been scheduled. The period with the lowest total conflict score was selected. As discussed above, we allowed people the option of specifying a *preferred* time slot and *alternate* time periods. We introduced a penalty for using the alternate time slots so that the algorithm would use the preferred times unless there was a *significant* reduction in student conflicts with the alternate time.

Every section is assigned to some period unless there is no feasible period! (Realistically, the only time this occurs is when too many "large" sections are all pre-assigned to the same time period, and we run out of large rooms.)

The 2-opt local improvement heuristic was quite effective in finding good solutions. At the end of the initial assignment phase, we would ensure that there is no single section that can move to a new feasible time slot and reduce the total student conflicts. We then ensure that there is no *pair* of sections that can switch time slots and reduce the total cost.

For the Fall 1985 session, approximately 60% of the meets were pre-assigned to fixed times by the departments. The remaining 40% (1 230 out of 3 052) had potential alternate times. The program used alternate time slots for only 55 of the meets. (All others were assigned to their preferred time. By assigning these 1.8% of the meets to an alternate time, we were able to reduce the total student conflicts from 19 229 to 17 109 (an 11% decrease). (Note that the previous system simply assigned all sections to their preferred times.)

6   Automated Classroom Assignment

In this phase, we assign classes to specific rooms across campus. At Waterloo, the central Registrar's Office controls the regular lecture rooms. No one is allowed to book these rooms until the classroom assignment procedure has been run. The laboratories are typically reserved for specific lab courses, and the room assignment is simple. However, labs could easily be accommodated as separate sub-problems.

The algorithm uses a lagrangian relaxation approach to solve the problem as a generalized assignment problem (see Carter [3] for complete details; Carter and Tovey [7] present a general discussion of the difficulty of classroom assignment).

$$
\begin{aligned}
\underset{x \in \{0,\,1\}}{\text{minimize}} \quad & \sum_{i=1}^{n} \sum_{j=1}^{m} c_{ij} x_{ij} \\
\text{subject to} \quad & \sum_{j=1}^{m} x_{ij} \quad \text{for all } i \\
& \sum_{i \in P_k} x_{ij} \le 1 \quad \text{for all } j, k
\end{aligned}
$$

where c_{ij} is the cost of assigning class i to room j, $x_{ij}=1$ if class i is assigned to room j and P_k is the set of all classes in period k. The problem is a tough mathematical problem in itself. However, it is complicated by the problem of trying to determine an appropriate set of costs. We assign a cost of zero if the room is a perfect match. There are a number of criteria that are used to measure the quality of the match between a class and a given room:

- Location: The University of Waterloo is spread out over a few dozen buildings over a square mile. Each instructor has preferred buildings, often based on the proximity to his/her office or to the department. However, it may be based on the predominant type of seating, facilities or other factors.
- Space Utilization: We want to make effective use of the available space. This includes having sufficient space, and not wasting too many seats. Moreover, when room assignment is run in early June, we do not know exactly how many students are likely to be taking the course; we only have partial pre-registration data. At this stage in 1985, less than 50% of the students had pre-registered. The user must enter a "minimum required seats" for each class. We assume that any room that is too small is not feasible. We penalize rooms that are too large based on the number of empty seats. Forecasting the minimum required number of seats is a major problem on its own. Professors may tend to over-estimate, and historical trends can be misleading as curricula are revised. At Waterloo, the Registrar's Office makes an educated guess based on discussions with the Faculty.
- Preferred Rooms: Many instructors have specific rooms that they like best, and we allow them to express a preference. We give a small penalty to all non-preferred rooms. Unfortunately, instructors tend to prefer the same rooms, so it was difficult to satisfy all requests.

- Special Facilities: Instructors may request special features such as audio-visual, computer support, television, special seating types, large blackboards, etc. Some of these are considered requirements while others can be substituted with a penalty.

- Fragmentation: When a class meets two or three times a week, professors at Waterloo liked to get the same room in all meetings. They felt that some people might get confused if meetings were assigned to more than one room. We tried to convince people that this might not be advisable. For example, a poor assignment on Monday (due to lack of proper space) would require the same space on Wednesday and Friday meetings! This was a particularly difficult constraint. It meant that each day could not be solved separately. In the end, we simply used the Monday assignment as a strong preference when we solved the Wednesday problem, etc.

- Standard Slots: The problem of timetabling is much easier if everyone is using a standard set of slot times for their courses. For example, classes at Waterloo tend to meet on the half hour commencing at 8:30 Monday, Wednesday and Friday for one hour. Similar slots occur beginning at 9:30, 10:30, etc. Daytime slots such as two 1.5 hours, two hours, three hours, etc are normally held on Tuesday and Thursday with the slot pattern beginning at 8:30. When someone uses a meeting time of Monday at 9:00 for two hours and Thursday at 2:00 for one hour, they create conflict times for all students in their class for a large number of other standard time slots. We allowed non-standard times, but we introduced a slight penalty: when two classes that are identical in all other respects are competing for the same (preferred) room, the class using standard times will win.

In order to keep the problem manageable, we divided the week into separate days, and divided each day into three parts. We introduced a rule that no class should straddle the 12:30PM time, and no class should go through 6:30PM. This was consistent with the standard time slots in use, and consistent with most logical schedules. The rule allowed us to solve each day as three separate problems: morning, afternoon and evening. Any classes that insisted on straddling the boundaries could either be pre-assigned to a room, or could look for a room after the automated assignment has been run.

The room assignment algorithm is based on the following simplification. Suppose all classes were one hour long (on the half hour). Then room assignment could be solved using a sequence of assignment problems. Using the above criteria, we could compute a *cost* for assigning each class to each room. Unfortunately, at Waterloo, classes have varying lengths. The assignment solution would have students in three~hour classes changing rooms at the end of each hour (potentially), which is not really acceptable. (In practice, most classes stayed in the same room.) In our algorithm, we solved the individual assignment problems, and then introduced "room jumping" constraints to stop classes from changing rooms every hour. These constraints were incorporated as penalty terms in the objective function, and the resulting problem was again solved as a sequence of assignment problems. In a sense, the hard part was coming up with a set of costs that would imitate the preferences and trade-off evaluations of the current schedulers.

The results were excellent. In a parallel comparison with the manual assignments, the algorithm satisfied 50% more preferred building requests (from 89% to 95.5% of the requests for first or second choice space). The automated system had fewer empty seats (from an average of ten empty seats a meeting down to seven). The implication is that the computer system leaves more large rooms empty. This produces more flexibility in the later stages when revisions and additions require late modifications.

The manual system was unable to satisfy 12% of special facility requirements, while the automated system met them all. For room preferences, the computer system satisfied 50% while the manual approach met 25%. The manual system had fewer examples of fragmentation (47 compared with 57), but this was deemed acceptable. Moreover, the manual assignment was estimated to take three people three weeks each term to perform (three terms a year). Therefore, the estimated savings from room assignment alone are 27 person-weeks per year.

In the Fall of 1988, the University of Waterloo opened the Davis Building. A total of 40 (smaller) classrooms across campus were replaced by 30 generally larger classrooms in the new building. The number of rooms decreased, but the total number of seats increased. The Scheduling Office was very concerned about how the new space would impact space allocation. The automated system effectively solved a potentially serious planning problem. The algorithm can be used as a fast "What if?" tool to evaluate proposed changes in space allocation.

7   Interactive Course Timetabling

At Waterloo, each of the 57 departments has a "timetable representative" who has access to on-line timetable information. After the initial timetable and room assignment has been run, timetable reps can log into the system and make manual improvements. The primary display for a selected course section shows a "timetable grid" with days of the week across the top, and times of the day down the right margin. Each square displays the number of student conflicts (required and elective) that would be created if this course section were moved to that day and time. It also displays room (R) and instructor (I) conflicts. The user can select any square in the matrix and drill down to find out what programs the conflicting students are in, or the actual students affected. The display in Figure 5 is intended only to illustrate the type of information in the grid. It is not an accurate representation of the actual screen.

The user can choose a specific course to investigate, or let the system select "problem" courses. For example, in one option the system searches for courses with a better available time. The system also identifies pairs of courses that could be interchanged to reduce total conflicts.

The timetable reps can also change room assignments. For example, some professors will not get their preferred rooms. They will have an opportunity to change the time for the class to a time when the room is available. Again, the timetable reps use the conflict matrix. They can search for an appropriate available room (at a specified time), display when a given room is available, and book a new room. Essentially, any room that is still open is available to them.

Typically, at this stage, if there is a better time for a class, the program did not find it because the original constraints prohibited some times. The timetable reps can now contact their counterparts in other departments to discuss mutual problems;

and/or go directly to the corresponding professors in "problem" courses, and recommend a new time slot. They are allowed to move any of their *own* classes to a new time; but only if the *total cost* improves. If the timetable reps want to move a course to a worse period, they need the approval of the Registrar's Office once approved by the Dean of the Faculty. The changes are done in real time, so that any revisions will immediately be available to all other users.

MATH 110 Lecture 01   9:30 MWF   87 students   Room MC 3012 Prof. Carter

Time	Monday		Tuesday		Wednesday		Thursday		Friday	
8:30	5		8	R	3		174		4	I
9:00	5		8	R	3		174		4	I
9:30	9		30		8		40		10	
10:00	8		30		8		42		10	
10:30	6		75		7	R	174	IR	6	
11:00	6		75		7	R	174	IR	6	
11:30	20	IR	174		22		20		15	
etc...										

Fig. 5. Conceptual timetable grid

8   Automated Student Scheduling (Final Section Assignments)

In the homogeneous sectioning algorithm, students were assigned to sections of courses based on similarity to other students in the group. In this routine, we now consider each student separately, and assign him or her to course sections in order to minimize conflicts. The routine also tries to balance the number of students assigned to each section. Each student is solved separately in the order that his or her course requests were posted electronically. The algorithm searches for a feasible section for each course in the order that the students entered them on the registration form. We do not schedule a course for a student unless all of the corresponding sections (lectures, labs, tutorials, etc.) have a feasible assignment.

We wanted to ensure that all students get a space in their required (core) courses. Otherwise, students who register early could fill up the available space in some courses and prevent later students from getting into their core courses. In many cases, we did not know which courses were core for the students, so we simply assumed that the students' first n choices were required. A table that governs the value of n by Program, Year & Term was developed to separately maintain the calculation of this value. So, we perform the algorithm in two passes. In the first pass, each student is assigned to sections of his or her required courses. We then repeat from the beginning on elective courses.

The routine uses an extension of the Hungarian method for constructing an "alternating path" in a graph. In this case, we construct an "alternating tree". Consider the graph in Figure 6. Each vertex represents a section of a course (for a single student). The vertices are clustered into groups for one course. Suppose that we have already assigned section 01 to course A, section 02 to course B, and we are now trying to find a section for a new course C. If there is a feasible section, we take the one with the maximum number of free seats. Otherwise, we construct a tree where each section of course C is connected to all sections of other courses that would have to be rescheduled (because of conflicts) if this section were assigned to the student. For each of these "conflicting" sections, we construct an edge to every alternate section that could be feasibly assigned. An alternating tree is a feasible set of reassignments to provide an opening for one of the sections of the new course.

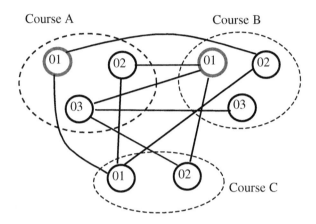

Fig. 6. Multi-section student scheduling

In addition, we want to construct *balanced* section sizes. Therefore, on the *first pass* of the algorithm, we only include sections with the largest number of empty spaces. If there is no feasible solution, we add the next smallest sections and repeat.

9   Student "Drop/Add" Process

In September, the students arrive at the University and go through a "drop/add" procedure. They may change some of their course selections because they did not get all of the courses they asked for, or they may simply show up at the first lecture and decide to switch courses. Department timetable reps are able to change the student schedules on-line. The routine contains a "re-optimize" facility that allows the student to juggle some or all of their existing course sections in order to find a way to insert a new course. The on-line routine allows pre-determined sections, preferred sections and permissible conflicts if approved by the instructors.

10   Conclusions

When the project was initiated, the authors believed that the key to a successful system would depend on the quality of the assignments produced by the automated algorithms. In retrospect, while these components have played a key role, we also realize that simply giving timetable reps the facility to make real time on-line changes was the single most important contribution. Timetabling is highly political, and you cannot hope to incorporate the complete spectrum of preferences and alternatives in a quantitative system.

The system was in use for 15 years. Unfortunately, like so many other schools, Waterloo found themselves with serious year 2000 problems. Moreover, the system was not portable – even within Waterloo. They have decided to purchase a commercial package from Peoplesoft to provide their administrative student system support, and they are in the process of buying other packages for optimization. Converting the old ones with a new database is to labour intensive.

Acknowledgements. There were a large number of people who contributed to the success of this project. At the risk of omitting someone, the author would like to mention Vivienne Ballantyne, Adam Belsky, Martha Dolan, Mike Esmond, Janice Foster, Dave Fritz, Gary Marx, Dave Mason, Cam McKay, Wayne Montgomery, Cheryl Myers, Lynn Tucker, Roy Wagler and Rudy Zigler

I am particularly grateful to Karen Ledrew, Associate Registrar, for her help in the design of the system and her constant support.

References

1.   Aubin, J. and Ferland, J.A., "A Large Scale Timetabling Problem", Computers & Operations Research 16, 67-77, 1989.
2.   E.K. Burke and J.P. Newall, "A Multi-Stage Evolutionary Algorithm for the Timetable Problem", the IEEE Transactions on Evolutionary Computation, Vol 3.1, 63-74, 1999
3.   Carter, M.W., "A Lagrangian Relaxation Approach to the Classroom Assignment Problem", INFOR 27, No. 2, pp. 230-246, 1989.
4.   Carter, M.W., "A Decomposition Algorithm for Practical Timetabling Problems", Working Paper, Mechanical and Industrial Engineering, University of Toronto, 1983 (available from www.mie.utoronto.ca/staff/profiles/carter.html)
5.   Carter, M.W., and Gendreau, M., "A Practical Algorithm to Find a Large Clique in a Graph", Working Paper, Mechanical and Industrial Engineering, University of Toronto, 1999.
6.   Carter, M.W. and Laporte, G., "Recent Developments in Practical Course Timetabling", in Practice and Theory of Automated Timetabling, Springer-Verlag Lecture Notes in Computer Science 1408, E.K. Burke and M.W. Carter, eds., Springer, 1998.
7.   Carter, M.W. and Tovey, C.A., "When Is the Classroom Assignment Problem Hard?", Operations Research 40, S28-S39, 1992.
8.   Hu, T.C., Integer Programming and Network Flows, Addison-Wesley, Reading, MA, 1969.

9.   Paechter, B., Cumming, A. and Luchian, H., "The Use of Local Search Suggestion Lists for Improving the Solution of Timetable Problems with Evolutionary Algorithms", In Proceedings of the AISB Workshop on Evolutionary Computing, (Sheffield, April 3~7, 1995.

10. Rankin, R.C., "Automatic Timetabling in Practice", in Practice and Theory of Automated Timetabling, Springer-Verlag Lecture Notes in Computer Science 1153, E.K. Burke and P. Ross, eds., Springer, 1996.

11. Sampson, S.E., Freeland, J.R. and Weiss, E.N., "Class Scheduling to Maximize Participant Satisfaction", Interfaces 25, No. 3, 30-41, 1995.

12. University of Waterloo web site http://www.uwaterloo.ca.

Examination Timetabling

Examination Timetables and Tabu Search with Longer-Term Memory

George M. White and Bill S. Xie

School of Information Technology and Engineering,
University of Ottawa,
Ottawa, K1N 6N5, Canada
white@site.uottawa.ca

Abstract. The examination scheduling problem has been examined and
a four-phase system using a tabu search algorithm, OTTABU, has been
implemented. This system uses both recency-based short-term memory
and move (or frequency)-based longer-term memory to improve the quality of the solutions found. The system was tested using real data obtained
from the University of Ottawa registrar's office and real examination
schedules were produced. It was found that the use of longer-term memory produced better schedules that those produced without such memory
– typically a 34% improvement was obtained due to this factor alone. The
length of the long term memory list was also found to be important. A
length that is too small can greatly reduce its effectiveness. A list that
is too long only reduces the effectiveness by a small amount. A quantitative analysis method is applied to estimate the appropriate length of
the longer-term tabu list and a controlled tabu relaxation technique is
used to improve the effectiveness.

1 Introduction

The examination timetabling problem is a difficult combinatorial exercise that
has been studied for many years, dating back to at least the early 1960s [1]. This
work has led to several implementations that have been used with much success
at universities [2], [3].

A review of most of the work published in this area can be found in the
comprehensive coverage of Carter and Laporte [4].

Much of the early work was based on bin packing algorithms with heuristic
rearrangement of the schedule based on some version of the travelling salesman's
algorithm to reduce the number of consecutive examinations taken by students.
More recent work is based on the observation that the examination timetabling
problem is an assignment-type problem and can be considered to be a graph
colouring problem. Some researchers have applied tabu search (TS) techniques
to solving examination timetabling problems in recent years. Hertz and de Werra
[5] used a tabu search algorithm to generate solutions for the problem modelled
as a graph colouring problem with great success. After this initial approach,
Hertz [6] applied the approach developed for graph colouring to construct and

E. Burke and W. Erben (Eds.): PATAT 2000, LNCS 2079, pp. 85–103, 2001.
© Springer-Verlag Berlin Heidelberg 2001

solve course timetabling and examination timetabling problems. Boufflet et al.
[7] modified this method to solve a particular practical examination timetabling
problem. They found that their TS techniques improved the solutions well.

In these works only recency-based short-term tabu techniques were used. The
move *(exam, original period)* is stored instead of the solution into the tabu list
in order to save space and time, and the tabu tenure is set to a fixed value 7.
Boufflet et al. also set *nbmax*, the maximum number of null iterations, to 200 but
Hertz did not indicate which value he used. Hertz generated the neighbourhood
of a current solution s only from the exams which create at least one conflict in s,
but Boufflet et al. considered all the feasible moves of exams from one feasible
set to another one.

In this paper, we describe an automated TS approach such that a frequency-
based longer-term memory mechanism combined with tabu relaxation technique
is used to optimize an examination timetable problem. The tabu relaxation technique can accelerate downhill movement and diversify the search space. We also
introduce a quantitative analysis method which investigates the examination
distribution and estimates the size of the lightest or the most active exams and
the size of the heaviest or the most inactive exams in the examination set, and
makes it possible to choose automatically the appropriate parameters for TS for
an individual examination timetabling problem.

Section 2 describes the examination timetabling problem formulated as a
graph colouring problem. The following sections describe the ideas behind TS
and show how a specific implementation of a TS algorithm can be used to cast
timetables with certain desirable qualities. Data taken from the Fall 1996 registration data at the University of Ottawa were used to test the system using real
data. Some comparisons with results obtained from other researchers conclude
this paper.

2   Graph Colouring and Examination Timetabling Models

The examination timetabling problem is an assignment-type problem and can
also be considered to be a graph colouring problem. The assignment-type problem can be described as follows [8]:

> Given n items (exams) and m resources (timeslots), determine an
> assignment of each item to a resource in order to optimize an objective
> function and satisfy s additional side constraints (classified as hard and
> soft constraints).

We use an undirected graph $G = G(V, E)$ to describe the examinations and
their relationships. Let V be the examination (or node) set, v be the number
of examinations, v_i be the ith examination and s_i be the number of students
taking exam v_i. If there are m students who must take both exam v_i and v_j, we
consider there is an edge e_{ij} with a weight m between the node v_i and v_j. Let
E be the edge set in the graph, e the total number of edges in the graph and w
the sum of the weights of edges in E.

Let T be a given set of consecutive timeslots, each one containing zero or more exams: i.e. $T = (T_1, T_2, \ldots, T_k)$, where k is the number of timeslots. C_i is the maximum number of seats available for T_i. Each exam in the graph is assigned into a timeslot T_i. For a timetable to be *feasible*, the sum of weights of edges having both endpoints in the same timeslot must be zero. In addition, the sum of weights of edges between the timeslots with a given distance (i.e. of adjacent timeslots) should be minimized. Such a timetable is called optimal. We must, of course, ensure that no timeslot requires more examination seats than are available.

Let $E_{i,0}$ be the set of edges having both endpoints in T_i $(i = 1, 2, \ldots, k)$, $E_{i,1}$ be the set of edges between T_i and T_{i+1} $(i = 1, 2, \ldots, k-1)$ and $E_{i,2}$ be the set of edges between T_i and T_{i+2} $(i = 1, 2, \ldots, k-2)$. Let W_0 be the sum of weights of edge set $E_0 = \cup E_{i,0}$ $(i = 1, 2, \ldots, k)$. W_1 is the sum of weights of edge set $E_1 = \cup E_{i,1}$ $(i = 1, 2, \ldots, k-1)$. W_2 is the sum of weights of the set $E_2 = \cup E_{i,2}$ $(i = 1, 2, \ldots, k-2)$. We define the objective function $f(s)$ for solution s as

$$f(s) = p_0 \times W_0 + p_1 \times W_1 + p_2 \times W_2$$

where s is subject to the condition that for all T_i, the sum of the enrolments $\leq C_i$ (the maximum number of seats available for T_i). p_0, p_1 and p_2 are the penalties chosen to weight the different conflict classes in s that are characterized by W_0, W_1 and W_2 respectively. For this problem we set $p_0 \gg p_1 \gg p_2 = 1$.

Hence, the problem of optimal examination scheduling is equivalent to minimizing the value of the objective function $f(s)$ of a solution set S.

3   Concepts of Tabu Search

The TS optimization schemes are derived from ideas proposed at various times in the 1960s by Fred Glover, and subsequently developed by him and by other researchers [9], [10]. As defined in [10],

> Tabu search is a meta-heuristic that guides a local heuristic search procedure to explore the solution space beyond local optimality.... The local procedure is a search that uses an operation called *move* to define the neighbourhood of any given solution.
>
> A meta-heuristic refers to a master strategy that guides and modifies other heuristics to produce solutions beyond those that are normally generated in a quest for local optimality. The heuristics guided by such a meta-strategy may be high level procedures or may embody nothing more than a description of available moves for transforming one solution into another, together with an associated evaluation rule.

The local search strategies are often memoryless ones that keep no record of their past moves or solutions. The TS meta-heuristic makes abundant use of memory in two ways. Different *adaptive memory structures* are often incorporated that remember some of the recent moves made by the algorithm and may record some of the more recent or more promising solutions found. This memory of the search history is used to incorporate a *responsive exploration* of the state space helping the search to find superior solutions faster.

Glover and Laguna [10] emphasize four important dimensions of the memory structures used in TS. These are

- *recency*: The TS memory keeps track of the solution attributes that have changed in the recent past. These attributes are parts of the solution that change in moving from one solution to another.
- *frequency*: TS calculates ratios that keep track of which attributes change the most and how often they move.
- *quality*: In principle, TS can distinguish a better solution from a worse one by using criteria other than a single objective function, i.e. TS can directly incorporate multiple-criteria decision capability.
- *influence*: TS uses the information it has in its memory to evaluate the choices it will make and the quality of the solutions it finds.

The strategies used by TS can be classified as using *intensification* and/or *diversification* [10]. The strategy of intensification is implemented by modifying the choice rules used to select moves in order to aid the location of solutions that have been found to be good in previous search areas. The basic idea here is that if certain regions have yielded good solutions in the past, they may well yield even better solutions in the future.

When diversification is used, the exact opposite behaviour is encouraged. This drives the search area into those regions that have not yet been well explored. Perhaps better solutions can be found there, as that area has not yet been well examined.

4   The OTTABU Algorithm

We have implemented a TS algorithm OTTABU and have used it in an attempt to provide a practical examination timetable using data provided by the University of Ottawa. The data were generated in the Fall of 1996 and are based on the course enrolments of that time. This section outlines the strategies used in developing the algorithm. Details are given in Section 6.

4.1   Initial Solution

An algorithm derived from a bin packing algorithm (largest enrolment first) is used to generate an initial solution. The main idea for this algorithm is presented in the next paragraph. Note that this initial solution may be either feasible or infeasible.

Let G be a set of examinations and T a set of timeslots. G consists of v exams with enrolments $s_1, s_2, \ldots, s_v$ and T consists of t timeslots with seating capacities $C_1, C_2, \ldots, C_t$. We assign the exam with the largest enrolment from G, denoted as A, into the first timeslot T_1. Then we construct a set of examinations

G^A (a subset of G) whose members have no conflicts with the exam A, and try to schedule the largest exam in G^A into T_1. If T_1 does not have the capacity to accept it, try the second largest one. ... Repeat this procedure until T_1 is full or no exam without conflict with exams assigned in T_1 is available. This procedure can guarantee timeslot T_1 to be conflict free. Apply the same algorithm to the unassigned exams for $T_2, T_3, \ldots, T_t$. If one or more of the timeslots is empty and all exams are assigned, we have a feasible timetable. If all t timeslots are used up and there are still some exams to be assigned, we assign them into these non-full timeslots regardless of the conflicts generated.

4.2   Atomic Move, Neighbourhood, and Local Search

Let a solution $s = (T_1, T_2, \ldots, T_t)$, where T_i is the set of exams assigned to timeslot T_i and t is the number of timeslots. We generate a new solution s' from a solution s by an *atomic move*. An atomic move is one such that exactly one node x in s is moved from a timeslot T_i to another timeslot T_j, denoted as (x, i, j). We call s' a neighbour of s and all the neighbours generated from s by an atomic move as the *neighbourhood* of s. The size of the neighbourhood depends on the size of candidate node lists and the number of timeslots. There are two useful types of candidate node lists: those composed of all nodes in the graph and those containing only those nodes related to the conflicts concerned [7]. These are used to generate the neighbours of the current solution. Obviously, the former is V, and the latter is the set of nodes, denoted by V^*, which contains the end-nodes of edges in E_0, E_1 and E_2.

Starting from an initial solution, the TS algorithm iteratively explores a subset $N^*(s)$ of the neighbourhood $N(s)$ of the current solution s. The member of $N^*(s)$ that gives the minimum value of the objective function becomes the new current solution independently of the whether its value is better or worse than the value corresponding to s (i.e. an uphill move is acceptable). If $N(s)$ is not too large, it is possible and convenient to take $N^*(s) = N(s)$.

4.3   Recency-Based Short-Term Memory

Whenever a node x is moved from timeslot T_i to T_j, a move denoted by (x, i, j), to get a new current solution s^*, (x, i) becomes *tabu* and is put into a tabu list TS with a given tenure (so-called short-term memory). The tenure of each entry already in the tabu list is decreased by 1 and those entries with zero tenure are dropped from the tabu list. We choose 9 as the maximum tenure for TS. If a move (x, i, j) creates the best solution so far, we will accept this move regardless of its tabu status in TS, and if (x, j) is in TS, (x, j) will be dropped from TS. We have found that if both long-term and short-term memory are used, the tenure of the short-term tabu list is not critical.

4.4   Transitional Frequency-Based Longer-Term Memory

In order to investigate the effect of a frequency-based long-term memory, we have incorporated a move frequency table (MFT) to store the move frequency of each node in the graph. When a node is moved, its move frequency is incremented by 1. We use a second, longer-term tabu list TL to dynamically forbid moving over-active nodes in order to get diversification and help to prevent cycling. If a node x has been moved more than two times and TL is not full, it will be put into TL. If it is full and if some node y already in TL has a lower move frequency than x, we drop y from TL and add x into TL. Hence, a node in TL will not be dropped unless TL is full and a new node with higher move frequency is added. If a node is in TL and if we move this node to get a better solution than the latest best solution, we will accept this move but will not drop this node from TL.

In order to choose an appropriate parameter for the length of TL, we need to analyse the properties of the examination set to estimate the size of the lightest and the heaviest exams in the graph. We call a node (exam) with lower enrolment, degree or weight a *light* node (exam). Similarly, a node (exam) with higher enrolment, degree or weight is a *heavy* node (exam).

From the point of view of local search (move), the lightest exams usually are the most active, and the heaviest are the most inactive and most influential if moved. In the process of search, the light nodes can act as "crack fillers" [10] to perform a fine-tuning function. This, of course, helps the generation of more optimal solutions. But, if there are too many light nodes in the graph, there will be a high likelihood that the light exams may repeatedly move from one timeslot to another timeslot. The values of the objective function of the solutions examined will then change by only small amounts or have no change at all. This increases the likelihood of cycling in the TS. The cycling may waste time and iterations, lure the search away from the optimal region, and cause the search to stop prematurely.

Hence, estimating the number of the potentially active nodes will help us to decide what is the appropriate length for TL, denoted as l_{TL}.

4.5   Tabu Relaxation

Another strategy used is the relaxation of tabu lists. If a given number of iterations (this number should be less than $nbmax$) has elapsed and TL is full since the last best solution was found, or if the current solution is much worse than the last best solution, we empty all entries in both TS and TL. Relaxation of the tabu lists will change the neighbourhood of the current solution dramatically, which may drive the search into a new region and increase the likelihood of finding a better solution.

In our algorithm, the maximum number of null iterations $nbmax$ is set at $3 \times l_{TL}$. If the search has passed $2 \times l_{TL}$ iterations since the latest best solution was found and TL is full, or if the value of the current solution is better than the value of the latest best solution by a predefined threshold r, the entries in TS and TL will be dropped automatically. That is, if $(f(s) - f(best))/f(best) > r$, where $f(s)$ is the value of objective function of the current solution s and $f(best)$ is the value of objective function of the latest best solution, both lists are emptied.

Experiments show that the threshold r should be set between 5% to 15%. It is suggested that at the early stage of the search, r should be set to the high end and gradually decreased as better and better solutions are uncovered. This is because it is found that a threshold r that is too high may lead to too much uphill movement such that the search may not be able to go downhill to the best solution after relaxation of tabu lists

It is found that if the number of iterations has passed from l_{TL} to $2 \times l_{TL}$ and TL is full, the relaxation will lead to a dramatic downhill movement, which may in turn lead to new regions or search spaces.

4.6   Intensification

For a large-scale examination timetable, the neighbourhood $N(s)$ of the current solution s generated with nodes in V has a much larger size than the neighbourhood $N^*(s)$ generated with the nodes in V^*. We also noticed that the V^* becomes smaller and smaller while the solution is getting better and better. Hence, we use a strategy such that a smaller $nbmax$ is chosen for the search over V and a larger $nbmax$ chosen for the search over V^*. It is shown that the four-pass search strategy presented below can generate better solutions and save search time.

Pass 1: We start the TS from a given initial solution. We generate neighbours of current solution s with the nodes in V^* and set a bigger value for $nbmax$ to allow the search to do more downhill–uphill movements which may increase the likelihood that more optimal solutions may be found, We store the best solution and the last solution. The last solution may be as good as, or worse than, the best solution.

Pass 2: We use the last solution obtained in pass 1 as the initial solution to start a new search round with smaller l_{TL} and $nbmax$ (compared with the size of V). The best solution obtained in pass 1 is saved as the default best solution. We erase both the short-term memory and longer-term memory before starting the search. The neighbourhood of the current solution is generated from V. We have found that this will change the search behaviour and solution space dramatically and not take much more time because the $nbmax$ is small.

Pass 3: We repeat pass 1 with the last solution and the best solution obtained in pass 2.

Pass 4: We repeat pass 2 with the last solution and the best solution obtained in pass 3. The best solution obtained in this pass is the final solution.

Obviously, this strategy is an intensification strategy in which each pass except Pass 1 intensifies the search in the region containing the best solution obtained in the previous pass. We found that this strategy works well.

5   Quantitative Analysis of Examination Graph

In order to choose an appropriate parameter for the length of TL, we need to analyse the graph of the examination set. We define the *degree* d_i of a node v_i

to be the number of edges connected with v_i directly, and the sum of weights of these edges w_i to be the *weight* of the node.

We calculate:

- the total number of edges, $e = \sum d_i/2$ $(i = 1, 2, \ldots, v)$
- the total enrolment, $s = \sum s_i$ $(i = 1, 2, \ldots, v)$
- the total weights of graph, $w = \sum w_i/2$ $(i = 1, 2, \ldots, v)$
- the density of matrix, $dom = 2e/v(v-1)$
- the average enrolment per exam, $\bar{e} = s/v$
- the average degree per exam, $\bar{d} = 2e/v$
- the average weight per exam, $\bar{w} = 2w/v$.

Experiments have shown that the nodes (exams) with lower enrolment, degree or weight have higher probability to be the most active nodes. This, of course, not only helps the generation of more optimal solutions but also increases the probability of cycling in the TS. Hence, estimating the number of the potentially active nodes may help to decide what is the appropriate length for the longer-term tabu list TL.

If the nodes $(v_1, v_2, \ldots, v_v)$ are sorted according to the number of students enrolled, by the degree of node, and by weight of node, we can create three ordered sets:

$$s' = (s'_1, s'_2, \ldots, s'_v) \qquad \text{where } s'_i \le s'_j \text{ for any } i < j,$$
$$d' = (d'_1, d'_2, \ldots, d'_v) \qquad \text{where } d'_i \le d'_j \text{ for any } i < j,$$
$$w' = (w'_1, w'_2, \ldots, w'_v) \qquad \text{where } w'_i \le w'_j \text{ for any } i < j.$$

Based on s', d' and w', we estimate how many light and heavy nodes there are in the graph by two methods. The first method is called mean points estimation and the second method is called accumulation percentage estimation.

5.1   Method 1: Mean Points Estimation

For a series $z = z_1, z_2, \ldots, z_m$, sorted in ascending order, we calculate the mean value of z, $\bar{z} = \sum z_i/m$. If the kth element z_k is the largest element $le\bar{z}$ then there are k elements with a value less than or equal to $\bar{z}$. We calculate five points $\bar{z}/3$, $\bar{z}/2$, $\bar{z}$, $2\bar{z}$ and $3\bar{z}$, and find the corresponding k. We apply this procedure to series s', d' and w' to get Table 1.

The numbers in the table are the number of nodes with values (enrolment, edge, weight) less than or equal to the corresponding mean points. For example, the value 311 means that there are 311 exams for which the enrolment is less than $1/2$ the average enrolment. We define nm to be the average of the number of exams below the mean point of the three series s', d' and w'. Thus, the entry $300 = (311 + 253 + 336)/3$. The number of nodes in the graph $v = 771$.

From this table, we estimate there are about 194 light exams (below the mean point $\bar{z}/3$) and 32 heavy exams (above the mean point $3\bar{z}$).

Table 1. Mean Points (data: University of Ottawa, 1996)

Mean point	s'	d'	w'	nm	nm/v
min	1	10	10	7	
$\bar{z}/3$	211	143	229	194	0.25
$\bar{z}/2$	311	253	336	300	0.38
$\bar{z}$	533	488	533	518	
$2\bar{z}$	681	687	680	683	
$3\bar{z}$	734	753	731	739	
max	771	771	771	771	

5.2   Method 2: Accumulation Percentages Estimation

From s', d' and w', we calculate ss, dd and ww respectively to get three new series:

$$ss = s'_1, s'_1 + s'_2, s'_1 + s'_2 + s'_3, \ldots, \sum s'_i \qquad (i = 1, 2, \ldots, v)$$

i.e. $ss = (ss_1, ss_2, \ldots, ss_v)$ where $ss_j = \sum s'_i$ $(i = 1, 2, \ldots, j)$, for $j = 1, 2, \ldots, v$, which indicates the accumulative enrolments of the j lightest exams ranked by their enrolment;

$$dd = d'_1, d'_1 + d'_2, d'_1 + d'_2 + d'_3, \ldots, \sum d'_i \qquad (i = 1, 2, \ldots, v)$$

i.e. $dd = (dd_1, dd_2, \ldots, dd_v)$ where $dd_j = \sum d'_i$ $(i = 1, 2, \ldots, j)$, for $j = 1, 2, \ldots, v$ which indicates the accumulative degrees of the j lightest exams ranked according to their degree;

$$ww = w'_1, w'_1 + w'_2, w'_1 + w'_2 + w'_3, \ldots, \sum w'_i \qquad (i = 1, 2, \ldots, v)$$

i.e. $ww = (ww_1, ww_2, \ldots, ww_v)$ where $ww_j = \sum w'_i$ $(i = 1, 2, \ldots, j)$, for $j = 1, 2, \ldots, v$, which indicates the accumulative weights of the j lightest exams ranked according to their weight.

Obviously, we have $s = ss_v$ (the total enrolment), $e = dd_v/2$ (the number of edges), and $w = ww_v/2$. ss_i/ss_v is the ratio of the accumulative enrolment of the i lightest nodes to the total enrolment. For the same reason, dd_i/dd_v (or ww_i/ww_v) are respectively the ratio of the accumulative degrees (or weights) of the i lightest nodes to $2e$ (or $2w$). We choose the percentages 5%, 10%, 15%, 50%, 85%, 90% and 95% and find the corresponding numbers k in the series s', d' and w'.

We define np to be the average of the number of exams below the percentages of the three series.

Note that 196/766 or 25% of the total nodes in the graph have only 5% of the total enrolment, degree, and weights. It is not difficult to conclude that the lowest 25% of the nodes (196 nodes) in the graph have a greater likelihood to be moved in the process of local search.

Table 2. Accumulation percentages (data: University of Ottawa, 1996)

Percentage	s'	d'	w'	np	np/v
5%	199	171	218	196	0.25
10%	299	258	316	291	0.38
15%	372	321	388	360	0.47
50%	648	597	653	631	
85%	757	738	757	751	
90%	763	752	763	759	
95%	768	763	768	766	

5.3   Estimating the Length of the Longer-Term Tabu List

We use the data calculated above to estimate the number of the lightest nodes in the graph. In general, the nodes located in the area $A = (0, \bar{z}/3) \cup (0, 5\%)$ are the most active nodes. The length of the long-term tabu list l_{TL} can be set to the number of elements in area A: for instance, the largest one among np or nm.

If A is too small for some individual application, we can choose area $B = (0, \bar{z}/2) \cup (0, 5\%)$ or something similar.

6   Details of the OTTABU Algorithm

Procedures

$/ * Initialization * /$
set np by analysing the examination set G with accumulation percentage method;
$s :=$ BinPackingWithLargestEnrolmentFirst(G);
set p_0, p_1, and p_2;
$l_{TS} := 9$; nbiter $:= 0$; bestiter $:= 0$; optiter $:= 0$; bestsol $:= s$;
fmin $:= 0$; NeighbourType $:= V^*$;

$/ * A\ Four\ Pass\ Tabu\ Search * /$
t $:= 0$;
while t < 4 do {
   ExamTimeTableTabuSearch();
   if NeighbourType $= V^*$ then NeighbourType $:= V$
   else NeighbourType $:= V^*$;
   t $:= $ t $+ 1$;
   } endwhile;
return;

$/ * ExamTimeTableTabuSearch * /$
MFT $:= \emptyset$; TS $:= \emptyset$; TL $:= \emptyset$; nv $:= np$; nbmax $:= 0$;
while $f(bestsol) > fmin$ and $(nbiter - bestiter) < nbmax$ do {
   nbiter $:= $ nbiter $+ 1$; optiter $:= $ optiter $+ 1$;

if NeighbourType $= V^*$ then VC $:= V^*$ else VC $:= V$;
if $|VC| \leq$ nv then nv $:= 2|VC|/3$;
if NeighbourType $= V^*$ then { $l_{TL} :=$ nv; nbmax $:= 3l_{TL}$; }
    else { $l_{TL} :=$ nv/3; nbmax $:= 3l_{TL}/2$;}
For any $x \leq VC$, generate NS$(s) := \{s_i | s_i := s \oplus m_i\}$ from all atomic
moves $m_i := (x, T_i, T_j)$, where either $((x, j)$ not in TS$) \wedge (x$ not in TL$))$ or
$f(s_i) < f(bestsol)$;
Choose a solution $s^* \in NS(s)$ with minimum f over NS(s) and if more
than one solution is minimal, choose one with lowest move frequency;
$s := s^*$;
update MFT(x); update TS(x, i, j); update TL(x);
if $f(s) < f(bestsol)$ then { bestsol $= s$; bestiter $:=$ nbiter; optiter $:= 0$}
 elseif optiter $> 2l_{TL}$ or $(f(s) - f(bestsol))/f(bestsol) > r$
  then { optiter $:= 0$; TS $:= \emptyset$; TL $:= \emptyset$;}
 } endwhile;
return;

Notes

If, in some iteration, we get a new current solution $s^* = s \oplus (x, T_i, T_j)$, then we
do

1. Update $TS(x, i, j)$: All tenures are decreased by 1. Drop the entries whose
   tenures are 0, and add (x, i). If s^* is the best solution so far, drop (x, j) if it
   exists;
2. Update $TL(x)$: add (x) into TL if TL is not full and $MFT(x) \geq 2$, or replace
   (y) in TL with (x) if TL is full and $MFT[y] < MFT[x]$;
3. Update $MFT(x)$: $MFT[x] := MFT[x] + 1$.

6.1 Symbols

TS, l_{TS}, a short-term tabu list and its maximum tenure;
TL, l_{TL}, a longer-term tabu list and its maximum length;
np, the estimated number of most active nodes in the examination set;
MFT, a node move frequency table, $MFT[x]$ is the move frequency of node x;
$nbiter$, the current iteration number;
$nbmax$, the maxmum number of null iterations;
$bestiter$, the number of iterations since the latest best solution was found;
$optiter$, the number of iterations since the latest tabu relaxation;
s, the current solution;
s^*, the minimal solution generated in some iteration from current solution s;
$bestsol$, the best solution so far;
$NS(s)$, a subset of neighbourhood $N(s)$ or $N^*(s)$ of solution s;
$f(s)$, the value of objective function of solution s. $f(s) = p_0 W_0 + p_1 W_1 + p_2 W_2$,
where p_0, p_1, p_2 are the penalties used to weight the classes of conflicts, and
W_0, W_1, W_2 are the numbers of first-order, second-order and third-order conflicts:

i.e. simultaneous exams, consecutive exams and exams with only one free period between them;

V^*, a set of nodes having at least one conflict in current solution s;

V, all nodes in graph (i.e. all examinations);

T_i, timeslot i (or period i);

m_i, any feasible atomic move (x, T_i, T_j) that moves a node x from T_i into T_j $(j \neq i)$;

r, a ratio used to limit the uphill movement during TS.

7   The Complete System

We have constructed a multi-phase system to optimize the examination timetabling procedure. This can find better solutions than a single-phase one.

Phase One:

- Analyse the examination set with the quantitative analysis methods described earlier in Section 5 and estimate the number of the lightest exams;
- Apply the bin packing algorithm to generate the initial solution (with a given number of timeslots);
- Evaluate this initial solution. If the initial solution is feasible (i.e. there are no first-order conflicts), go to phase three; otherwise go to phase two.

Phase Two:

- Use the solution generated in phase one as the initial solution;
- Construct an objective function using only first-order conflicts: $f(s) = W_0$;
- Apply OTTABU to the infeasible solution to generate a feasible solution; if OTTABU cannot generate a feasible solution, then add one more period to the timetable and re-run phase two;
- If a feasible solution is generated then proceed to phase three.

Phase Three:

- Use the solution generated in phase one or phase two as the initial solution. Construct an objective function using first-order and second-order conflicts: $f(s) = p_0 \times W_0 + W_1, \quad p_0 \gg 1$;
- Apply OTTABU to the feasible solution to optimize the solution;
- Go to phase four.

Phase Four:

- Use the solution generated in phase three as the initial solution. Construct an objective function using first-order, second-order and third-order conflicts: $f(s) = p_0 \times W_0 + p_1 \times W_1 + W_2, \quad p_0 \gg p_1 \gg 1$;
- Apply OTTABU to the feasible solution to optimize the solution.

8   Results

There are 771 exams to be assigned into 36 timeslots and each timeslot has 2200 seats available. In order to avoid seating large numbers of students at the end of the examination session (where there are no following exterior edges) the seating for these three timeslots is reduced to 1100. Some quantitative data about the examinations are

- The total number of periods, 36,
- The total number of examinations, 771,
- The total number of students, 14 032,
- The average number of exams per period, 21,
- The number of seats for per period, 2200,
- The total number of seats available, 75 900,
- The total number of student-exam enrolments, 46 899,
- The average number of exams per student, 3.34,
- The average number of student-exams per period, 1303,
- The ratio of seat utilization, 61.79%,
- The average number of students per exam, 60.82,
- The total number of edges in the graph, 18 522,
- The average degree per node, 48,
- The density of conflict matrix, 0.06,
- The total weights of edges in the graph, 140 704,
- The average weights of edges per exam, 182.

The number of lightest exams is estimated as $np = 196$ at the 5% accumulation percentage (see Table 2).

In phase one, we apply a bin packing algorithm to get a feasible solution s_0 such that $(W_0/W_1/W_2) = (0/8686/7844)$. This means that no student needs to take two exams at the same time (first-order conflict), 8686 students have to take two consecutive exams (second-order conflicts) and 7844 students have to take exams having third-order conflicts.

We use the TS technique with both short-term tabu list and longer-term tabu list to optimize this timetable. Since a feasible solution was achieved in phase one, we go directly to phase three, apply OTTABU to optimize W_0/W_1 and obtain the result shown in Table 3. We set $p_0 = 500$, $p_1 = 1$, $np = 196$, and $l_{TS} = 9$. l_{TL} was set to 190 in pass 1 and pass 3. The initial solution is characterized by 0/8686/7844 obtained by the bin packing algorithm of phase one ($f(s) = 8686$).

Then, going to phase four, we use the solution s obtained above as an initial value. We set $p_0 = 250\,000$, $p_1 = 500$, $p_2 = 1$, $np = 196$ (at the 5% accumulation percentage), $l_{TS} = 9$, and the initial solution has 0/496/5643, obtained in phase three ($f(s) = 253\,643$).

The best solution is *bestsol* where $(W_0/W_1/W_2) = (0/494/4354)$ which had an execution time of approx. 86 min on a Pentium 100 MHz PC. In this solution, 494 students (approx. 3% of the total enrolment) have to take two consecutive exams. This is the best solution obtained by the system with these data.

Table 3. Case 1: both longer-term and short-term memory applied (phase three)

Pass	Neighbourhood	l_{TL}	$nbmax$	Iterations	Time	bestsol	$f(bestsol)$
1	$N^*(s)$	190	570	2595	0:16:00	0/776	776
2	$N(s)$	63	94	761	0:12:06	0/592	592
3	$N^*(s)$	190	570	500*	0:02:14	0/590	590
4	$N(s)$	63	94	1224	0:19:36	0/496	496
Total				5080	0:49:56		

* Because V^* in pass 3 is less than l_{TL}, both l_{TL} and $nbmax$ were changed automatically.

Table 4. Case 2: both longer-term and short-term memory applied (phase four)

Pass	Neighbourhood	l_{TL}	$nbmax$	Iterations	Time	bestsol	$f(bestsol)$
1	$N^*(s)$	190	570	772	0:12:15	0/496/4546	252546
2	$N(s)$	63	94	94	0:02:18	0/496/4546	252546
3	$N^*(s)$	190	570	664	0:19:47	0/494/4354	251354
4	$N(s)$	63	94	94	0:02:17	0/494/4354	251354
Total				1624	0:36:37		

9   Comparison

In order to examine the usefulness of longer-term frequency-based memory, we have done some experiments that attempt to explore the behaviour of OTTABU when the longer-term memory is disabled or set with different maximum lengths.

First, the longer-term tabu list in OTTABU was disabled; the penalties p_0, p_1 and p_2 were set to the same values as in cases 1 and 2; $nbmax$ was fixed at 270. We set $p_0 = 500$, $p_1 = 1$, $l_{TS} = 7$ and the initial solution is 0/8686/7844. The results are shown in Table 5.

Table 5. Case 3: short-term memory only. I

Pass	Neighbourhood	$nbmax$	Iteration	Time	bestsol	$f(bestsol)$
1	$N^*(s)$	270	785	0:06:00	0/892	892
2	$N(s)$	270	494	0:07:00	0/772	772
3	$N^*(s)$	270	307	0:02:00	0/759	759
4	$N(s)$	270	286	0:04:38	0/748	748
Total			1772	0:19:38		

We have repeated the search, using the same conditions of case 3 with the sole difference that the order of applying neighbourhood definitions is reversed, resulting in a solution initially slower but better than previously obtained. But in the end, the solution quality is about the same, as is the time required to obtain it. We set $p_0 = 500$, $p_1 = 1$, $l_{TS} = 7$ and the initial solution is 0/8686/7844. The results are shown in Table 6.

Table 6. Case 4: short-term memory only. II

Pass	Neighbourhood	$nbmax$	Iterations	Time	$bestsol$	$f(bestsol)$
1	$N^*(s)$	270	767	0:12:16	0/801	801
2	$N(s)$	270	342	0:02:01	0/776	776
3	$N^*(s)$	270	291	0:04:44	0/760	760
4	$N(s)$	270	390	0:02:20	0/749	749
total			1790	0:21:21		

The results show that:

1. By using $N(s)$, we can get a much better solution than by using $N^*(s)$ in pass 1 even though the former took twice as much time as does the latter;
2. The four-pass algorithm can greatly improve the timetable. The best solution obtained in pass 1 is improved in the later passes. The value of the objective function is minimized by about 16% in case 3, and about 7% in case 4;
3. Even though we set a bigger value for $nbmax$, the search ended much earlier than we expected because of cycling. In fact, we have later used a smaller $nbmax$ to save time without having a negative impact on the search.

Comparing cases 3 and 4 with cases 1 and 2, we notice that the use of longer-term memory had a greater effect on the quality of the final result. The use of longer-term memory resulted in a solution having a value of 496 whereas without longer-term memory, the best obtainable solution had a value of 748. The reduction of $(748 - 496)/748 = 0.34$ is due entirely to the use of longer-term memory.

We have also set different values for the maximum length of the longer-term tabu list, and find the right l_{TL} not only to get the best solution and but also to reduce search time.

It can be seen from Table 7 that reducing l_{TL} to 30 results in solutions that are much worse than those shown in Table 4, (solution value is 326 668 as opposed to 251 354). We set $p_0 = 250\,000$, $p_1 = 500$, $p_2 = 1$, $l_{TL} = 30$, $l_{TS} = 9$, and the initial solution is 0/8686/7844.

We set $p_0 = 500$, $p_1 = 1$ and $np = 354$ (at 15% accumulation percentage), $l_{TS} = 9$, and the initial solution is 0/8686/7844 obtained by a bin packing algorithm ($f(s) = 8686$). The result is shown in Table 8.

Thus far we have examined the influence of the longer-term tabu list on the Ottawa University data used. A similar study was performed on some data available in Carter's repository [11]. These data sets, called CAR-F-92 and UTA-S-92 in Table 1 of [11], were used as the input data of OTTABU. In both cases, when the longer-term list was removed from the program the results were not

Table 7. A longer-term tabu list that is too small

Pass	Neighbourhood	l_{TL}	$nbmax$	Iterations	Time	bestsol	$f(bestsol)$
1	$N^*(s)$	30	270	2513	0:39:47	0/812/4745	470745
2	$N(s)$	30	270	768	0:22:06	0/659/4644	334144
3	$N^*(s)$	30	270	579	0:08:40	0/651/4817	330317
4	$N(s)$	30	270	839	0:19:33	0/644/4688	326668
Total				4699	1:29:05		

Table 8. A longer-term tabu list that is too large

Pass	Neighbourhood	l_{TL}	$nbmax$	Iterations	Time	bestsol	$f(bestsol)$
1	$N^*(s)$	354	1062	2595*	0:16:32	0/776	776
2	$N(s)$	118	177	844	0:14:12	0/592	592
3	$N^*(s)$	354	1062	502*	0:02:29	0/590	590
4	$N(s)$	118	177	1425	0:23:49	0/509	509
Total				5080	0:55:52		

* Both l_{TL} and $nbmax$ are changed automatically when the size of V^* is less than np.

as satisfactory as when it was present. For the CAR-F-92 data set, the use of a longer-term list reduces the ultimate cost by about 15% and for the UTA-F-92 data set, the figure is about 7%. Both these values are less than the value calculated in the previous section for the Ottawa data.

We believe that the main reason for this is related to the matrix density. Both the CAR and UTA data sets have a matrix density that is relatively high, 0.14 and 0.13 respectively. The density of the Ottawa data was calculated to be 0.06. Thus there are less than half as many edges per node on average for the Ottawa data as there are for the others.

Thus, when a move is made using the Ottawa data a large reduction in the penalty function is possible because the matrix density is smaller, increasing the likelihood that a move of a node to the target timeslot will result in a smaller (or zero) penalty. Study of this effect is continuing.

10   Comparison of OTTABU with Other Results

The comparison of various strategies within an environment in which everything else is held constant is relatively easy. This is the approach followed in the preceding sections in which it became evident that incorporating constraint relaxation and a longer-term tabu list into a TS led to better-quality solutions than could be obtained otherwise, at least in the environment used. It is something else to decide whether these techniques will also work in other environments.

This question is complicated by the fact that many of the other algorithms have been written to solve certain problems particular to some institution. As Carter et al. [11] point out, "As all real timetabling problems have different side constraints with them, for all practical purposes each school has a unique problem."

Nonetheless, in order to see whether the improvements demonstrated by OTTABU in the Ottawa University environment would continue to be shown in other environments, some changes were made to the algorithm and the data in such a way that some comparisons could be made. These included the following modifications:

1. Data from two other universities were downloaded from Carter's repository [11]. The data sets, CAR-F-92 and UTA-S-92, were used as the input data of OTTABU.
2. The value of the objective function was changed to include farther proximities or higher-order conflicts. The cost weights were taken as $w_1 = 16$, $w_2 = 8$, $w_3 = 4$, $w_4 = 2$, $w_5 = 1$. The possibility of simultaneous examinations was taken as a hard constraint and was forbidden.
3. The common metric "cost", taken to be the total penalty using the weights listed above divided by the number of students involved, was used as the basis of comparison.
4. OTTABU was modified to incorporate a random tenure in its short-term tabu list so that it could be restarted with the same parameters several times to get a number of final solutions and a selection of best solutions and costs. This also makes the short-term tabu list part of OTTABU more like the algorithm of Di Gaspero and Schaerf [12].

With these modifications several runs were made and the final solutions obtained by OTTABU could be compared directly with the results published by Carter et al. [11] and by Di Gaspero and Schaerf [12]. The program EXAMINE, reported by Carter et al. [11] uses 40 different strategies overall, incorporating five basic sorting rules, using cliques or not, using costs or not and using backtracking or not. The results published in their Table 5 have associated costs ranging from 6.2 to 8.2 for the data set CAR-F-92 and ranging from 3.5 to 6.4 for data set UTA-S-92.

The program of Di Gaspero and Schaerf uses a short-term memory with tenure varying randomly between 15 and 25. The stopping criteria is based on the number of iterations from the last improvement (idle iterations), the number of iterations depending of the instance, varying from 2000 to 20 000. Their average cost for CAR-F-92 was 5.6 and for UTA-S-92 was 4.5.

In comparison, the results obtained by OTTABU as modified above were an average of 4.7 for CAR-F-92 and 4.0 for UTA-S-92. These results are shown in Table 9.

The average cost of the OTTABU solution is less than that reported by Di Gaspero and Schaerf in both cases and is less than all but one of the values reported by Carter et al. for these two data sets.

The overall conclusion resulting from this comparison is that OTTABU's use of longer-term tabu lists has a demonstrated beneficial effect in all the cases tested and that OTTABU's results compare favourably with the other algorithms examined.

Table 9. Comparison of results

Data set	Number of timeslots	OTTABU average	Di Gaspero and Schaerf average	Carter et al.
CAR-F-92	32	4.7	5.6	6.2–8.2
UTA-S-92	35	4.0	4.5	3.5–6.4

11   Conclusion

In the OTTABU algorithm, we use both a recency-based short-term tabu list and a move (or transition)-frequency-based longer-term tabu list which prevent cycling and diversify the search space effectively to help get a better solution. We have developed a four-pass TS technique to intensively search the region containing the best solution in order to search for a better solution. Based on our experimental tests using real data from three institutions we conclude:

1. For a large-scale examination timetabling problem, TS techniques, based on short-term memory only, make progress quickly from a non-optimal initial solution but cannot generate a really high-quality solution. We have implemented an algorithm in which both longer-term and short-term memory are used to generate better solutions than a short-term memory alone. This is observed to be true on every data set we have used.
2. It is shown in our study that the move- or transition-frequency-based long-term memory is a technique that is easy to understand and implement. The TL tabu list was observed to forbid the movement of potential overactive exams, thus preventing cycling and diversifying the search space.
3. We successfully applied a quantitative analysis procedure to estimate the number of the lightest exams (i.e. the potentially movable exams). This makes it possible to automatically determine the appropriate size of the longer-term tabu list.
4. The relaxation of tabu lists is an effective way of avoiding potential over-uphill movements, reduce the search time and search space and increase the likelihood of finding optimal solutions.
5. The four-pass algorithm OTTABU is an effective mechanism to control the balance between diversification and intensification in the solution search.

Acknowledgements. The former registrar of the University of Ottawa, Mr George H. von Schoenberg, was instrumental in initiating the research that led to this study. The Manager, Registration and Records, Ms Pauline Bélanger, helped greatly in all facets of the implementation and data collection.

References

1. Peck, J.E.L., Williams, M.R.: Algorithm 286 – Examination Scheduling. Commun. A.C.M. **9** (1966) 433–434

2. White, G.M. Chan, P.-W.: Towards the Construction of Optimal Examination Schedules. INFOR **17** (1979) 219–229
3. Carter, M.W., Laporte, G., Chinneck, J.W.: A General Examination Scheduling System. Interfaces **24** (1994) 109–120
4. Carter, M.W., Laporte, G.: Recent Developements in Practical Examination Timetabling. Lecture Notes in Computer Science, Vol. 1153. Springer-Verlag, Berlin Heidelberg New York (1996) 3–21
5. de Werra, D., Hertz, A.: Tabu Search Techniques – a Tutorial and Applications to Neural Networks. OR Spektrum **11** (1989) 131–141
6. Hertz, A.: Tabu Search for Large Scale Timetabling Problems. Eur. J. Oper. Res. **54** (1991) 39–47
7. Boufflet, J.P., Negre, S.: Three Methods Used to Solve an Examination Timetable Problem. Lecture Notes in Computer Science, Vol. 1153. Springer-Verlag, Berlin Heidelberg New York (1996) 3–21
8. Ferland, J.A.: Generalized Assignment-Type Problems: a Powerful Model Scheme. Lecture Notes in Computer Science, Vol. 1408. Springer-Verlag, Berlin Heidelberg New York (1998) 53–77
9. Glover, F: Tabu Search: a Tutorial. Interfaces **20** (1990) 74–94
10. Glover, F., Laguna, M.: Tabu Search. Kluwer, Dordrecht (1997)
11. Carter, M.W., Laporte, G., Lee, S.Y.: Examination Timetabling: Algorithmic Strategies and Applications. J. Oper. Res. Soc. **47** (1996) 373–383
12. Di Gaspero, L., Schaerf, A.: Proc. 3rd Int. Conf. Practice and Theory of Automated Timetabling (Konstanz, Germany) (2000) 176–179

Tabu Search Techniques for Examination Timetabling

Luca Di Gaspero[1] and Andrea Schaerf[2]

[1] Dipartimento di Matematica e Informatica
Università degli Studi di Udine
via delle Scienze 206, I-33100, Udine, Italy
digasper@dimi.uniud.it
[2] Dipartimento di Ingegneria Elettrica, Gestionale e Meccanica
Università degli Studi di Udine
via delle Scienze 208, I-33100, Udine, Italy
schaerf@uniud.it

Abstract. The EXAMINATION TIMETABLING problem regards the scheduling for the exams of a set of university courses, avoiding the overlapping of exams having students in common, fairly spreading the exams for the students, and satisfying room capacity constraints.

We present a family of solution algorithms for a set of variants of the EXAMINATION TIMETABLING problem. The algorithms are based on tabu search, and they import several features from the research on the GRAPH COLOURING problem.

Our algorithms are tested on both public benchmarks and random instances, and they are compared with previous results in the literature. The comparison shows that the presented algorithms performs as well as constructive methods and memetic algorithms, and only a decomposition based approach outperforms them in most cases.

1 Introduction

The EXAMINATION TIMETABLING problem is a combinatorial problem that commonly arises in universities: schedule a certain number of examinations in a given number of time slots in such a way that no student is involved in more than one exam at a time. The assignment of exams to days and to time slots within the day is also subject to constraints on availabilities, fair spreading of the students workload, and room capacity. Different variants of the timetabling problem have been proposed in the literature, which differ from each other based on the type of constraints and objectives (see [4], [17] for recent surveys). Constraints involve room capacity and teacher availability, whereas objectives mainly regard student workload.

In this paper, we present an ongoing research on the development of a family of solution algorithms for a set of variants of the EXAMINATION TIMETABLING problem. Our algorithms are based on tabu search [11], and they make use of several features imported from the literature on the GRAPH COLOURING problem [9, Problem GT4, page 191].

E. Burke and W. Erben (Eds.): PATAT 2000, LNCS 2079, pp. 104–117, 2001.
© Springer-Verlag Berlin Heidelberg 2001

The investigation of different versions of the problem allows us not only to obtain a more flexible application, but also to understand the general structure of the problem family. As a consequence, we should be able to perform a more robust parameter tuning mechanisms and to compare our results with most of previous ones on different versions of the problem.

We tested our algorithms on the popular Toronto benchmarks [6] and on the Nottingham instance, which are the only publicly available ones at present[1] for EXAMINATION TIMETABLING. In this way, we could compare the outcomes of our algorithms with previous results obtained in [1], [2], [6] using different techniques.

This paper is organized as follows. In Section 2 we introduce several formulations of the EXAMINATION TIMETABLING problem, which take into account different combinations of hard and soft constraints. Section 3 contains a presentation of the general local search framework and, in particular, of the tabu search meta-heuristic and of the tandem strategy. The family of algorithms employed in this study are described in detail in Section 4, whereas in the following section we compare their results with those obtained in previously published works. In Section 6 we draw some conclusions and consider some future lines of research that we intend to pursue.

2   Examination Timetabling Problem

We introduce the EXAMINATION TIMETABLING problem in stages. In Section 2.1 we present the basic search problem. In Section 2.2 we consider further (hard) constraints, and we describe various components of the objective function (soft constraints). Finally, in Section 2.3 we describe other versions of the problem not considered in this work.

2.1   Basic Search Problem

Given is a set of n exams $E = \{e_1, \ldots, e_n\}$, a set of q students $S = \{s_1, \ldots, s_q\}$, and a set of p time slots (or periods) $P = \{1, \ldots, p\}$.

Consecutive time slots lie one unit apart; however, the time distance between periods is not homogeneous due to lunch breaks and nights. This fact is taken into account in second-order constraints as explained below.

There is a binary *enrolment* matrix $C_{n \times q}$, which tells which exams the students plan to attend: i.e. $c_{ij} = 1$ if and only if student s_j wants to attend exam e_i.

The basic version of EXAMINATION TIMETABLING is the problem of assigning examinations to time slots avoiding exam overlapping.

The assignment is represented by a binary matrix $Y_{n \times p}$ such that $y_{ik} = 1$ if and only if exam e_i is assigned to period k. The corresponding mathematical formulation is the following.

---

[1] At `ftp://ie.utoronto.ca/pub/carter/testprob` and `ftp://ftp.cs.nott.ac.uk/ttp/Data`, respectively.

$$\text{find} \quad y_{ik} \quad (i = 1, \dots, n; k = 1, \dots, p)$$

$$\text{s.t.} \quad \sum_{k=1}^{p} y_{ik} = 1 \quad (i = 1, \dots, n) \tag{1}$$

$$\sum_{h=1}^{q} y_{ik} y_{jk} c_{ih} c_{jh} = 0 \quad (k = 1, \dots, p; i, j = 1, \dots, n; i \neq j) \tag{2}$$

$$y_{ik} = 0 \text{ or } 1 \quad (i = 1, \dots, n; k = 1, \dots, p). \tag{3}$$

Constraints (1) state that an exam must be assigned exactly to one time slot; Constraints (2) state that no student shall attend two exams scheduled at the same time slot.

It is easy to recognize that this basic version of the EXAMINATION TIMETABLING problem is a variant of the well-known NP-complete GRAPH COLOURING problem. In particular, colours represent time slots, each node represents an exam, and there is an (undirected) edge between two nodes i and j if at least one student is enrolled in both exams e_i and e_j.

2.2 Additional Constraints and Objectives

Many different types of hard and soft constraints have been considered in the literature on EXAMINATION TIMETABLING. The hard ones that we take into account are the following.

Capacity: Based on the availability of rooms, we have a *capacity* array $L = (l_1, \dots, l_p)$, which represents the number of available seats. For each time slot k, the value l_k is an upper bound of the total number of students that can be examined at period k. The capacity constraints can be expressed as follows:

$$\sum_{i=1}^{n} \sum_{h=1}^{q} c_{ih} y_{ik} \leq l_k \qquad k = 1, \dots, p. \tag{4}$$

Notice that we do not take into account the number of rooms, but only the total number of seats available in that period. This is reasonable under the assumption that more than one exam can take place in the same room. Alternative formulations that assign one exam per room are discussed in Section 2.3.

Preassignments and Unavailabilities: An exam e_i may have to be scheduled necessarily in a given time slot k, or, conversely, may not be scheduled in such a time slot. These constraints are added to the formulation by simply imposing y_{ik} to be 1 or 0 respectively.

We now describe the soft constraints, which contribute, with their associated weights, to the objective function to be minimized. For the sake of brevity, we do not provide the mathematical formulation of the objective function.

Second-Order Conflicts: A student should not take two exams in consecutive periods. To this aim, we include in the objective function a component that counts the number of times a student has to take a pair of exams scheduled at adjacent periods. Many versions of this constraint type have been considered in the literature, according to the actual time distance between periods:
   I) Penalize conflicting exams equally.
  II) Penalize *overnight* (last one of the evening and the first of the morning after) adjacent periods less than all others [1].
 III) Penalize less the exams just before and just after lunch, do not penalize overnight conflicts [7].

Higher-Order Conflicts: This constraint penalizes also the fact that a student takes two exams in periods at distance three, four, or five. Specifically, it assigns a proximity cost $pc(i)$ whenever a student has to attend two exams scheduled within i time slots. The cost of each conflict is thus multiplied by the number of students involved in both examinations. As in [6], we use a cost function that decreases from 16 to 1 as follows: $pc(1) = 16$, $pc(2) = 8$, $pc(3) = 4$, $pc(4) = 2$, $pc(5) = 1$.

Preferences: Preferences can be given by teachers and student for scheduling exams to given periods. This is the soft version of preassignments and unavailability.

We have considered many versions of the problem, which differ from each other based on which of the above hard and soft constraints are taken into account.

2.3   Other Variants of the Problem

We briefly discuss variants of the problem and different constraint types not considered in our work.

Room assignment: Some authors (see, e.g., [5]) allow only one exam per room in a given period. In this case, then exams must be assigned not only to periods, but also to rooms. The assignment must be done based on the number of students taking the exams and capacity of rooms.

Special rooms: Some other authors (see, e.g., [14]) consider also different types of rooms, and exams that may only be held in certain types of rooms.

Exams of variable length: Exams may have length that do not fit in one single time slot. In this case consecutive time slots must be assigned to them.

Minimize the length of the session: We have assumed that the session has a fixed length. However, we may also want to minimize the number of periods required to accomplish all the exams. In that case, the number of periods p becomes part of the objective function.

Other higher-order conflicts: Carter et al. [5] generalize the higher-order constraints and consider a penalty for the fact that a student is forced to take x exams in y consecutive periods.

3   Local Search

Local search is a family of general-purpose techniques for search and optimization problems. Local search techniques are *non-exhaustive* in the sense that they do not guarantee to find a feasible (or optimal) solution, but they search non-systematically until a specific stop criterion is satisfied.

3.1   Basics

Given an instance π of a problem Π, we associate a *search space* S with it. Each element $s \in S$ corresponds to a potential solution of π, and is called a *state* of π. Local search relies on a function N, depending on the structure of Π, which assigns to each $s \in S$ its *neighbourhood* $N(s) \subseteq S$. Each state $s' \in N(s)$ is called a *neighbour* of s.

A local search algorithm starts from an initial state s_0 (which can be obtained with some other technique or generated randomly) and enters a loop that *navigates* the search space, stepping from one state s_i to one of its neighbours s_{i+1}. The neighbourhood is usually composed by the states that are obtained by some local change (called *move*) from the current one.

Local search techniques differ from each other according to the strategy they use to select the move in each state and to stop the search. In all techniques, the search is driven by a *cost function* f that estimates the quality of each state. For optimization problems, f generally accounts for the number of violated constraints and for the objective function of the problem.

The most common local search techniques are *hill climbing*, *simulated annealing*, and *tabu search* (TS). We now describe in more detail TS, which is the technique that we use in our application. However, a full description of TS is beyond the scope of this paper (see, e.g., [11] for a detailed review): we only present the formulation of the technique which has been used in this work.

3.2   Tabu Search

At each state s_i, TS explores a subset V of the current neighbourhood $N(s_i)$. Among the elements in V, the one that gives the minimum value of the cost function becomes the new current state s_{i+1}, independently of the fact that $f(s_{i+1})$ is less or greater than $f(s_i)$.

Such a choice allows the algorithm to *escape* from local minima, but creates the risk of cycling among a set of states. In order to prevent cycling, the so-called *tabu list* is used, which determines the forbidden moves. This list stores the most recently accepted moves. The *inverses* of the moves in the list are forbidden.

The simplest way to run the tabu list is as a queue of fixed size k. That is, when a new move is added to the list, the oldest one is discarded. We employ a more general mechanism which assigns to each move that enters the list a random number of moves, between two values $k_{\min}$ and $k_{\max}$ (where $k_{\min}$ and $k_{\max}$ are parameters of the method), that it should be kept in the tabu list.

When its tabu period is expired, a move is removed from the list. In this way the size on the list is not fixed, but varies dynamically in the interval k_{min}–k_{max}.

There is also a mechanism, called *aspiration*, that overrides the tabu status: If a move m leads to a state whose cost function value is better than the current best, then its tabu status is dropped and the resulting state is acceptable as the new current one.

The stop criterion is based on the so-called *idle iterations*: The search terminates when it reaches a given number of iterations elapsed from the last improvement of the current best state.

3.3   Tandem Search

One of the attractive properties of the local search framework is that different techniques can be combined and alternated to give rise to complex algorithms.

In particular, we explore what we call the *tandem* strategy, which is a simple mechanism for combining two different local search techniques and/or two different neighbourhood relations. Given an initial state s_0 and two basic local search techniques t_1 and t_2, that we call *runners*, the tandem search alternates a run of each t_i, always starting from the best solution found by the other one.

The full process stops when it performs a round without an improvement by any of the two runners, whereas the component runners stop according to their specific criteria.

The effectiveness of tandem search has been stressed by several authors (see [11]). In particular, when one of the two runners, say t_2, is not used with the aim of improving the cost function, but rather for diversifying the search region, this idea falls under the name of *iterated* local search (see, e.g., [18]). In this case the run with t_2 is normally called the *mutation* operator or the *kick* move.

4   Tabu Search for Examination Timetabling

We propose TS algorithms for EXAMINATION TIMETABLING, along the lines of the TS algorithm for GRAPH COLOURING proposed by Hertz and de Werra [12].

As already mentioned, EXAMINATION TIMETABLING is an extension of the GRAPH COLOURING problem. In order to represent the additional constraints, we extend the graph with an edge-weight function that represents the number of students involved in two conflicting examinations and a node-weight function that indicates the number of students enrolled in each examination.

4.1   Search Space and Cost Function

As in [12], the search space is composed by all complete assignments including the infeasible ones. The only constraints that we impose to be satisfied in all states of the search space are the unavailabilities and preassignments. This can be easily obtained generating initial solutions that satisfy them, and forbidding moves that lead to states that violate them.

The cost function that guides the search is a hierarchical one, in the sense that it is a linear combination of hard and soft constraints with the weight for hard constraints larger than the sum of all weights of the soft ones. For many cases, though, this simple strategy of assigning fixed weights to the hard and the soft components does not work well. Therefore, during the search the weight w of each component (either hard or soft) is let to vary according to the so-called *shifting penalty* mechanism (see, e.g., [10]):

- If for K consecutive iterations all constraints of that component are satisfied, then w is divided by a factor γ randomly chosen between 1.5 and 2.
- If for H consecutive iterations all constraints of that component are satisfied, then the corresponding weight is multiplied by a random factor in the same range.
- Otherwise, the weight is left unchanged.

The values H and K are parameters of the algorithm (and their values are usually between 2 and 20).

This mechanism changes continuously the shape of the cost function in an adaptive way, thus causing TS to visit solutions that have a different structure than the previously visited ones.

4.2   Neighbourhood Relation

In our definition, two states are *neighbours* if they differ for the period assigned to a single course. Therefore, a move corresponds to changing the period of one course, and it is identified by a triple ($\langle course \rangle$, $\langle old\_period \rangle$, $\langle new\_period \rangle$).

Regarding the notion of inverse of a move, we experimented with various definitions, and the one that has given the best results is the one that considers inverse of (e, k_1, k_2) any move of the form $(e, \_, \_)$. That is, the period of the course cannot be changed again to any new one.

In order to identify the most promising moves at each iteration, we maintain the so-called *violations list* VL, which contains the exams that are involved in at least one violation (either hard or soft). A second (shorter) list HVL contains only the exams that are involved in violations of hard constraints. In different stages of the search (as explained in Section 4.4), exams are selected either from VL or from HVL. Exams not in the lists are never analysed.

For the selection of the move among the exams in the list (either VL or HVL), we experimented with two different strategies:

Sampling: Examine a sample of candidate exams selected based on a dynamic random-variate probability distribution biased on the exams with higher influence in the cost function.

Exhaustive: Examine systematically all exams.

In both cases, the selection of the new period for the chosen exam u is exhaustive, and the new period is assigned in such a way that leads to a smallest value of the cost function, arbitrarily tie breaking.

More complex kinds of neighbourhood relations (see [16] for a review) are currently under investigation.

4.3   Initial Solution Selection

Many authors (see, e.g., [12,13]) suggest for GRAPH COLOURING to start local search from an initial solution obtained with an *ad hoc* algorithm, rather than from a random state. We experimentally observe that, indeed, giving a good initial state saves significant computational time, which can be usefully exploited for a more complete exploration of the search space. Therefore, we use a greedy algorithm that builds p independent sets (feasible period classes) and assigns all the remaining exams randomly.

4.4   Search Techniques

We implemented two main TS-based solvers. The first one is a single runner that uses an adaptive combination of VL and HVL. In detail, it selects from HVL when there are some hard violations, and resorts to VL in any iteration in which HVL is empty.

Our second solver is a tandem that alternates the above-described runner with a second one that always selects the exams from the violations list VL. Intuitively, the first runner focuses on hard constraints, favouring the moves that affect them, whereas the second one searches for any kind of improvement. The former, however, once it has found a feasible solution, automatically expands the neighbourhood to also include moves that deal with soft constraints.

Both runners use the shifting penalty mechanism. In addition, they both use the exhaustive exploration of the violation list, because it "blends" well with the shifting penalty mechanism. In fact, in the presence of a continuous change of the cost function, the use of a more accurate selection of the best move is experimentally proven to be more effective.

5   Experimental Results

Our code has been written in C++ using the framework EASYLOCAL++ [8]. We use also the LEDA libraries [15] that provide a toolbox for data structures and algorithms. Our algorithms have been coded in C++ using the GNU g++ compiler version 2.95.1. The tests reported in the following sections were conducted on an AMD Athlon 650MHz PC running Linux.

5.1   Benchmarks and Experimental Setting

Up to now, the real-world data sets available to the timetabling community are the 12 Toronto instances and the single Nottingham instance. We tested our algorithms with these and with random instances. Unavailabilities, preassignments, and preferences are considered only in the random instances, while all other components are present in the benchmarks, too.

Our best results have been obtained by the solver that uses one single runner. The setting of the parameters for each instance has been made after an extensive tuning session, and the best performances are obtained with a tabu tenure varying randomly between 15 and 25. The stop criterion is based on the number of iterations from the last improvement (idle iterations); it depends on the instance and it is set, varying from 2000 to 20 000, so that the duration of the runs is normally kept below 300 s.

In the following sections, we present the results on the benchmark instances and compare them with previous ones obtained using different methods. For the sake of brevity, we omit the results on random instances. Due to the difference in the computing power of the machines involved in the comparison, it seems unfair to compare running times. For this reason, we decided to report only the quality of the solution found.

5.2   Comparison with Carter, Laport, and Lee

Carter et al. [6] present some results about the application of a variety of constructive algorithms for all the Toronto instances. The algorithms are based on backtracking and exploit several well-known GRAPH COLOURING heuristics. They consider the formulation of the problem with second-order conflicts (version I) and higher-order conflicts, but no capacity constraints.

The objective function is normalized based on the total number of students. In this way they obtain a measure of the number of violations "per student", which allows them to compare results for instances of different size.

Table 1 summarizes the performances of our first algorithm with respect to Carter's results. The last column shows the best and the worst result among the set of techniques experimented by Carter et al.

The table shows that our results are comparable with Carter's in many cases, though we perform better than all constructive techniques in four cases (boldface).

5.3   Comparison with Burke, Newall, and Weare

Burke et al. [2] consider the problem with capacity constraints and second-order conflicts (version I), and they solve it using a memetic algorithm (MA1). The algorithm interleaves evolutionary steps (mutation and cross-over) with a local search operator. This way, the genetic component is able to focus on a restricted search space made up of local optima only.

Table 1. Comparison with results of Carter et al. [6]

Data set	Exams	Time slots	TS solver		Carter et al.
			Best	Avg	Costs
CAR-F-92	543	32	**5.2**	5.6	6.2–7.6
CAR-S-91	682	35	**6.2**	6.5	7.1–7.9
EAR-F-83	189	24	45.7	46.7	36.4–46.5
HEC-S-92	80	18	12.4	12.6	10.8–15.9
KFU-S-93	461	20	18.0	19.5	14.0–20.8
LSE-F-91	381	18	15.5	15.9	10.5–13.1
STA-F-83	138	13	**160.8**	166.8	161.5–165.7
TRE-S-92	261	23	10.0	10.5	9.6–11.0
UTA-S-92	638	35	4.2	4.5	3.5–4.5
UTE-S-92	184	10	29.0	31.3	25.8–38.3
YOR-F-83	180	21	**41.0**	42.1	41.7–49.9

Table 2. Comparison with results of Burke et al. [2]

Data set	Exams	Time slots	TS solver		Burke et al.
			Best cost	Avg cost	MA1
CAR-F-92	543	40	424	443	331
CAR-S-91	543	51	88	98	81
KFU-S-93	461	20	**512**	597	974
TRE-S-92	261	35	4	5	3
NOTT	800	26	**11**	13	53
NOTT	800	23	**123**	134	269
UTA-S-93	638	38	**554**	625	772

Table 2 presents the comparison of our TS solver with the best results obtained by Burke *et al.* [2]. The table shows that our results are superior in many cases.

5.4   Comparison with Burke and Newall

The results of Burke et al. [2] are refined by Burke and Newall [1], who consider the problem with capacity constraints and second-order conflicts, but in this case with version II. They present results about a subset of the Toronto instances and on the Nottingham one.

They propose a new version of the above memetic algorithm. This version uses a multistage procedure that decomposes the instances into smaller ones and combines the partial assignments. The decomposition is performed along the lines proposed by Carter in [3]. For comparison, they also implement a constructive method.

Table 3. Comparison with results of Burke and Newall [1]

Data set	Exams	Time slots	TS solver		Burke and Newall		
			Best	Avg	MA2	MA2+D	Con
CAR-F-92	543	36	3048	3377	12167	1765	2915
KFU-S-93	461	21	1733	1845	3883	1608	2700
NOTT	800	23	751	810	1168	736	918
PUR-S-93	2419	30	123 935	126 046	219 371	65 461	97 521

Table 3 shows the comparison of their best[2] results with our TS solver. In the table, we call MA2 the memetic algorithm which uses decomposition only at the coarse-grain level, MA2+D the one with a strong use of decomposition (into groups of 50–100 exams), and Con the constructive method.

The table shows that our solver works better than the pure memetic algorithm and the constructive one. Only the approach based on the decomposition performs better.

Decompositions are independent of the technique used, and we are currently working on exploiting such idea also in our TS algorithms. Unfortunately, preliminary experiments do not show any improvement with respect to the results presented in this paper.

5.5   Discussion of the Algorithms

To conclude the experiments, we want to show the relative importance of the various features of our TS algorithm. To this aim we compare our regular algorithm with some modified versions that miss one feature at a time.

In detail, we consider the following modifications of our algorithm:

1. Disable the shifting penalty mechanism. Penalties are fixed to their original values throughout the run. Hard constraints are all assigned the same value α, which is larger than the sum of all soft ones.
2. Make the selection of the best neighbour always based on the full violation list VL. In the regular algorithm the selection is performed on the sole HVL when hard-conflicts are present.
3. Explore the whole set of examinations at each search step, instead of focusing on the conflicting examination only.
4. Set a fixed value for the tabu list, rather than letting it vary within a given range.
5. Start from a random initial state instead of using the heuristics that searches for p independent sets.

We perform five runs on each version of the algorithm, recording the best, the worst and the average value, and their computing time. Table 4 shows the

---

[2] The best combination of heuristic and size of decomposition; the results are the average on five runs.

results of such experiments for the instance KFU-S-93 (21 periods and 1955 seats per period). We use the Burke and Newall formulation, and a parameter setting as follows: tabu list length 10–30, idle iterations 10 000.

The results show that the key features of our algorithm are the shifting penalty mechanism and the management of the conflict set. Removing these features, on the average, the quality of the solution degrades more than 60%. In fact, both these features prevent the algorithm from wasting time on large plateaux rather then making worsening moves that diversify the search toward more promising regions.

The intuition that the landscape of the cost function is made up of large plateaux is confirmed from a modified version of the algorithm which explores the whole set of examinations at each step of the search. This algorithm is not even able to find a feasible solution, and uses all the time at its disposal in exploring such regions.

Regarding the selection of the initial state, the loss of starting from a random state is relatively small on regular runs. However, the random initial state sometimes leads to extremely poor results, as shown by the maximum cost obtained. In addition, as previously observed, starting from a good state saves computation time.

The use of a fixed-length tabu list also affects the performance of the algorithm. Furthermore, additional experiments show that the fixed length makes the selection of the single value much more critical (the value of 20 moves used in the reported experiment has been chosen after a long trial-and-error session). Conversely, the variable-length case is more robust with respect to the specific values in use, and gives good results for a large variety of values.

Table 4. Relative influence of the features of the algorithm: time in seconds

	Modified feature	Minimum cost		Maximum cost		Average cost	
		cost	time	cost	time	cost	time
	None (regular algorithm)	1793	127.65	2463	189.95	2100.4	179.47
1	Fixed penalties	3391	55.9	5827	80.95	4395.2	64.63
2	Selection on full VL	2662	110.22	4233	210.04	3507.6	3547
3	Extended neighbourhood	23943	172.55	35105	171.59	29774.8	174.07
4	Fixed tabu list length	2024	116.30	3333	68.95	2382.4	125.58
5	Random initial state	2142	182.62	8017	64.82	3377.8	187.24

6   Conclusions and Future Work

We have implemented different TS-based algorithms for the EXAMINATION TI-METABLING problem and we have compared them with the existing literature on the problem. The most effective algorithm makes use of a shifting penalty mechanism, a variable-size tabu list, a dynamic neighbourhood selection, and a heuristic initial state. All these features are shown experimentally to be necessary for obtaining good results.

Unfortunately, the experimental analysis shows that the results of our algorithms are not satisfactory on all benchmark instances. Nevertheless, we consider these preliminary results quite encouraging, and in our opinion they provide a good basis for future improvements. To this aim we plan to extend our application in the following ways:

- Use decomposition techniques as Burke and Newall for large instances.
- Implement and possibly interleave other local search techniques, different from TS.
- Implement more complex neighbourhoods relations. In fact, many relations have been proposed inside the GRAPH COLOURING community, which could be profitably adapted for our problem.

The long-term goal of this research is twofold: On the one hand we want to assess the effectiveness of local search techniques for GRAPH COLOURING to EXAMINATION TIMETABLING. On the other hand, we aim at drawing a comprehensive picture of the structure and the hardness of the numerous variants of the EXAMINATION TIMETABLING problem. For this purpose, we are going to consider further versions of the problems, as briefly discussed in Section 2.3.

References

1. Burke, E., Newall, J.: A Multi-stage Evolutionary Algorithm for the Timetable Problem. IEEE Trans. Evol. Comput. **3** (1999) 63–74
2. Burke, E., Newall, J., Weare R.: A Memetic Algorithm for University Exam Timetabling. In: Proc. 1st Int. Conf. on the Practice and Theory of Automated Timetabling (1995) 241–250
3. Carter, M.W.: A Decomposition Algorithm for Practical Timetabling Problems. Working Paper 83-06. Industrial Engineering, University of Toronto (1983)
4. Carter, M.W., Laporte, G.: Recent Developments in Practical Examination Timetabling. In: Proc. 1st Int. Conf. on the Practice and Theory of Automated Timetabling (1996) 3–21
5. Carter, M.W., Laporte, G., Chinneck, J.W.: A General Examination Scheduling System. Interfaces **24** (1994) 109–120
6. Carter, M.W., Laporte, G., Lee, S.Y.: Examination Timetabling: Algorithmic Strategies and Applications. J. Oper. Res. Soc. **74** (1996) 373–383
7. Corne, D., Fang, H.-L., Mellish, C.: Solving the Modular Exam Scheduling Problem with Genetic Algorithms. Technical Report 622. Department of Artificial Intelligence, University of Edinburgh (1993)
8. Di Gaspero, L., Schaerf, A.: EASYLOCAL++: An Object-Oriented Framework for Flexible Design of Local Search Algorithms. Technical Report UDMI/13/2000/RR. Dipartimento di Matematica e Informatica, Università di Udine (2000). Available at http://www.diegm.uniud.it/ aschaerf/projects/local++
9. Garey, M.R., Johnson, D.S.: Computers and Intractability – A Guide to NP-completeness. Freeman, San Francisco (1979)

10. Gendreau, M., Hertz, A., Laporte, G.: A Tabu Search Heuristic for the Vehicle Routing Problem. Manage. Sci. **40** (1994) 1276–1290
11. Glover, F., Laguna, M.: Tabu Search. Kluwer, Dordrecht (1997)
12. Hertz, A., de Werra, D.: Using Tabu Search Techniques for Graph Coloring. Computing **39** (1987) 345–351
13. Johnson, D.S., Aragon, C.R., McGeoch, L.A., Schevon, C.: Optimization by Simulated Annealing: an Experimental Evaluation. Part II: Graph Coloring and Number Partitioning. Oper. Res. **39** (1991) 378–406
14. Laporte, G. Desroches, S.: Examination Timetabling by Computer. Comput. Oper. Res. **11** (1984) 351–360
15. Mehlhorn, K., Näher, S., Seel, M., Uhrig, C.: The LEDA User Manual. Max Plank Institute, Saarbrücken, Germany (1999). Version 4.0
16. Morgenstern, C., Shapiro, H.: Coloration Neighborhood Structures for General Graph Coloring. In: 1st Ann. ACM–SIAM Symp. on Discrete Algorithms (1990) 226–235
17. Schaerf, A.: A Survey of Automated Timetabling. Artif. Intell. Rev. **13** (1999) 87–127
18. Stützle, T.: Iterated Local Search for the Quadratic Assignment Problem. Technical Report AIDA-99-03. FG Intellektik, TU Darmstadt (1998)

A Multicriteria Approach to Examination Timetabling

Edmund Burke, Yuri Bykov, and Sanja Petrovic

University of Nottingham, School of Computer Science and Information Technology,
Wollaton Road,
Nottingham NG8 1BB, UK
{ekb,yxb,sxp}@cs.nott.ac.uk

Abstract. The main aim of this paper is to consider university examination timetabling problems as multicriteria decision problems. A new multicriteria approach to solving such problems is presented. A number of criteria will be defined with respect to a number of exam timetabling constraints. The criteria considered in this research concern room capacities, the proximity of the exams for the students, the order and locations of events, etc. Of course, the criteria have different levels of importance in different situations and for different institutions.

The approach that we adopt is divided into two phases. The goal of the first phase is to find high-quality timetables with respect to each criterion separately. In the second phase, trade-offs between criteria values are carried out in order to find a compromised solution with respect to all the criteria simultaneously. This approach involves considering an ideal point in the criteria space which optimises all criteria at once. It is, of course, generally the case that a solution that corresponds to such a point does not exist. The heuristic search of the criteria space starts from the timetables obtained in the first phase with the aim of finding a solution that is as close as possible to this ideal point with respect to a certain defined distance measure.

The developed methodology is validated, tested and discussed using real world examination data from various universities.

1   Introduction

Timetabling problems arise in a wide variety of domains including education (e.g., school and university timetabling), transport (e.g., train and bus driver timetabling), healthcare institutions (e.g., nurse and surgeon timetabling), etc. This research is focused particularly on university examination timetabling problems. The problem of automated university timetabling has attracted the attention of scientists from a number of differing disciplines, including operations research and artificial intelligence, since the 1950s. Over this period, several different approaches have been developed which can be roughly classified into four groups: sequential methods, cluster methods, constraint-based methods, and meta-heuristics (Carter, 1996), (Carter, 1998).

E. Burke and W. Erben (Eds.): PATAT 2000, LNCS 2079, pp. 118–131, 2001.

1. Sequential methods order the events for scheduling using heuristics (often graph colouring heuristics). They assign the ordered events to valid time periods so that no events in the period are in conflict with each other, i.e. two events which require the same resource are not scheduled in the same time period (Carter, 1986).
2. Cluster methods split the set of events into groups which are conflict-free and then assign the groups to time periods to fulfil other constraints imposed on the timetabling problem (Fisher and Shier, 1983), (Balakrishnan et al., 1992). A multi-phase exam-scheduling package described in Arani and Lotfi (1989) and Lotfi and Cerveny (1991) consists of three phases. In the first phase, clusters of exams are formed with the aim of minimising the number of students with simultaneous exams. In the second phase, these clusters are assigned to exam days while minimising the number of students with two or more exams per day. Finally, the exam days and clusters are arranged to minimise the number of students with consecutive exams.
3. Constraint-based approaches model a timetabling problem as a set of variables (i.e. events) to which values (i.e. resources such as rooms and time periods) have to be assigned to satisfy a number of constraints (Brailsford et al., 1999), (White, 2000).
4. Meta-heuristic methods such as simulated annealing, tabu search and genetic algorithms begin with one or more initial timetables and employ search strategies which try to avoid local optima (see, for example, Burke and Newall (1999), Erben and Kappler (1996), Paechter et al. (1998), Ross et al. (1998), Elmohamed et al. (1998), White and Zhang (1998), Dowsland (1998)).

Different categories of people have differing priorities in the timetabling process and are affected differently by its outcome. For example, in the exam timetabling process there are three main groups of people who are affected by the results of the process (Romero, 1982):

1. Administration of the university: usually sets the minimum standards to which the timetables must confirm.
2. Departments: may have demands for specific classrooms for examinations, may prefer to place large exams early in the timetables, etc.
3. Students: usually prefer exams to be spread out in time and to have a break between consecutive exams, may be concerned about the order in which exams are scheduled, etc.

Consequently, the quality of a timetable can be assessed from various points of view and the importance of a particular constraint can very much depend upon the priorities of the three categories presented above.   However, algorithms for timetabling usually address only subsets of the constraints. Very often, single-cost functions which penalise violations of the selected constraints are defined to assess the quality of the timetable. A reasonable computational cost is achieved at the expense of flexibility and "comprehensiveness". Usually, single-cost functions are additive. This means that a single value is usually obtained that is a (weighted) sum of the penalties obtained for each individual constraint. However, values which are similar may well have been obtained by very different means. For example, one evaluation may be the result of adding a good result for one constraint to a bad result for another. A very similar result might be obtained by adding two mediocre results for both constraints. It might well be that one of these evaluations is much more desirable than the other, in the context of the particular problem in hand, even though these two cost functions have similar results.

This observation motivated the development of the research described in this paper, which considers examination timetabling problems as multicriteria decision problems. However, we believe that a similar methodology can be used to model and solve other timetabling problems.

A number of criteria is defined to evaluate the quality of the timetables from different points of view with respect to different constraints imposed on the problem. These fall into three categories: room capacity, proximity of exams and time and order of exams (see below). A multicriteria method that attempts to consider all criteria simultaneously is employed in the timetabling process. This paper is organised in the following way. Section 2 presents the exam timetabling problem as a multicriteria problem. Our approach is described in Section 3. In Section 4 we discuss the results obtained on real-world exam timetabling problems and in Section 5 we present a few concluding remarks.

2   Exam Timetabling Problems as Multicriteria Problems

The exam timetabling problem can be defined to be the problem of assigning a number of exams into a limited number of time periods (Carter, 1996). The assignment is subject to a number of constraints. Usually, they are split into two categories: hard and soft constraints. Hard constraints must not be violated under any circumstances. Soft constraints are desirable but can be violated if necessary. In a real-world situation it is (almost always) not possible to find a solution without violating some soft constraints. If we assume that solutions that satisfy the hard constraints exist, then the quality of the solutions can be assessed on the basis of how well they satisfy the soft constraints.

The sets of hard and soft constraints differ significantly from university to university. We introduce one hard constraint which is present in all exam timetabling problems: that exams which are in conflict (i.e. have common students) must not be scheduled in the same time period. Criteria are defined with respect to the other constraints that are imposed on specific problems. Each criterion expresses a measure of the violation of the corresponding constraint. We introduce nine criteria $\{C_1,..., C_9\}$ which are split into three groups related to room capacities, proximity of exams, and the time and order of exams. However, the list of the criteria can be relatively easily extended and included in the multicriteria timetabling process. The three groups of criteria are presented below.

Room capacities. C_1 represents the number of times that room capacities are exceeded. It may, of course, be the case that the same student contributes to several capacity violations.

Proximity of exams. These criteria express the number of conflicts where exams are not adequately spread out in time so that students do not have enough free time between two exams:

- C_2 represents the number of conflicts where students have exams in adjacent periods on the same day.
- C_3 represents the number of conflicts where students have two or more exams in the same day.
- C_4 represents the number of conflicts where students have exams in adjacent days.
- C_5 represents the number of conflicts where students have exams in overnight adjacent periods.

Time and order of exams. These criteria express the number of times that students are affected by inappropriate times of, or order of, exams. As was the case with (1) above, one particular student may be affected at several different points in one timetable:

- C_6 represents the number of times that a student has an exam that is not scheduled in a time period of the proper duration.
- C_7 represents the number of times that a student has an exam that is not scheduled in the required time period.
- C_8 represents the number of times that a student has an exam that is not scheduled before/after another specified exam.
- C_9 represents the number of times that a student has an exam that is not scheduled immediately before/after another specified exam.

Our main goal is to minimise each of these nine criteria. However, it is very clear that the criteria are of a very different nature. They are incommensurable due to their different units of measure with different scales. In addition, they are partially or totally conflicting. For example, two exams which should be scheduled immediately before/after each other may have common students and consequently, in the timetable construction, improvements of the value of criterion C_9 will lead to the degradation of the value of criterion C_2.

Universities will assign different priorities to the constraints imposed on their timetabling problems. A Multicriteria approach to timetabling enables timetabling officers to assign different relative importances (weights) to each of the criteria to reflect different institutional regulations and requirements. The weights are not related to how easy it is to satisfy the constraints. Also, the criteria which correspond to constraints that are rigidly enforced by the university (i.e. additional hard constraints that the university may set) will be assigned relatively much higher weights than those assigned to criteria that are related to the soft constraints. For example, the constraint that the resources used at any time must not exceed the resources available (represented by our criterion C_1) is often considered to be hard. So we would assign it a (much) higher weight than other criteria.

2.1   A Multicriteria Statement of the Exam Timetabling Problem

We introduce the following notation:

N is the number of exams.
P is the number of time periods.
K is the number of criteria.
$C = [\, c_{nm}\,]_{N \times N}$ is the symmetrical matrix which represents the conflicts of exams.
c_{nm} is the number of students taking both exam n and exam m where $n \in \{1,..., N\}$ and $m \in \{1,..., N\}$.
$T = [\, t_{np}\,]_{N \times P}$ is the matrix which represents assignments of the exams into time periods.

$$t_{np} = \begin{cases} 1, & \text{if exam } n \text{ is scheduled in time period } p \\ 0, & \text{otherwise} \end{cases}$$

where $n \in \{1,..., N\}$ and $p \in \{1,..., P\}$,
$f_k\,(T)$ is the value of criterion C_k where $k \in \{1,..., K\}$,
w_k is the weight of criterion C_k where $k \in \{1,..., K\}$.

The multicriteria exam timetabling problem can be stated analytically in the following way. Given the number of exams N and the number of time periods P, determine the timetable T which makes all the elements of the vector $WF = (w_1 f_1(T),\ ...,\ w_k f_k(T)\,,\ ...,\ w_K f_K(T))$ as small as possible, subject to

$$\sum_{p=1}^{P} t_{np} = 1 \qquad \text{where } n \in \{1,..., N\} \tag{1}$$

$$\sum_{n=1}^{N-1} \sum_{m=n+1}^{N} \sum_{p=1}^{P} t_{np}\, t_{mp}\, c_{nm} = 0 . \tag{2}$$

3   A Method for Multicriteria Timetabling

A mathematical apparatus has been introduced which enables different criteria to be taken into consideration simultaneously in the construction of a timetable. A criteria space is defined whose dimension is equal to the number of criteria. Each timetable is represented as a point in the criteria space. An ideal point, which optimises all criteria simultaneously, is defined in the criteria space as the vector $I = (f_1^*,..., f_k^*,..., f_K^*)$ where f_k^* (for $k \in \{1,..., K\}$) denotes the best value of the criterion C_k. As it is not possible to actually find these single-criterion optimal solutions, we can talk only about the approximate ideal point. It is generally the case that the solution that corresponds to the ideal point does not exist. The notion of an approximate anti-ideal point is used to define the vector $A = (f_{1*},..., f_{k*},..., f_{K*})$, where f_{k*} (for $k \in \{1,..., K\}$) is the worst value of criteria C_k.

3.1   Preference Space

In order to overcome the difficulties caused by different units of measure for the criteria and their different scales, we introduce a preference space and a mapping of the criteria space into the preference space (Petrovic and Petrovic, 1995). For each criterion C_k, (where $k \in \{1,..., K\}$), a linear preference function is defined which maps the values of the criterion to the real interval $[0, w_k]$. The linear preference function, s_k, for the criterion C_k is defined below. Note that the worst value of the criterion f_{k*} is mapped on to the value 0 and the best value of the criterion f_k^* is mapped onto w_k:

$$s_k = w_k \frac{f_{k*} - f_k}{f_{k*} - f_k^*} \tag{3}$$

where f_k and s_k denote the values of the criterion C_k in the criteria space and the preference space, respectively.

The best value of the criterion is achieved when there are no violations of the corresponding constraint in the timetable (i.e. when $f_k^*=0$ for $k \in \{1,..., K\}$). Of course, we need to calculate the values of f_{k*}, for $k \in \{1,..., K\}$.

As an example of such a calculation, we will consider f_{2*} which expresses the maximum number of conflicts when students have exams in adjacent periods. First, the maximum number of conflicts is calculated for each student. The worst possible situation for the student is taken into consideration when all of the student's exams are scheduled in adjacent periods. Table 1 shows the maximum number of conflicts for a student as a function of the number of the student's exams when there are three time periods per day.

Table 1. Maximum number of conflicts when a student has exams in adjacent periods

No. of exams E	Period 1	Period 2	Period 3	Period 4	Period 5	Period 6	No. of conflicts adj_s
1							0
2							1
3							2
4							2
5							3
...							...

The value adj_s can be calculated using the following formula:

$$adj_s = \begin{cases} (E \text{ div } 3)*2 & if \quad E \bmod 3 = 0 \\ ((E-1) \text{ div } 3)*2 & if \quad E \bmod 3 = 1 \\ ((E+1) \text{ div } 3)*2-1 & if \quad E \bmod 3 = 2 \end{cases} \qquad (4)$$

where E is the number of the student's exams. The operators div and mod denote the quotient and remainder of two integer values, respectively.

Now f_2^* is calculated as follows:

$$f_2^* = \sum_{s=1}^{S} adj_s \qquad (5)$$

where S is the total number of students.

A mapping from the criteria space into the preference space is based on the linear preference functions for all criteria. As an illustration, Figure 1 depicts the mapping of the ideal point $I = (0, 0)$ from the criteria space on the point $W = (w_1, w_2)$ in the preference space, when $K = 2$.

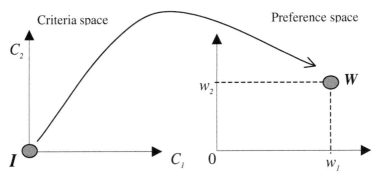

Fig. 1. Mapping from the criteria space into the preference space

3.2   An Algorithm for Heuristic Search of the Preference Space

A new algorithm for heuristic search of the preference space has been developed. It is based on compromise programming – a multicriteria decision making method that attempts to determine so-called compromise solutions which are closest to the ideal point (Zeleny, 1974, 1973). A family of L_p metrics is used to measure the distance between the solutions and the ideal point W in the preference space. L_p metrics are defined by the following formula:

$$L_p(S,W) = \left\{ \sum_{k=1}^{K} [s_k - w_k]^p \right\}^{1/p} \qquad 1 \le p \le \infty \tag{6}$$

where $L_p(S,W)$ is the distance between the solution S with coordinates s_k in the preference space and the ideal point W with coordinates w_k. Three values of p are usually of particular interest in L_p: $p = 1$, $p = 2$, and $p = \infty$. It can be shown that

$$L_\infty = (S,W) = \max_k [s_k, w_k]. \tag{7}$$

L_2 is a well known Euclidean norm which gives the compromise solution geometrically closest to W. Smaller values of p imply compensations of criteria values (i.e. a weak value of one criterion can be offset by a strong value of another). Higher values of p lead to solutions where each criterion stands on its own and there are no trade-offs between the criteria.

The algorithm consists of two phases. The goal of the first phase is to find a set of high-quality timetables in terms of each criterion separately. These timetables will be used as starting points in the search of the preference space in the second phase of the algorithm.

We use sequential methods to construct timetables that are good with respect to each criterion separately. The graph colouring heuristic *saturation degree* is used to order the exams. In each step of the timetable construction, the heuristic selects the exam which has the smallest number of remaining valid periods in the timetable constructed so far. The chosen exam is scheduled in the time period which causes the lowest increase of the criterion value. Ties are broken in the following way: for each exam and for each of its valid periods the increase of the criterion value is calculated and the exam with the highest average increase is selected and scheduled in the best valid period in terms of the criterion value. The motivation is that such an exam should be scheduled early because it contributes more significantly to the criterion value.

These timetables, which are good with respect to each criterion separately, are used as the initial points in the search of the preference space. The algorithm iteratively searches the neighbourhood of each of these timetables trying to improve the other criteria values with the aim of getting close to the ideal point. Two operators are used to explore the neighbourhood of the timetables. They are based on hill climbing and the heavy-mutation operators employed in the memetic algorithm described in Burke and Newall (1999), Burke et al. (1996). These operators are modified to take into account the distance of the timetables from the ideal point.

The hill climbing operator randomly chooses an exam from the timetable and reschedules it in the valid time period which yields a maximum decrease in the distance from the ideal point. The purpose of the hill climbing operator is to direct the search toward the local optima. The hill climbing operator is applied until its application does not decrease the distance of the solution from the ideal point for a predefined number of times.

The mutation operator reschedules the exams (from a number of time periods) which contribute the most to the distance from the ideal point. The exams from these time periods are rescheduled with respect to the distance from the ideal point. The purpose of the mutation operator is to direct the search away from local optima and to explore new areas of the preference space.

In each step of the preference space search, multiple applications of the hill climbing operator are followed by an application of the mutation operator until the distance between the solution and the ideal point has not decreased for a predefined number of steps.

The search of the neighbourhood of each of the initial solutions yields one timetable. The final solution is the timetable chosen from the set of obtained timetables which has the minimum distance from the ideal point.

4   A Real Timetabling Problem: Results and Discussion

We discuss here the results obtained for the exam timetabling problem of the University of Nottingham (1994) whose description is given in Table 2. The data are publicly available from ftp://ftp.cs.nott.ac.uk/ttp/Data.

Table 2. Characteristics of the examination data (University of Nottingham, 1994)

Characteristics of the examination data	Value
No. of exams	800
No. of students	7896
No. of enrolments	33997
No. of time periods per day	3
Duration of time periods (hours)	3, 2, 2
No. of seats per period	1630
No. of exams which require particular time periods	9
No. of times that exams should be scheduled before/after another	2
No. of times that exams should be scheduled immediately before/after another	1

The timetable officer may express his/her preference in two ways: (1) by assigning different weights to constraints and consequently to corresponding criteria, and (2) by favouring solutions which permit or do not permit compensation for weak criteria values. Two parameters are changed: w_k, for $k \in \{1,..., K\}$ and p in the L_p metrics. We consider four different cases with respect to the total number of time periods (i.e. 23 periods, 26 periods, 29 periods and 32 periods).

The results obtained for the case when all criteria are of the same importance and for L_2 are given in Table 3, which presents the values of criteria and the distance of the solutions from the ideal point. Of course, when more time periods are available,

Table 3. Solutions when all criteria are of the same importance

	23 periods	26 periods	29 periods	32 periods
C_1(Rooms)	1038 (4.29%)	137 (0.57%)	139 (0.58%)	25 (0.10%)
C_2(Adj per.)	1111 (5.02%)	655 (2.96%)	513 (2.32%)	314 (1.42%)
C_3(Same day)	3518 (8.49%)	2814 (6.79%)	2239 (5.40%)	1546 (3.73%)
C_4(Adj days)	4804 (12.16%)	2759 (6.98%)	2172 (5.50%)	1646 (4.17%)
C_5(Overnight)	405 (2.70%)	265 (1.77%)	231 (1.54%)	174 (1.16%)
C_6(Duration)	4 (0.28%)	0 (0.00%)	0 (0.00%)	0 (0.00%)
C_7(Time per)	0 (0.00%)	0 (0.00%)	0 (0.00%)	0 (0.00%)
C_8(bef/after)	0 (0.00%)	0 (0.00%)	0 (0.00%)	0 (0.00%)
C_9(Im.bef./after)	0 (0.00%)	0 (0.00%)	0 (0.00%)	0 (0.00%)
L_2	0.164559	0.103448	0.082128	0.058842

higher-quality solutions can be obtained (i.e. ones which are closer to the ideal point). The values given in parentheses are the percentages of the anti-ideal values of the criteria that are achieved in the solution. For example, in the solution with 23 time periods there is 5.02% of the total number of possible conflicts where students have two exams in adjacent periods. In most of the cases the final solution usually significantly improves the values of the other criteria (which are not taken into consideration in the initial solution). However, this is, of course, at the expense of the light degradation of the criterion from the initial solution.

Let us now suppose that the timetable officer prefers timetables in which the number of conflicts where students have exams in adjacent periods is as small as possible. Therefore, the criterion C_2 is assigned a higher importance than the other criteria. In fact, three cases were investigated where $w_2= 2$, 5, and 10 (while all the other weights were assigned value 1). This leads to solutions with a lower number of violations of the corresponding constraint (see Table 4). The timetable officer may choose the timetable which best satisfies his/her preferences concerning the other constraints.

The values of the parameter p significantly influence the solutions. The solutions presented in Table 5 are obtained when all criteria have the same weights. For $p=1$, and $p=2$, the eventual weak value of one criterion is compensated by the stronger values of all the other criteria. On the other hand for $p=\infty$ all the criteria values are reasonably strong, although in some cases not as strong as in the solutions for $p=1$ and $p=2$.

5 Conclusion

This paper presents a new method for multicriteria exam timetabling. It employs the concept of an ideal point that corresponds to the solution which does not violate any of the stated constraints. Such a solution does not usually exist in real-world timetabling problems but the aim is to try to approach it. The method evaluates timetables according to their distances from the approximate ideal point taking the relative importance of the constraints into account. The timetable officer may express

Table 4. Solutions when C_2 is of higher importance than the other criteria

$W = (1, 2, 1, 1, 1, 1, 1, 1, 1)$				
	23 periods	26 periods	29 periods	32 periods
C_1(Rooms)	982 (4.06%)	301 (1.25%)	187 (0.77%)	85 (0.35%)
C_2(Adj per.)	713 (3.22%)	378 (1.71%)	330 (1.49%)	230 (1.04%)
C_3(Same day)	3511 (8.47%)	2922 (7.05%)	2083 (5.02%)	1562 (3.77%)
C_4(Adj days)	5216 (13.20%)	3291 (8.33%)	2358 (5.97%)	1740 (4.40%)
C_5(Overnight)	486 (3.24%)	351 (2.34%)	221 (1.47%)	171 (1.14%)
C_6(Duration)	39 (2.70%)	0 (0.00%)	0 (0.00%)	0 (0.00%)
C_7(Time per)	0 (0.00%)	0 (0.00%)	0 (0.00%)	0 (0.00%)
C_8(bef/after)	0 (0.00%)	0 (0.00%)	0 (0.00%)	0 (0.00%)
C_9(Im.bef/after)	0 (0.00%)	0 (0.00%)	0 (0.00%)	0 (0.00%)
L_2	0.179373	0.117357	0.085153	0.062710
$W = (1, 5, 1, 1, 1, 1, 1, 1, 1)$				
	23 periods	26 periods	29 periods	32 periods
C_1(Rooms)	1358 (5.62%)	418 (1.73%)	274 (1.13%)	78 (0.32%)
C_2(Adj per.)	435 (1.97%)	220 (0.99%)	149 (0.67%)	131 (0.59%)
C_3(Same day)	3262 (7.87%)	2925 (7.05%)	2068 (4.99%)	1564 (3.77%)
C_4(Adj days)	5701 (14.43%)	3768 (9.53%)	3219 (8.15%)	1982 (5.02%)
C_5(Overnight)	565 (3.77%)	463 (3.09%)	347 (2.32%)	221 (1.47%)
C_6(Duration)	4 (0.28%)	0 (0.00%)	0 (0.00%)	0 (0.00%)
C_7(Time per)	0 (0.00%)	0 (0.00%)	0 (0.00%)	0 (0.00%)
C_8(bef/after)	0 (0.00%)	0 (0.00%)	0 (0.00%)	0 (0.00%)
C_9(Im.bef/after)	0 (0.00%)	0 (0.00%)	0 (0.00%)	0 (0.00%)
L_2	0.203092	0.133388	0.104502	0.071009
$W = (1, 10, 1, 1, 1, 1, 1, 1, 1)$				
	23 periods	26 periods	29 periods	32 periods
C_1(Rooms)	1275 (5.28%)	321 (1.33%)	192 (0.79%)	96 (0.40%)
C_2(Adj per.)	212 (0.96%)	106 (0.48%)	90 (0.41%)	44 (0.20%)
C_3(Same day)	3525 (8.50%)	2810 (6.78%)	2240 (5.40%)	1660 (4.00%)
C_4(Adj days)	6584 (16.66%)	4564 (11.55%)	3306 (8.37%)	2382 (6.03%)
C_5(Overnight)	855 (5.71%)	574 (3.83%)	420 (2.80%)	321 (2.14%)
C_6(Duration)	100 (6.92%)	49 (3.39%)	0 (0.00%)	0 (0.00%)
C_7(Time per)	0 (0.00%)	0 (0.00%)	0 (0.00%)	0 (0.00%)
C_8(bef/after)	0 (0.00%)	0 (0.00%)	0 (0.00%)	0 (0.00%)
C_9(Im.bef/after)	0 (0.00%)	0 (0.00%)	0 (0.00%)	0 (0.00%)
L_2	0.234499	0.151720	0.111444	0.078141

Table 5. Solutions for different distance measures, $W = (1, 1, 1, 1, 1, 1, 1, 1, 1)$

	23 periods	26 periods	29 periods	32 periods
C_1(Rooms)	0 (0.00%)	0 (0.00%)	0 (0.00%)	0 (0.00%)
C_2(Adj per.)	879 (3.97%)	604 (2.73%)	393 (1.78%)	316 (1.43%)
C_3(Same day)	3623 (8.74%)	2544 (6.14%)	1957 (4.72%)	1332 (3.21%)
C_4(Adj days)	6381 (16.15%)	4571 (11.57%)	3438 (8.70%)	2482 (6.28%)
C_5(Overnight)	264 (1.76%)	164 (1.09%)	151 (1.01%)	53 (0.35%)
C_6(Duration)	0 (0.00%)	0 (0.00%)	0 (0.00%)	0 (0.00%)
C_7(Time per)	0 (0.00%)	0 (0.00%)	0 (0.00%)	0 (0.00%)
C_8(bef/after)	0 (0.00%)	0 (0.00%)	0 (0.00%)	0 (0.00%)
C_9(Im.bef/after)	0 (0.00%)	0 (0.00%)	0 (0.00%)	0 (0.00%)
L_1	0.306187	0.215262	0.162033	0.112748
C_1(Rooms)	2848 (11.78%)	2044 (8.46%)	1559 (6.45%)	1243 (5.14%)
C_2(Adj per.)	2608 (11.78%)	1872 (8.46%)	1435 (6.48%)	1138 (5.14%)
C_3(Same day)	4886 (11.78%)	3507 (8.46%)	2688 (6.48%)	2132 (5.14%)
C_4(Adj days)	4658 (11.79%)	3343 (8.46%)	2563 (6.49%)	2033 (5.14%)
C_5(Overnight)	807 (5.39%)	475 (3.17%)	441 (2.94%)	334 (2.23%)
C_6(Duration)	170 (11.76%)		89 (6.16%)	74 (5.12%)
C_7(Time per)	40 (9.90%)	24 (5.94%)	24 (5.94%)	18 (4.46%)
C_8(bef/after)	0 (0.00%)	0 (0.00%)	0 (0.00%)	0 (0.00%)
C_9(Im.bef/after)	0 (0.00%)	0 (0.00%)	0 (0.00%)	0 (0.00%)
L_∞	0.117867	0.084592	0.064855	0.051444

his/her preference by altering the weights of the criteria (which correspond to the relative importance of the constraints) and by choosing a distance measure which enables/prevents the compensation of weak criteria values.

The initial results presented here have confirmed that such methods can give an insight into timetabling problems that is not provided by existing approaches. We do not compare the proposed multicriteria approach to other timetabling approaches based on single-cost functions because the measures used to evaluate the quality of solutions are incomparable. However, we can conclude that a multicriteria approach provides a flexibility in the handling of different types of constraints which is not possible when using a single-objective function. It enables constraints of a fundamentally different nature to be handled together and makes an appropriate compromise between them according to the regulations and requirements of particular universities.

Our future research work will include an investigation into the possibility of a multicriteria approach to meta-heuristic methods for timetabling such as genetic algorithms, tabu search and simulated annealing.

References

Arani, T., Lotfi, V.: A Three Phased Approach to Final Exam Scheduling. IIE Trans. **21** (1989) 86–96

Balakrishnan, N., Lucena, A., Wong, R.T.: Scheduling Examinations to Reduce Second-Order Conflicts. Comput. Oper. Res. **19** (1992) 353–361

Brailsford, S.C., Potts, C.N., Smith, B.M.: Constraint Satisfaction Problems: Algorithms and Applications. Eur. J. Oper. Res. **119** (1999) 557–581

Burke, E.K., Newall, J.P.: A Multi-stage Evolutionary Algorithm for the Timetable Problem, IEEE Trans. Evol. Comput. **3** (1999) 63–74

Burke, E., Carter, M. (eds.): The Practice and Theory of Automated Timetabling II: Selected Papers (PATAT '97, University of Toronto). Lecture Notes in Computer Science, Vol. 1408. Springer-Verlag, Berlin Heidelberg New York (1998)

Burke, E., Ross, P. (eds.): The Practice and Theory of Automated Timetabling: Selected Papers (PATAT'95, Napier University). Lecture Notes in Computer Science, Vol. 1153. Springer-Verlag, Berlin Heidelberg New York (1996)

Burke, E.K., Newall, J.P., Weare, R.F., 1996b. A Memetic Algorithm for University Exam Timetabling. In: Burke, E., Ross, P. (eds.): The Practice and Theory of Automated Timetabling: Selected Papers (PATAT'95, Napier University). Lecture Notes in Computer Science, Vol. 1153. Springer-Verlag, Berlin Heidelberg New York (1996) 241–250

Carter M.W., and Laporte G., 1998. Recent Developments in Practical Course Timetabling. In: Burke, E., Carter, M. (eds.): The Practice and Theory of Automated Timetabling II: Selected Papers (PATAT'97, University of Toronto). Lecture Notes in Computer Science, Vol. 1408. Springer-Verlag, Berlin Heidelberg New York (1998) 3–19

Carter, M.W, Laporte, G., 1996. Recent Developments in Practical Examination Timetabling. In: Burke, E., Ross, P. (eds.): The Practice and Theory of Automated Timetabling: Selected Papers (PATAT'95, Napier University). Lecture Notes in Computer Science, Vol. 1153. Springer-Verlag, Berlin Heidelberg New York (1996) 3–21.

Carter, M.W, Laporte, G., Lee, S.Y.: 1996. Examination Timetabling: Algorithmic Strategies and Applications. J. Oper. Res. Soc. **47** (1996) 373–383

Carter, M.W.: A Survey of Practical Applications of Examination Timetabling Algorithms, Oper. Res. **34** (1986) 193–202

Dowsland, K.A.: Off the Peg or Made to Measure. In: Burke, E., Carter, M. (eds.): The Practice and Theory of Automated Timetabling II: Selected Papers (PATAT '97, University of Toronto). Lecture Notes in Computer Science, Vol. 1408. Springer-Verlag, Berlin Heidelberg New York (1998) 37–52

Elmohamed, S., Coddington, P., Fox., F.A.: Comparison of Annealing Techniques for Academic Course Scheduling. In: Burke, E., Carter, M. (eds.): The Practice and Theory of Automated Timetabling II: Selected Papers (PATAT'97, University of Toronto). Lecture Notes in Computer Science, Vol. 1408. Springer-Verlag, Berlin Heidelberg New York (1998) 92–112

Erben, W., Keppler, J.: 1996. A Genetic Algorithm Solving a Weekly Course-Timetabling Problem. In: Burke, E., Ross, P. (eds.): The Practice and Theory of Automated Timetabling: Selected Papers (PATAT'95, Napier University). Lecture Notes in Computer Science, Vol. 1153, Springer-Verlag, Berlin Heidelberg New York (1996) 198–211

Fisher, J.G., Shier, D.R.: A Heuristic Procedure for Large-Scale Examination Scheduling Problems. Technical Report 417. Department of Mathematical Sciences, Clemson University (1983)

Lotfi, V., Cerveny, R.: A Final Exam-Scheduling Package. J. Oper. Res. Soc. **42** (1991) 205–216

Paechter, B., Rankin, R.C., Cumming, A.: Improving a Lecture Timetabling System for University-Wide Use. In: Burke, E., Carter, M. (eds.): The Practice and Theory of Automated Timetabling II: Selected Papers (PATAT'97, University of Toronto). Lecture Notes in Computer Science, Vol. 1408. Springer-Verlag, Berlin Heidelberg New York (1998) 156–165

Petrovic, S., Petrovic, R.: Eco-Ecodispatch: DSS for Multicriteria Loading of Thermal Power Generators. J. Decis. Syst. **4** (1995) 279–295

Romero B.P.: 1982. Examination Scheduling in a Large Engineering School: A Computer-Assisted Participative Procedure. Interfaces **12** (1982) 17–23

Ross, P., Hart, E., Corne, D.: Some Observations about GA based Timetabling. In: Burke, E., Carter, M. (eds.): The Practice and Theory of Automated Timetabling II: Selected Papers (PATAT '97, University of Toronto). Lecture Notes in Computer Science, Vol. 1408. Springer-Verlag, Berlin Heidelberg New York (1998) 115–129

White, G.M., 2000. Constrained Satisfaction, Not So Constrained Satisfaction and the Timetabling Problem. A Plenary Talk. In: Burke, E., Erben, W. (eds.): Proc. 3rd Int. Conf. on the Practice and Theory of Automated Timetabling (University of Applied Sciences, Konstanz, August 16–18). Lecture Notes in Computer Science. Springer-Verlag, Berlin Heidelberg New York (2000) 32–47

White, G.M., Zhang, J. 1998. Generating Complete University Timetables by Combining Tabu Search with Constraint Logic. In: Burke, E., Carter, M. (eds.): The Practice and Theory of Automated Timetabling II: Selected Papers (PATAT'97, University of Toronto). Lecture Notes in Computer Science, Vol. 1408. Springer-Verlag, Berlin Heidelberg New York (1998) 187–210

Zeleny, M., 1974. A Concept of Compromise Solutions and the Method of Displaced Ideal. Comput. Oper. Res. **1** 479–496.

Zeleny, M., 1973. Compromise programming. In: J.L.Cochrane and M.Zeleny (eds.): Multiple Criteria Decision Making, University of South Carolina Press, Columbia, 262–301.

A Grouping Genetic Algorithm for Graph Colouring and Exam Timetabling

Wilhelm Erben

Department of Computer Science, FH Konstanz – University of Applied Sciences,
78462 Konstanz, Germany
erben@fh-konstanz.de

Abstract. It has frequently been reported that pure genetic algorithms for graph colouring are in general outperformed by more conventional methods. There is every reason to believe that this is mainly due to the choice of an unsuitable encoding of solutions. Therefore, an alternative representation, based on the *grouping* character of the graph colouring problem, was chosen. Furthermore, a fitness function defined on the set of partitions of vertices, guiding the Grouping Genetic Algorithm well in the search, was developed. This algorithm has been applied to a choice of hard-to-colour graph examples, with good results. It has also been extended to the application to real-world timetabling problems. As a by-product, phase transition regions of a class of randomly generated graphs have been located.

1 Introduction

1.1 Graph Colouring

A *k-colouring* of a undirected graph G is a partition of its set of nodes into k subsets ("colours") such that no two nodes connected by an edge belong to the same subset. In a graph colouring problem, the goal is to find a k-colouring for either a given k, or with k as small as possible. The first variant is a pure constraint satisfaction problem involving binary constraints that forbid particular pairs of nodes to have the same colour. The other also includes an optimisation, the search for the *chromatic number* K of graph G.

The graph colouring decision and the optimisation problem are known to be NP-complete and NP-hard, respectively. Therefore, techniques based on heuristics have to be used for large-sized problems. Among them, the *greedy algorithm* is the most simple. Taking the nodes of the graph one by one in given order, the first colour used so far that does not cause a conflict is assigned; if such a colour cannot be found, a new one is requested. The greedy algorithm becomes more powerful if the nodes are first decreasingly ordered by their *vertex degree* (the number of neighbours), before the

E. Burke and W. Erben (Eds.): PATAT 2000, LNCS 2079, pp. 132–156, 2001.
© Springer-Verlag Berlin Heidelberg 2001

greedy method is applied. A much better ordering heuristic is used in Brelaz' *Dsatur algorithm* [1]: The next node to be coloured is chosen first among those with the largest number of already (differently) coloured neighbours (the nodes with a maximum *saturation degree*); whenever two of the yet uncoloured nodes have the same saturation degree, the one with higher vertex degree in the uncoloured subgraph is taken. The *Iterated Greedy algorithm* of Culberson et al. [8] uses the greedy algorithm repeatedly; on each iteration, a new permutation of the nodes based on the previously found colouring is produced and presented to the greedy algorithm.

Several other (probabilistic) techniques have been applied to graph colouring problems, such as tabu search (see [14], for example), simulated annealing [15], genetic and other evolutionary algorithms (including hybrids; see [10], [11], [13], for example).

1.2 Timetabling

In a simple timetabling problem, given events have to be scheduled in a number of *periods* in such a way that binary constraints of the form

"Event A must not be assigned to the same period as event B"   (∗)

are satisfied. In exam timetabling, for instance, no student may be scheduled to take two exams at the same time.

This is easily mapped into a graph colouring problem. The events are represented as the nodes of a graph, and an edge between two nodes corresponds to a binary constraint of the type (∗). A valid k-colouring is at the same time a feasible timetable using k periods.

Real-world problems, however, usually involve additional, *non-binary* constraints. Capacity constraints such as seating limitations must be obeyed, for instance. *Near-clashes* should be avoided, if possible: events with an edge constraint should be assigned not only to different, but also non-consecutive periods. Some events must or must not take place at a specific period. Ordering conditions stipulating that an event A must precede a set of other events, or take place at the same time as event B, may occur.... Numerous other examples of constraints could be listed here, where *hard constraints*, fundamental for a solution to be valid, should be distinguished from *soft constraints,* stipulating desirable, though not essential, properties of a timetable.

As in graph colouring, two variants of the timetabling problem can be considered: events have to be scheduled either in a fixed number of periods, or in as few periods as possible, leaving the number of violated soft constraints within allowed limits.

Some graph colouring techniques can be extended to deal also with non-binary constraints. In [18], for instance, a modified Dsatur algorithm that takes seating limitations into account and tries to reduce near-clashes was introduced. Similarly, Carter et al. [4] apply graph colouring heuristics to real-world exam timetabling problems.

1.3 Genetic Algorithms for Graph Colouring/Timetabling

It has frequently been reported that pure genetic algorithms for graph colouring are in general outperformed by more conventional methods (see [20], for instance). This is why various hybrid schemes, incorporating local search procedures into a genetic algorithm, have been proposed in order to improve results (see [13]). Alternatively, adaptive schemes, dynamically changing control parameters of the evolutionary search process depending on the achieved progress, have been developed:

In the evolutionary algorithm presented by Eiben et al. [11], chromosomes are penalised according to the number of uncoloured nodes they contain. In their *Stepwise Adaptation of Weights* (SAW) mechanism, nodes that could not be coloured after a certain number of search steps without violating an edge constraint are considered "hard", and their weights in the penalty function are increased in order to focus the attention of the algorithm on colouring them, as things develop. Results reported in [11] are very good, even a backtracking version of the Dsatur algorithm is mostly outperformed. Another approach yielding similar results is presented by Ross and Hart [19]. In their *adaptive mutation* scheme the mutation probability of a gene is dynamically changed in such a way that wrongly coloured nodes receive higher chances to be mutated and get better colour assignments. A similar approach using *targeted mutation* had been described before by Paechter et al. [16].

For timetabling problems, hybrids seem to yield better results than pure genetic algorithms (see the memetic algorithm of Burke et al. [2] for an example). However, it is, in general, only those genetic algorithms which use a simple direct representation that have been compared with other methods (see [18], for instance, but also [21] for a totally different approach using non-direct representation).

1.4 An Alternative Approach: Grouping Genetic Algorithms

Graph colouring is an instance of the larger family of *grouping problems* (cf. Falkenauer [12]) that consist of partitioning, due to some hard constraints, a set of entities into k mutually disjoint subsets (*groups*). In graph colouring, it is required that connected nodes are placed into distinct groups, whereas the hard constraints for the bin packing problem – the most prominent member of the grouping family – stipulate that items are "packed" into k "bins" without exceeding the capacity limits of these partition sets. Task allocation problems, piling problems, and some vehicle scheduling problems can also be mentioned as members of the grouping family. The grouping decision problems give rise to associated optimisation problems: a cost function, which in most cases is proportional to the number of groups used for the partition, is to be minimised.

In a standard genetic algorithm for a grouping problem, a straightforward direct representation of a solution is used: If N entities (nodes, exams, items, etc.) are to be placed in groups, a chromosome is a string of N genes, and the value of the i-th gene indicates the number of the group that contains the i-th entity. As Falkenauer convincingly argues, this *standard encoding* is not the best choice: It is highly redundant because two distinct chromosomes using different numberings of the groups may encode the same solution. Worse, combined with the classic one- or two-point crossover, good schemata are likely to be disrupted, and "while the schemata are well transmitted with respect to the *chromosomes* under the standard encoding/crossover, their meaning with respect to the *problem* to solve...is lost in the process of recombination" (see [12]).

In an *order-based* (indirect) representation, solutions are encoded as permutations of the entities, and a decoder is applied to retrieve corresponding solutions. In graph colouring, for instance, the greedy algorithm could be used to construct a valid colouring from a permutation of the nodes. Falkenauer gives similar reasons as in the case of direct representation as to why he believes that also this approach is not suitable for grouping problems.

In a *Grouping Genetic Algorithm* (GGA) that he proposes, a chromosome is made up of *groups* as genes. In a graph colouring problem a chromosome could, for instance, consist of the groups {1, 2, 5}, {3, 7}, {4} and {6, 8}, indicating that nodes 1, 2 and 5 have the same colour (no matter which one!) and that another colour has been assigned to nodes 3 and 7, etc. Genetic operators working on these groups have to be defined: In crossover, groups from one parent are injected into the other chromosome while some existing groups are deleted or their contents somehow re-allocated; a mutation could essentially consist of the elimination of a group, or the creation of a new one. In this way it is taken into account that the value of the fitness function actually depends on the *grouping* of the entities, rather than on the single entities themselves, and the genetic operators make sure that whole groups, forming the building blocks used by the genetic algorithm, are processed.

1.5 The Aims of This Paper

A GGA has been developed and tested using some "hard" graph colouring problems and a real-world exam timetabling problem.

A similar algorithm has been implemented before by Eiben et al. [11]. However, results reported by them were quite disappointing: while their algorithm seemed to perform better than a genetic algorithm using standard encoding, it was outperformed by the SAW-ing evolutionary algorithm, which is also presented in [11]. It is assumed that the main reason for this is that the fitness function they chose leads to an inhospitable landscape of the search space, and cannot guide the search for an optimal solution satisfactorily. This is detailed in Sections 2.2 and 3.1.

In Section 2.2, an alternative fitness function is proposed which takes into account, as do many colouring heuristics, that graph vertices of high degree are usually the "hardest" ones. As the test results presented in the subsequent sections show, the GGA equipped with this fitness function, and with the other features given in Section 2, compares well with other powerful techniques.

Graph colouring is known to be NP-hard, even though most graphs are easily coloured [22]. There are, however, hard problem instances, arising at a certain *critical value* of the edge connectivity of a graph [6]. In Section 3, such hard K-colourable (randomly generated) graphs are investigated and taken as a benchmark for the GGA. The test results are compared with those in [11]. Furthermore, a new approximation of the critical value – the point where the *phase transition* occurs – is given for the special class of *equipartite* graphs, showing that this value strongly depends on the size of the graph and its chromatic number. We hope that investigations of such kind will help to develop a good understanding of *real* timetabling problems as well and will make it possible to predict whether a particular problem is difficult to solve or not.

In Section 4, finally, the GGA is applied to a real-world timetabling problem. First test results are presented.

2   The Grouping Genetic Algorithm

The GGA has essentially been implemented following Falkenauer's concepts [12]. It is a steady-state algorithm using tournament selection.

As applications, Falkenauer choses grouping problems such as bin packing and line balancing, but not graph colouring problems. In the following section, therefore, the basic ideas of the GGA are only roughly outlined, whereas its graph colouring specifics are depicted in more detail.

2.1   The Encoding and the Genetic Operators

Given the set $V=\{1,2,....,N\}$ of vertices, a k-colouring of a graph is (most naturally) represented as a set of k mutually disjoint *groups* $G_i \subseteq V$, where

$$\bigcup_{i=1}^{k} G_i = V .$$

Due to the hard constraints, two nodes connected by an edge must be assigned to distinct groups (cf. Section 1.4).

Note that the length of the chromosomes, which is the number of groups (the number of colours used), is variable. Also, as long as pure colouring problems are considered, a chromosome may be regarded as a *set* of groups rather than an *array*. For timetabling problems, however, the groups must be ordered according to increasing time.

In the group-oriented encoding, groups are the genes, and hence the genetic operators will work on these groups of equally coloured nodes. The mutation operator selects – according to a given probability (the mutation rate) – groups of the population at random and eliminates them. The nodes of these eliminated groups have to be reallocated in order to receive valid colourings again. To do this, the existing groups of a chromosome are randomly ordered, and the nodes are greedily coloured using the first group that does not cause a conflict; if such a group cannot be found, a new one is created.

For timetabling problems, another mutation operator makes sense: The positions of two randomly chosen groups in a chromosome are exchanged, i.e. the entities originally assigned to period i are scheduled in period j, and vice versa. Other operators, involving more than two periods in the exchange, or inverting the order of groups scheduled to consecutive periods, have been tested, although without showing significant performance improvement.

The crossover operator has been implemented following a general pattern proposed by Falkenauer [12]. Suppose the parent chromosomes are denoted by $P_1 = \{G_1, G_2, ..., G_k\}$ and $P_2 = \{H_1, H_2, ..., H_m\}$, respectively. A non-empty subset $T \subseteq P_1$ of groups is randomly selected from the first parent, and injected into the second one. However, $P_2 \cup T$ is not yet a valid partition of the vertex set, in general, because it will contain groups of the second parent, $H_{i_1}, H_{i_2}, ..., H_{i_r}$, say, which have elements in common with some injected groups. $H_{i_1}, H_{i_2}, ..., H_{i_r}$ are therefore eliminated, and those nodes that now no longer belong to a group are re-inserted in the same way as described above for the mutation operator. The resulting chromosome, containing, in general, groups from both parents as well as some new groups, represents a first offspring. The second one is produced in the same way, with exchanged roles of both parents.

For the initialisation of a population either the greedy algorithm – applied to a random order of the vertices – is used, or else a random colouring producing bad solutions in the beginning, as a challenge for the GGA.

2.2   The Fitness Function

The GGA is intended to be applied to the graph colouring optimisation problem, i.e. to find the minimum number K of colours needed for a colouring. If this chromatic number K is known beforehand, the cost function proposed by Eiben et al. [11] can be used and minimised: In their approach, colourings using more than K colours are simply penalised by counting the number of extra colours, and adding the number of nodes assigned to such unnecessary colours.

$$\text{cost}(P) = k - K + \sum_{j=K+1}^{k} |G_j|, \qquad (2.1)$$

where $P=\{G_1,G_2,....,G_k\}$ is a chromosome (a partition) consisting of k groups ($k \geq K$), and $G_{K+1},G_{K+2},....,G_k$ are assumed to be the $k - K$ smallest groups.

In most real-world problems, however, the chromatic number K is not known. Yet more serious, though, is the fact that cost function (2.1) cannot guide the search for an optimal solution satisfactorily: in general, there will be a huge number of sub-optimal (K+1)-colourings, all sharing the same cost value, and it will be difficult to escape from such a plateau. This will be further illustrated in Section 3.1.

For these reasons, an alternative fitness function has been created for the GGA: Given a chromosome $P=\{G_1,G_2,....,G_k\}$, for each group j the *total degree*

$$D_j = \sum_{i \in G_j} d_i ,$$

with d_i denoting the vertex degree of the i-th node, is determined first. The fitness of P is then defined by

$$f(P) = \frac{1}{k} \cdot \sum_{j=1}^{k} D_j^2 \qquad (2.2)$$

Obviously, f is to be maximised. Groups containing nodes of high degrees are preferred to groups having the same cardinality but a lower total degree. The idea behind this is nothing else than the well-known heuristic recommending colouring as many nodes of high degree as possible with the same colour. Simultaneously, the number k of used colours will reduce.

Because it is possible to distinguish between two colourings using the same number of colours, the fitness function f provides the algorithm with a search space landscape that is almost free of plains, and thus offers more information to be exploited. The test results presented in the subsequent sections show that, even for hard problems this fitness function is usually able to guide the search to an optimum colouring within a reasonable number of search steps.

3   Results for Graph Colouring Problems

For a number of "hard" graph colouring problem instances, test runs have been performed. In all cases, the chromatic number K is known beforehand, and the task is to find this minimum number of colours and a K-colouring (in short: an optimal solution). The results are compared with those reported in [18], [19] or [11].

3.1 Pyramidal Chain Problems

As a first object of investigation an "artificial counter-example" introduced by Ross, Hart and Corne [18] (see also [19]) has been chosen. It was constructed in order to reveal weaknesses of genetic algorithms with standard encoding (cf. Section 1.4). The inspiration for this class of problems came from studies in [17].

A *pyramidal chain* is a graph arranged as a ring made up of C cliques, each of size K, numbered 1...C where C is an even number (see Figure 1). Each pair of "pyramidal" cliques $2i-1$ and $2i$ ($i \in \{1,....,C/2\}$) overlaps by $K-1$ nodes, and thus takes the shape of a diamond. Two adjacent "diamonds" overlap by exactly 1 node.

It is clear that the graph is K-colourable, and all the "single" nodes which link successive "diamonds" must have the same colour. Brelaz' Dsatur algorithm, or any heuristic based on colouring nodes with maximal degree (which are the single nodes, here) first, will easily find an optimal solution. However, genetic algorithms using the standard representation of solutions fail to solve even small-sized instances of pyramidal chains (e.g., with $K=6$ and $C=20$, hence just $N=60$ nodes and $E=200$ edges), as reported in [18]. This is because "nothing is pushing the GA towards ensuring that all these single nodes have the same colour".

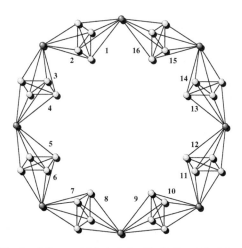

Fig. 1. A "pyramidal chain" with K=5 and C=16

But exactly this "pushing" is done in our grouping oriented approach.

Tables 1–4 present – for the pyramidal chain of Figure 1 – sample calculations of fitness values in different conceivable stages of the evolutionary search process. In an optimal solution using 5 colours (Table 1), all 8 single nodes must belong to the same group j. (In the table, this j is assumed to be 1.) Because each single node has degree 8, and there are no other nodes in group 1, its total degree is $D_1 = 8 \cdot 8 = 64$. Each of the other 4 groups contains exactly 1 (non-single) node from each "diamond", and thus the total degree of such a group amounts to 8•5 = 40. According to (2.2), the fitness value of an optimal solution P is now

$$f(P) = \frac{1}{5} \cdot \sum_{j=1}^{5} D_j^2 = 2{,}099 \; .$$

Tables 2–4 show three examples of colourings, all using exactly 1 colour more than necessary. A solution with just 6 of the 8 single nodes assigned to the same group is compared with situations where even 3 or 4 single nodes are wrongly coloured. Although the same number of colours is used in all three solutions, the sums of squared total degrees, and hence the fitness values, differ. The larger the fitness, the closer the optimal solution is approached. Note that the cost function (2.1), in contrast, yields the same value for all three sub-optimal instances!

Tables 1–4. Fitness values for sample colourings of the pyramidal chain of Figure 1

All single nodes have the same colour.

group j	single nodes	other nodes	total degree D_j	D_j^2
1	8	0	64	4,096
2	0	8	40	1,600
3	0	8	40	1,600
4	0	8	40	1,600
5	0	8	40	1,600
6	-	-	-	-
total	8	32	224	10,496

fitness function (2.2)	2,099
cost function (2.1)	0

Just 6 single nodes have the same colour.

group j	single nodes	other nodes	total degree D_j	D_j^2
1	6	0	48	2,304
2	2	4	36	1,296
3	0	8	40	1,600
4	0	8	40	1,600
5	0	8	40	1,600
6	0	4	20	400
total	8	32	224	8,800

fitness function (2.2)	1,467
cost function (2.1)	5

Just 5 single nodes have the same colour.

group j	single nodes	other nodes	total degree D_j	D_j^2
1	5	0	40	1,600
2	3	4	44	1,936
3	0	8	40	1,600
4	0	8	40	1,600
5	0	8	40	1,600
6	0	4	20	400
total	8	32	224	8,736

fitness function (2.2)	1,456
cost function (2.1)	5

Just 4 single nodes have the same colour.

group j	single nodes	other nodes	total degree D_j	D_j^2
1	4	0	32	1,024
2	4	3	47	2,209
3	0	8	40	1,600
4	0	8	40	1,600
5	0	8	40	1,600
6	0	5	25	625
total	8	32	224	8,658

fitness function (2.2)	1,443
cost function (2.1)	5

In fact, the GGA equipped with fitness function (2.2) easily solved all test instances of pyramidal chain problems (with between a few hundred and 5,000 evaluation steps, depending on the problem size), whereas cost function (2.1) often failed to find the optimum within the given limit of 10,000 search steps. Note that in order to make the search for an optimal solution more difficult, a simple initialisation procedure yielding random colourings (using, for instance, more than 70 colours for 6-colourable graphs) was used in all test runs, deliberately. For each problem instance several test runs with different seeds for the random generator have been performed.

3.2 "Sequences of Cliques" and Their Phase Transitions

This class of problems was also presented by Ross et al [18]. A quasi-random K-colourable graph is constructed by a graph generator as a sequence of C distinct cliques, each of size K, which are connected by a number of randomly generated edges. Some of the edges between nodes from different cliques, however, are forbidden in order to ensure that the graph-colouring problem is solvable with K colours: Suppose that the vertices of each clique are already assigned the colours 1...K, then no edge should join a node from one clique with the node of the same colour from another clique (cf. figure 2). In other words: The vertices are partitioned into K groups, each group containing exactly one node from each clique.

Hence, only $K(K-1)$ of the K^2 possible edges between two cliques are permissible, and, because there are $C(C-1)/2$ pairs of cliques, at most

$$\frac{C(C-1)}{2} \cdot K(K-1)$$

edges are allowed to be created. Each of these permissible edges is assigned by our graph generator with probability q.

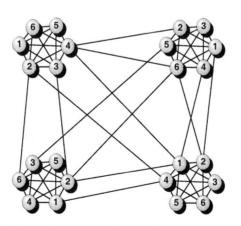

Fig. 2. A "sequence of cliques" with K=6 and C=4

A Preliminary Example. As initial test instances, 6-colourable graphs consisting of a sequence of 20 cliques (K=6 , C=20 , N=120 nodes) have been chosen. For each edge probability q, ranging from 0.05 to 0.25 in steps of 0.01, 10 problems were randomly generated. For each problem 10 runs were performed. The GGA used a population size of 100, crossover rate 0.2 and mutation rate 0.05. The initial populations were randomly created without using any known heuristic, leaving behind solutions with 70 or more colours at the very beginning. As a limit, at most 150,000 fitness evaluations were allowed before the search process was stopped.

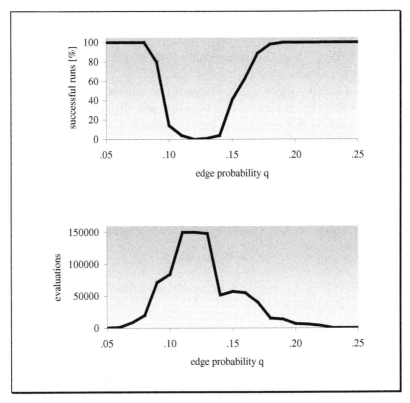

Fig. 3. "Sequences of cliques" with K=6 and C=20 (N=120 nodes):performance at the phase transition

Figure 3 shows the performance of the GGA – in dependence on the edge probability q – with regard to the number of successful runs within the given limit of steps and the average number of fitness evaluations in successful runs.

Obviously, problems with a low edge probability (q < 0.08) are easily solved. They belong to an under-constrained region in the space of problems; there are many optimal solutions around, and it is quite easy to find one of them. Problems with a large edge probability q are over-constrained and likely to contain only a few optimal solutions (or just one); such colourings are also found by the GGA with ease because they belong to prominent maxima in the fitness landscape. The hard problems occur at the boundary between the two regions. In our example, this *phase transition* appears to be at $q=0.12$, and the more the edge probability approaches this *critical value,* the more often the algorithm fails to solve the corresponding graph-colouring problems within a reasonable amount of search steps. As observed, adapting the parameter settings of the GGA to the hardness of the problem (for instance, by increasing the mutation rate) would help slightly. The overall behaviour, however, wouldn't change much.

Phase transitions for graph colouring and other NP-complete constraint satisfaction problems have been widely discussed: see for instance Cheeseman et al. [6] or Clearwater and Hogg [7]. With respect to the "sequence of cliques" class, Ross et al. [18] also report on the existence of the "fallible region" in which their genetic algorithm (using standard encoding) as well as other algorithms (among them the Dsatur algorithm) often fail or need a large number of search steps: For the problem instance with $K=4$ and $C=10$ they found that the *critical value* q_{crit} must be close to 0.20.

Their and our observations are supported by the following theoretical considerations. It will be described how the critical values depend on the parameters K and C.

Locating the Phase Transitions. Let us suppose that colours 1...K have been assigned arbitrarily to the K nodes of each clique (cf. figure 2). In order for the colouring to be valid, no nodes of different cliques sharing the same colour should be joined by an edge. For each $i \in \{1,2,....,K\}$ there exist $C(C-1)/2$ pairs of nodes with colour i. The graph generator does not create an edge between two such nodes with probability $1-\tilde{q}$, where $\tilde{q}=(K-1) \cdot q/K$ (recall that for each node only $K-1$ of K possible edges leading to another clique are permissible). Hence, the chosen colouring of the whole graph does not violate any constraints with probability

$$(1-\tfrac{K-1}{K}q)^{K \cdot \frac{C(C-1)}{2}} . \qquad (3.1)$$

On the other hand, knowing about the special structure of the graph, there are $K!^{C-1}$ ways to assign colours to its nodes (without obeying the constraints): having assigned K different colours to the nodes of a first clique, for every other clique there exist K! permutations of the colour set.

Now, taking into account that the underlying probability space suffers from maximal uncertainty (or entropy) if and only if all possible outcomes occur with equal probabilities, we conclude that the most difficult problems arise around an edge probability q_{crit} satisfying the equation

$$(1-\tfrac{K-1}{K}q_{crit})^{K \cdot \frac{C(C-1)}{2}} = \frac{1}{K!^{C-1}}$$

which implies

$$q_{crit} = \frac{K}{K-1} \cdot \left(1 - \left(\frac{1}{K!}\right)^{\frac{2}{N}} \right) .$$ (3.2)

According to this formula, the phase transition has to occur at an edge probability of $q_{crit} = 12.5\%$ in the preliminary example above ($K=6$ and $C=20$). Looking at our test results, this seems to be quite a good estimate. Also the findings of Ross et al [18] for $K=4$ and $C=10$ could have been predicted: $q_{crit} = 19.6\% \approx 0.20$.

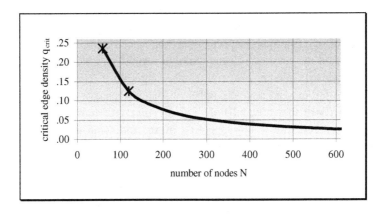

Fig. 4. Critical values $q_{crit} = q_{crit}(N)$ for 6-colourable "sequences of cliques"

Figure 4 shows, for fixed chromatic number $K=6$, how the critical edge probabilities q_{crit} decrease with increasing graph size $N = C \cdot K$. Test runs with the GGA for different values of N (and also K) empirically confirmed the predictions of phase transition regions.

Further Test Results. Figure 5 presents the results for small 6-colourable graphs with a sequence of just 10 cliques. The same parameter settings as before (population size 100, crossover rate 0.2 and mutation rate 0.05) were used. For each edge probability q, ranging from 0.18 to 0.30 in steps of 0.01, 10 problems were generated at random, and for each problem 10 runs were performed.

According to (3.2), the phase transition should occur at $q_{crit} = 23.6\%$. In fact, for some problems the success rate was below 100%, but this occurred only in a region around q_{crit}. The lowest percentage of successful runs was achieved for an edge probability of exactly 24%, and at this point the peak of the average number of evaluations, which are needed to find an optimal solution, is also found.

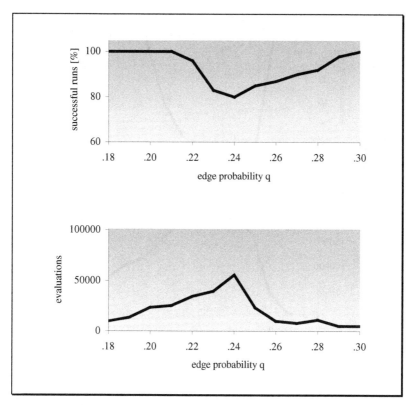

Fig. 5. "Sequences of cliques" with K=6 and C=10 (N=60 nodes): performance at the phase transition

The test results shown in Figure 6 belong to 6-colourable "sequences of cliques" with C=50. These graphs have N=300 nodes, and for an edge probability q in the phase transition region (which should be around $q_{crit}=5.1\%$, according to (3.2)), there are approximately $q \cdot C(C-1)/2 \cdot K(K-1) \approx 2,000$ edges connecting the cliques.

Again, several edge probabilities q around the phase transition region were chosen: for each $q \in \{0.01,0.02,....,0.15\}$, ten problems were generated at random. As a result, a broad ridge between 0.04 and 0.08 in which only very few runs were successful has been found. For q between 0.05 and 0.07, the GGA even failed consistently for all problems within the given limit of 200,000 evaluation steps.

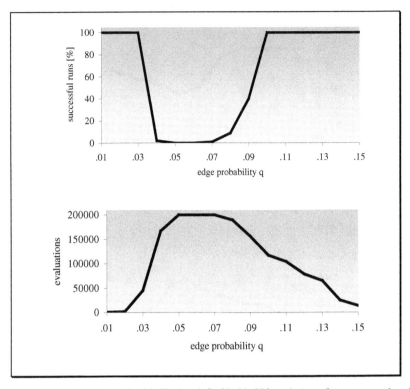

Fig. 6. "Sequences of cliques" with K=6 and C=50 (N=300 nodes) performance at the phase transition

3.3 Equipartite Graphs and Their Phase Transitions

In equipartite graphs the N nodes are partitioned into K almost equal-sized independent sets. Only edges between nodes from distinct partition sets are permitted. Hence, these graphs are K-colourable. (Note that the "sequences of cliques" of the preceding section are special cases of equipartite graphs: There, each partition set contains exactly one node from each of the C cliques.)

In order to produce test instances of equipartite graphs, Culberson's graph generator [9] was used. This generator assigns permissible edges with probability p. Again we conducted studies on phase transition regions depending on this *edge connectivity* p. (Note that for "sequences of cliques" the *edge probability* q refers to the edge density in the set of permissible edges between cliques, whereas Culberson's parameter p also includes the edges belonging to the cliques.)

Results for Equipartite 3-Colourable Graphs. For 3-colourable graphs produced with Culberson's graph generator, our findings were compared with the results obtained by Eiben et al. [11]. In all test runs, the GGA used population size 20, crossover rate 0.2 and mutation rate 0.05. Other settings were also tried, in particular larger population sizes, but hitherto no combinations of parameters yielding significantly better performance could be found. Initial populations were created by the greedy algorithm, which was applied to a random order of the nodes (Section 1.1) and typically assigned 40 or more colours to these initial solutions. The number of allowed fitness evaluations depended on the size of the graphs.

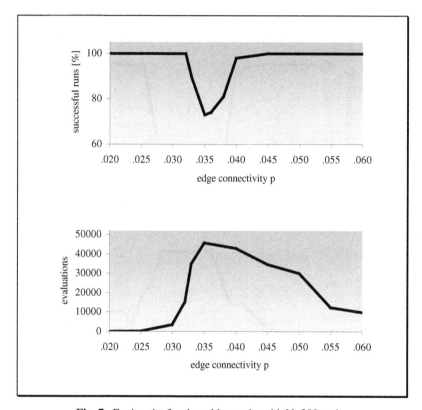

Fig. 7. Equipartite 3-colourable graphs with N=200 nodes

Figure 7 describes the phase transition region for equipartite 3-colourable graphs with N=200 nodes. Problem instances were generated for p from 0.02 to 0.06 in steps of at most 0.005. Obviously, the critical value p_{crit} is close to 0.035, and constitutes quite a sharp border to the easy problems to the left. The average number of evaluations needed to find an optimal colouring ranges from 60 at p=0.02 to 45,700 at p_{crit} =0.035, and about 9,850 evaluations were necessary for p=0.06. These results compare well with those reported by Eiben et al. [11]; only in a region to the right of the critical value p_{crit} is our GGA outperformed a little by their SAW-ing algorithm (cf. Section 1.3) with regard to the numbers of evaluations. The GGA they presented, however, shows quite a low performance, and comes off badly in comparison to ours: for $p \geq 0.045$ we always had successful runs, whereas the GGA in [11] leaves the "fallible" region only at 0.09.

The behaviour for larger edge densities p, outside the phase transition region, cannot be presented in Figure 7 due to the scaling of the diagram. To give an indication: for graphs with an edge connectivity of 0.1 and above, under 900 evaluations were needed on average; for $p=0.2$ just less than 100 were required.

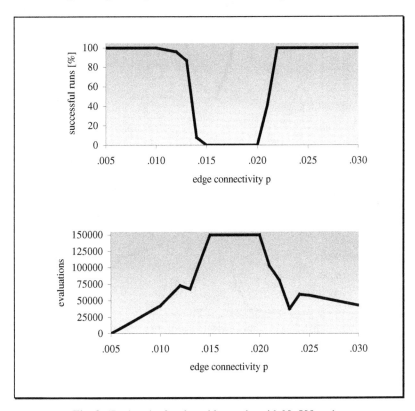

Fig. 8. Equipartite 3-colourable graphs with N=500 nodes

The performance of the GGA for equipartite 3-colourable graphs with $N=500$ nodes is depicted in Figure 8. Here, the edge connectivity p of the problem instances varies from 0.005 to 0.03 in steps of 0.001.

There is a broad ridge from $p=0.015$ to $p=0.02$, in which the GGA was unable to find a 3-colouring within the given limit of 150,000 evaluations. This region of phase transition is similar to what Eiben et al. [11] observed, although their SAW-ing algorithm was again more successful in the immediate right neighbourhood of the critical value p_{crit} (which seems to be at an edge connectivity of above 1.5%). Their GGA, however, is outperformed by far: at $p=0.025$, for instance, it could not find any optimal solution whereas ours successfully terminated all runs within 58,000 evaluations on average. (The peak at $p=0.023$ is certainly due to a sampling error.)

Further results, not shown in the diagrams of Figure 8, are, for instance: for $p=0.05$ still around 3,800 evaluations are necessary on average before an optimal colouring is found, but this number goes down to 500 at $p=0.1$, and under 90 for $p=0.20$.

Locating the Phase Transitions. As indicated before, graphs made up of "sequences of cliques" are special cases of equipartite graphs as defined by Culberson [8] and [9]. The difference is that in graphs produced with his generator, a node needs not be connected by an edge with any node of another partition set, let alone be a member of a clique consisting of K nodes of distinct partition sets. Hence, quite often the independent sets in the generated graphs are not maximal, and sometimes – in particular for very low edge densities – even ($K-1$)-colourable graphs may be produced.

Neglecting this difference for a moment, we assume that all our test instances of equipartite graphs are sequences of cliques, each made up of K nodes of distinct partition sets. This ensures that the chromatic number is K, and the independent partition sets are maximal and really equally sized (as might be the actual intention behind the term "equipartite"). Let us denote this size by C, as in Section 3.2, because $C=N/K$ is also the number of cliques.

Then we can approximate the critical edge connectivity p_{crit} by combining formula (3.2) with the following transformation between the edge probability q of Section 3.2 and Culberson's edge connectivity p:

$$p \cdot C^2 \cdot \frac{K(K-1)}{2} = 1 \cdot C \cdot \frac{K(K-1)}{2} + q \cdot \frac{C(C-1)}{2} \cdot K(K-1) . \tag{3.3}$$

Here, $p \cdot C^2 \cdot \frac{K(K-1)}{2}$ is the expected number of permissible edges, i.e. edges between pairs of the K partition sets, in a random equipartite graph. $C \cdot \frac{K(K-1)}{2}$ refers to the number of edges in the C cliques, and $q \cdot \frac{C(C-1)}{2} \cdot K(K-1)$ is the expected number of (permissible) edges between cliques.

Hence, $p = \frac{1}{C} + \frac{C-1}{C} \cdot q = \frac{K}{N} + \frac{N-K}{N} \cdot q$, and the sought approximation is

$$p_{crit} \approx \frac{K}{N} + \frac{N-K}{N} \cdot \frac{K}{K-1} \cdot \left(1 - \left(\frac{1}{K!}\right)^{\frac{2}{N}}\right) . \tag{3.4}$$

Using formula (3.4), for $K=3$ and $N=200$, we obtain $p_{crit} \approx 0.041$. As expected with the premise we made, this estimated value seems to be a bit too large compared with the empirical results described in Figure 7, although it is still contained in the critical region. However, the approximation is much better for larger graphs because then the number of edges within (fictitious) cliques is low in comparison with the number of other permissible edges. To illustrate, for the graphs with $N=500$ nodes (Figure 8) we calculate $p_{crit} \approx 0.017$ using (3.4), which fits perfectly with the empirical findings.

A brief calculation also shows that for $K=3$

$$\frac{3}{N} + \frac{3}{2}\frac{N-3}{N} \cdot \left(1 - \left(\frac{1}{6}\right)^{\frac{2}{N}}\right) \approx \frac{8}{N}.$$

This coincides with the estimation of the critical value for the special case of equipartite 3-colourable graphs in Eiben et al. [11].

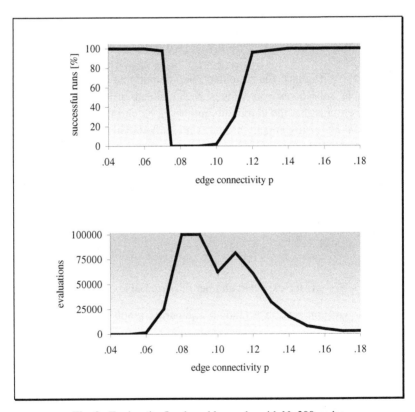

Fig. 9. Equipartite 5-colourable graphs with N=200 nodes

Results for other Equipartite Graphs. As an example for equipartite graphs with a chromatic number greater than 3, and in order to further confirm formula (3.4), the results for 5-colourable problem instances with $N=200$ nodes are presented in Figure 9. The edge probabilities p range from 0.04 to 0.18 in steps of 0.01 or lower. This region includes the phase transition: it was found by using the theoretical considerations above which are apparently again empirically supported. According to (3.4) with $N=200$ and $K=5$, p_{crit} should be about 0.082. In fact, the success rate decreased from 98% at p=0.07 to zero at p=0.075. Compared with the 3-colourable problems for graphs of the same size (see figure 7), the phase transition region is broader, and it takes longer to find an optimal colouring for $K=5$. For that reason, the maximal number of fitness evaluations was increased to 100,000. In the area at the right end of the phase transition region, the low success rates obviously give rise to sampling errors with regard to the measured evaluation numbers. Starting with p=0.12, however, optimal solutions are found in (almost) all runs, and the numbers of evaluations significantly dropped very fast with increasing edge connectivity p.

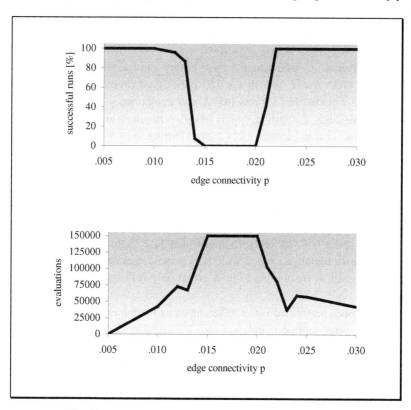

Fig. 10. Equipartite 8-colourable graphs with N=200 nodes

Similar results were obtained for other classes of equipartite graphs; just those for N=200 and K=8 shall be presented here (see Figure 10): Again, the "theoretical" value $p_{crit} \approx 0.15$ and the findings of the test runs go well together.

4   Adapting the GGA to Exam Timetabling

With some minor changes and extensions, the GGA described in the previous sections can also be applied to timetabling problems. A group represents a set of exams (nodes) scheduled for the same period (equipped with the same colour). A chromosome should reflect the chronological sequence of periods. Therefore, a solution (a timetable) is now represented as a *sequence* (rather than a *set*) of groups, although this is only essential as soon as constraints with regard to consecutive exams are introduced.

As a first test example, a data set from Carter's collection of real-world exam timetabling problems [5] has been chosen: in the tre-s-92 problem, which contains data from Trent University, N=261 exams have to be scheduled taking into account E=6,131 possible student conflicts (i.e. edges in the graph of the problem).

Binary Constraints. Considering the mere binary-constrained – the graph colouring – problem, the GGA easily finds that k=20 periods are enough. This result compares well with those reported in [18] and [4]: A non-evolutionary backtracking algorithm by Carter et al. finds a solution which likewise requires just 20 colours, whereas Brelaz' Dsatur algorithm only produces a 23-colouring.

Additional Fundamental Constraints. In the tre-s-92 problem, 4,360 students are involved, and there are 14,901 different enrolments. The maximum number of seats, however, is 655. Taking this limitation into consideration, a feasible chromosome in the GGA must not contain any group for which the sum of students taking an exam at this period exceeds 655. In order to implement this "hard" constraint, the greedy algorithm used for initialisation and within the genetic operators was modified in such a way that an exam is (re-)inserted into a group only if the seat capacity is still large enough.

Tests have been performed using different seeds for the random generator and the following parameter setting: population size 20, crossover rate 0.3 and mutation rate 0.06. As a result, a solution using k=25 periods is found by the GGA in less than 1,000 evaluation steps. After about 2,500 evaluations a timetable consisting of just 24 periods, and still meeting the seating limitation constraints, is produced. In this solution, most periods are at almost full capacity, and a short calculation shows that this seems to be the absolute optimum of the embedded bin packing problem.

"Soft" Constraints. It is desirable, though not fundamental, for a timetable that constraints like the following are also satisfied (cf. [2], for instance):

> "If a student is scheduled to take two exams in any one day
> there must be a complete period between the two exams."

It is not likely that such near-clashes could be avoided for all students completely. In the tre-s-92 problem, it is assumed that there are three periods a day. If an exam with a large degree (taken as a node in the problem graph) is scheduled to the middle period of a day it most probably causes near-clashes. Worse, if there are a large number of students taking not only this exam but also the same other exam (i.e., there are many *instances* of the same edge), a large number of violations of the condition may be produced at once. A good timetable, however, should have as few violations of the constraint as possible.

For the GGA this means that the number of near-clashes has to be incorporated into the fitness function, and also the genetic operators should assist the evolutionary process in finding better and better solutions. The fitness function (2.2) was slightly modified:

$$f(P)=\frac{1}{\alpha\cdot|k-k_{input}|+\beta\cdot(n_{nc}+1)}\cdot\sum_{j=1}^{k}D_j^2 \ .$$

Here, k_{input} is the constant number of available periods given as an input, n_{nc} is the number of near-clashes, and α and β are weights. Timetables using more or less than k_{input} "colours" receive lower fitness, and also each violation of the near-clashes constraint is penalised. Experiments showed that α and β should be in the ratio of at least 10 to 1.

A new mutation operator has been introduced, aimed at making the algorithm faster: it exchanges the positions (periods) of two randomly chosen groups in a chromosome. This operator was added to the existing mutation (which takes a group at random and reschedules their exams).

The modified GGA has been tested with the data of the tre-s-92 problem, obeying the seat capacity limitation. Several runs using different seeds for the random generator have been performed, and a limit of 10,000 evaluation steps was set in each run. As a result, the solutions using 35 periods (i.e. almost 12 days) contained 54 violations of the constraint, on average, and when all exams were scheduled into just 34 periods some 110 near-clashes remained.

At first glance, this seems to be unimpressive compared with the results reported in [2] or [21]: Burke et al. added a local search to the operators in their genetic algorithm, and found a solution for the tre-s-92 problem using 35 periods containing just 3 violations of the soft constraint. A completely different approach was followed up by Terashima-Marín et al. They developed a genetic algorithm searching for the best combination of strategies and heuristics to solve the timetabling problem, rather than using a direct representation of a timetable itself. The best solution obtained in this way for the tre-s-92 problem used only 27 slots and contained 586 near-clashes.

But taking into account that the GGA – originally developed for the mere graph colouring problem – has only slightly been changed for the application to the time-tabling problem, the results are quite promising. Moreover, the computational expense of an evaluation in the GGA is comparatively low. Further refinements with regard to the genetic operators and the parameter settings will certainly lead to noticeable improvements.

5   Conclusion

Graph colouring and timetabling problems have been considered as special instances of grouping problems, and a GGA based on Falkenauer's [12] concepts has been developed. First test results give grounds for the supposition that this idea deserves to be pursued further.

Comparing our findings with the results reported by Eiben et al. [11], it has been confirmed that not only must an adequate representation of individuals be devised for a genetic algorithm, but – perhaps even more important – the fitness function has to convey as much information about the quality of a solution as possible.

Although counter-examples could be generated, experience shows that partitions of the set of vertices that contain a few groups of high total degree are in general better than solutions where colours are almost equally spread out over the nodes, regardless of the fact that some nodes are "harder" than others. Thus, fitness function (2.2) seems to be a good measure reflecting progress towards improved colourings. This has successfully been tested, applying the GGA not only to some artificial graphs for which other genetic algorithms often fail to find an optimal solution, but also to hard-to-colour instances of randomly generated graphs.

As a by-product, the phase transition regions of this class of random graphs have been located (by a formula which has been empirically confirmed) more precisely than known before. The results could be used to generate further hard colouring or timetabling problems serving as benchmarks for the performance of different algorithms. Further investigations of such kind could also help to develop a good understanding of *real* timetabling problems. The structure of their underlying graphs is often similar to the graph instances used in the test runs described in this paper: Typically there is a number of clusters (each made up, for instance, of the exams within a particular faculty) which are more or less separated, but with high edge connectivity within a cluster. If the location of the phase transition were known it would be possible to decide if the introduction of additional edges between the clusters might make the timetabling problem easier.

As for all genetic algorithms, it turned out to be quite difficult to find an optimal parameter setting for the GGA. Further experiments have to be carried out in this respect. According to our present findings, a population size exceeding 20 individuals does not seem to improve performance significantly. To hard problem instances, in particular, large populations are rather detrimental. With regard to crossover and mutation rate settings, the GGA seems to be quite robust. However, best results have been achieved with a crossover rate between 0.2 and 0.3; for lower rates, performance declines dramatically. Thus, the crossover operator contributes well to the search.

The fitness function prompts further investigation. It has been defined as the average, over all colours, of *squared* total degrees. The square function therein could be replaced by another monotone function in such a way that the weight of colours with high total degree is either strengthened or weakened. In particular for the application to exam timetabling problems, more subtle modifications than hitherto performed should be made in order to improve results. The fitness function should, for instance, prevent some groups – like the middle periods of a day – from containing too many nodes of high degree, because otherwise they are likely to cause a lot of near-clashes.

Further research will also be directed at applying the GGA to more general timetabling problems, incorporating other soft constraints.

References

1. Brelaz, D.: New Methods to Color the Vertices of a Graph. Commun. of the ACM **22** (1979) 251–256
2. Burke, E., Newall, J., Weare, R.: A Memetic Algorithm for University Exam Timetabling. In: Burke, E., Ross, P. (eds.): Practice and Theory of Automated Timetabling, 1st International Conference (Edinburgh, August/September 1995), Selected Papers. Lecture Notes in Computer Science, Vol. 1153. Springer-Verlag, Berlin Heidelberg New York (1996) 241–250
3. Burke, E., Newall, J.: A Multi-stage Evolutionary Algorithm for the Timetable Problem. IEEE Trans. Evol. Comput. **3** (1999) 63–74
4. Carter, M., Laporte, G., Lee, S.Y.: Examination Timetabling: Algorithmic Strategies and Applications. J. Oper. Res. Soc. **47** (1996) 373–383
5. Carter, M.: ftp://ie.utoronto.ca/pub/carter/testprob
6. Cheeseman, P., Kanefsky, B., Taylor, W.: Where the Really Hard Problems Are. In: Mylopoulos, J., Reiter, R. (eds.): Proc. 12th IJCAI-91, Vol. 1. Kaufmann, San Francisco, CA (1991) 331–337
7. Clearwater, S., Hogg, T.: Problem Structure Heuristics and Scaling Behavior for Genetic Algorithms. Artif. Intell. **81** (1996) 327–347
8. Culberson, J., Luo, F.: Exploring the *k*-colorable Landscape with Iterated Greedy. In: Trick, M., Johnson, D. (eds.): Cliques, Colors, and Satisfiability: 2nd DIMACS Implementation Challenge, DIMACS Series in Discrete Mathematics and Theoretical Computer Science, Vol. 26. American Mathematical Society (1996) 245–284
9. Culberson, J.: Graph Coloring: http://www.cs.ualberta.ca/~joe/Coloring/index.html (1995)

10. Dorne, R., Hao, J.: A New Genetic Local Search Algorithm for Graph Coloring. In: Eiben, A., Bäck, T., Schoenauer, M., Schwefel, H.-P. (eds.): Parallel Problem Solving from Nature – PPSN V, Proc. 5th Int. Conf. (Amsterdam, September 1998). Lecture Notes in Computer Science, Vol. 1498. Springer-Verlag, Berlin Heidelberg New York (1998) 745–754

11. Eiben, A., Van der Hauw, J., Van Hemert, J.: Graph Coloring with Adaptive Evolutionary Algorithms. J. Heuristics **4** (1998) 25–46

12. Falkenauer, E.: Genetic Algorithms and Grouping Problems. Wiley, Chichester (1998)

13. Fleurent, Ch.; Ferland, J.: Genetic and Hybrid Algorithms for Graph Coloring. Ann. Oper. Res. **63** (1995) 437–463

14. Hertz, A., De Werra, D.: Using Tabu Search Techniques for Graph Coloring. Computing **39** (1987) 345–351

15. Johnson, D., Aragon, C., McGeoch, L., Schevon, C.: Optimization by Simulated Annealing: An Experimental Evaluation; Part II, Graph Coloring and Number Partitioning. Oper. Res. **39** (1991) 378–406

16. Paechter, B., Cumming, A., Norman, M., Luchian, H.: Extensions to a Memetic Timetabling System. In: Burke, E., Ross, P. (eds.): Practice and Theory of Automated Timetabling, 1st International Conference (Edinburgh, August/September 1995), Selected Papers. Lecture Notes in Computer Science, Vol. 1153. Springer-Verlag, Berlin Heidelberg New York (1996) 251–265

17. Ross, P., Corne, D., Terashima-Marín, H.: The Phase-Transition Niche for Evolutionary Algorithms in Timetabling. In: Burke, E., Ross, P. (eds.): Practice and Theory of Automated Timetabling, 1st International Conference (Edinburgh, August/September 1995), Selected Papers. Lecture Notes in Computer Science, Vol. 1153. Springer-Verlag, Berlin Heidelberg New York (1996) 309–324

18. Ross, P., Hart, E., Corne, D.: Some Observations about GA-Based Exam Timetabling. In: Burke, E., Carter, M. (eds.): Practice and Theory of Automated Timetabling II, 2nd International Conference (PATAT'97, Toronto, August 1997), Selected Papers. Lecture Notes in Computer Science, Vol. 1408. Springer-Verlag, Berlin Heidelberg New York (1998) 115–129

19. Ross, P., Hart, E.: An Adaptive Mutation Scheme for a Penalty-Based Graph-Colouring GA. In: Eiben, A., Bäck, T., Schoenauer, M., Schwefel, H.-P. (eds.): Parallel Problem Solving from Nature – PPSN V, Proc. 5th International Conference (Amsterdam, September 1998). Lecture Notes in Computer Science, Vol. 1498. Springer-Verlag, Berlin Heidelberg New York (1998) 795–802

20. Terashima-Marín, H.: A Comparison of GA-Based Methods and Graph-Colouring Methods for Solving the Timetabling Problem. Master's Thesis, Department of AI, University of Edinburgh (1994)

21. Terashima-Marín, H., Ross, P., Valenzuela-Rendón, M.: Evolution of Constraint Satisfaction Strategies in Examination Timetabling. Proc. Genetic and Evolutionary Conference 1999, Orlando, FL., July 13–17 (1999) 635–642

22. Turner, J.: Almost All k-colorable Graphs are Easy to Color. J. Algorithms **9** (1988) 63–82

Employee Timetabling

Cyclical Staff Scheduling Using Constraint Logic Programming

Peter Chan [1] and Georges Weil [2] [*]

[1] COSYTEC S.A., 4 Rue Jean Rostand, Parc Club Orsay-Université,
F-91893 Orsay-Cedex, France
[2] Laboratory TIMC-IMAG, Université J. Fourier, Faculté de Médecine de Grenoble,
F-38700 Grenoble, France
{pchan|gweil}@equitime.com
http::://www.equitime.com

Abstract. In organizations where duty is around the clock, seven days a week and every week of the year, timetabling is a very difficult task, juggling between the workload and the constraints to be respected. Our work concerns cyclical timetabling. This is not just duplicating a fixed sequence of assignments, but has to consider fixed annual leave, and various regulations on assignments on successive days. In some cases, the cycle sequence has to be *relaxed* and cycle length shortened or extended. In other cases, a *small* change in leave dates is allowed, except in summer. This paper describes the context and the use of work cycles in the real world, proposes an abstract model to take into account the various constraints, and finally shows how to implement an effective solution using constraint logic programming (in particular, CHIP V5) to produce timetables of up to 150 people over a yearly horizon.

1 Introduction

Employee timetabling addresses the management problem of scheduling human resources every day to meet estimated requirements such that goals and constraints of the management, labor union and personnel are satisfied. In establishments which operate around the clock and all around the year, this is a very highly combinatorial task. Hospitals have an additional dimension of qualification or grade (full diploma nurses, health workers, etc.). In most other cases, the staff is polyvalent and there is no distinction on qualification: police, prison guards, firemen, postal centers or public utility servicemen in the public sector. Private establishments include facility management centers, telephone call centers, tollbooths, bus drivers, etc. Employee timetabling does not simultaneously consider the planning of resources like classrooms or specific equipment.

---

[*] Both authors are currently at EQUITIME S.A., 4 Ave. de l'Obiou, F-38700 La Tronche, France.

E. Burke and W. Erben (Eds.): PATAT 2000, LNCS 2079, pp. 159–175, 2001.

1.1 Motivation

Timetables can either be cyclic or not. Non-cyclic timetables are created to meet daily requirements that may change every day of the month, yet cope with staff unavailability due to sickness, while satisfying management goals and individual preferences.

Cyclic timetables are based on predefined work cycles also called rotating schedules. They are designed to provide the ideal balance between daily work and rest, and are applied systematically to assure both equality among workers and the best service quality. When the same assignments are made in turn, everybody shares both popular and unpopular work equally.

The work reported in this paper is aimed at generating timetables for groups of some 150 people over one year, based on the choice of a work cycle and usual annual leave patterns. This simulation allows workers to choose their work cycle, after studying in detail resulting timetables, instead of some "theoretical" cycle properties which may be meaningless to workers. The simulator had to produce timetables that satisfy management requirements.

This work follows earlier work of the authors [2], [10] on non-cyclical timetables. Although non-cyclical timetables are outside the scope of this paper, a comparison between the two types of timetables will be given at the end of the paper.

1.2 Contribution

The contribution of this paper is to (a) describe the constraints applicable in timetables based on work cycles, (b) use a model unifying daily and weekly cycles and (c) demonstrate the application of constraint programming (CP) techniques to solve practical problem instances involving 150 people over a period of one year.

Much attention has been concentrated on the generation of work cycles [3], [7]; they are simply duplicated identically over the time horizon. In our users' establishments, this practice is unacceptable since absences are planned in advance. In addition to meeting daily manpower requirements, work schedules have to be adapted around fixed annual leave or professional training, extended or shortened to allow days off while ensuring rest after night duty. Similar to many practical applications, the real challenge is to introduce relaxation in an acceptable way. This is why CP was used for this application.

The paper continues with details of the cyclic timetable problem in Section 2. Our approach is given in Section 3, followed by a comparison and conclusion in Section 4.

As an anecdote, in one particular institution where a 12 week cycle was proposed, a four month timetable for 15 persons (with some leave periods) was produced manually. It was found that the cycle was disrupted for at least two persons every week, not to mention the severe headache it gave to the planning staff. Today, our application can generate annual timetables without breaking the cycle and in less than 20 s on an Intel Pentium 233 MHz PC.

In this paper, a resource can either be an individual worker or a team of workers who share the same timetable (we will not discuss here the situations where the individual timetable differs from the team timetable in some 20% of the days). The code examples in this paper use the logic programming syntax of CHIP. For more details, see [1].

2 Problem Definition

Our work is applied to the case where the workers are flexible to work all shifts. The problem is to schedule resources to meet requirements while respecting different constraints (outlined in Section 2.1). The institution is split into groups. Each group has the same kind of work and a set of requirements (numbers of workers) over each period of the given horizon.

2.1 Organization with Work Cycles

This section describes the various constraints of the problem. The last two constraints (Sections 2.1.5 and 2.1.6) are specific to the user's institution.

2.1.1 Cycle Constraint

In this section we discuss both *daily* cycles (where each work code lasts one day) and *weekly* cycles (work codes of one week). N denotes the number of work codes in the cycle; it is not a multiple of 7 in daily cycles. Both cycle types are defined in a uniform manner, so only weekly cycles are discussed in the rest of the paper. Figure 1 shows both a 5 day and a 4 week cycle. Each column represents one day, where S_i are daily work shifts, taken to be a set of legal working hours per day. We do not have to know what S_i represents (e.g., morning, afternoon, night or rest). The only condition is that the weekly day off is included.

			M	T	W	T	F	S	S
D_1	S_1	W_1	S_1	S_1	S_5	S_5	S_3	S_3	S_3
D_2	S_2	W_2	S_5	S_5	S_1	S_1	S_5	S_5	S_5
D_3	S_3	W_3	S_3	S_3	S_5	S_5	S_1	S_1	S_1
D_4	S_4	W_4	S_5	S_5	S_3	S_3	S_5	S_5	S_5
D_5	S_5								

Fig. 1. Definition of 5 day and 4 week cycles

The simple application of these cycles over 10 days and 8 weeks is shown in Figure 2 for N resources (R_1 to R_N), without taking leave into account. Real timetables are shown in Figure 3.

	1	2	3	4	5	6	7	8	9	10
R_1	D_5	D_1	D_2	D_3	D_4	D_5	D_1	D_2	D_1	D_4
R_2	D_4	D_5	D_1	D_2	D_3	D_4	D_5	D_1	D_2	D_1
R_3	D_3	D_4	D_5	D_1	D_2	D_3	D_4	D_5	D_1	D_2
R_4	D_2	D_3	D_4	D_5	D_1	D_2	D_3	D_4	D_5	D_1
R_5	D_1	D_2	D_3	D_4	D_5	D_1	D_2	D_3	D_4	D_5

1	2	3	4	5	6	7	8
W_4	W_1	W_2	W_3	W_4	W_1	W_2	W_3
W_3	W_4	W_1	W_2	W_3	W_4	W_1	W_2
W_2	W_3	W_4	W_1	W_2	W_3	W_4	W_1
W_1	W_2	W_3	W_4	W_1	W_2	W_3	W_4

Fig. 2. Simple application of highlighted cycles over N resources ($N = 4$ and $N = 5$)

The *daily work cycle* is defined by a sequence of N work codes D_i, each D_i being a daily shift composed of a given set of working hours. Any one resource, say R_j, is assigned to code D_i on one day then to D_{i+1} the following day and back to D_1 at the end of the cycle. Here, the duration of the work code is one day.

The *weekly work cycle* is defined by a sequence of N work codes W_i, each code being a sequence of shifts for each day of the week. Any one resource, say R_j, is assigned to shifts on W_i then to W_{i+1} the following week, and back to W_1 at the end of the cycle. Here, the duration of the work code is seven days.

In our application, N day cycles were translated into weekly cycles in which the duration of the work code is $N - 1$ days. This process allowed us to reduce the number of days to be handled and, most importantly, reduce the irregularity of the resulting timetable, since this can occur only between two consecutive *work codes*. Implementation is discussed in Section 3.2.

CYCLE CONSTRAINT: Each resource is assigned successively to codes X_i then X_{i+1}, and X_N to X_1, where X_i is a day code D_i, in the case of daily cycles, and W_i in the case of weekly cycles.

2.1.2 Requirement Constraint

Taking into account the operating mode of work cycles, each code of the cycle is assigned to a resource at every period. That is, during each period, the sum of the codes of the cycle must be sufficient to cover estimated needs. For daily cycles, the sum of all shifts must cover the needs of the day. For weekly cycles, the sum of weekly shifts cover the needs over the week: i.e. for each day of the week, the sum of shifts for each day covers the needs of that day.

REQUIREMENT CONSTRAINT: At each period of the horizon, each cycle *code* is assigned to at least one resource. If two resources are assigned to the same code, then there is surplus.

2.1.3 Leave Constraint

Annual leave is considered fixed because it is planned over several years to allow workers to have summer leave in August. Many workers take leave starting on a Monday and benefit from the preceding weekend. The equivalent rule in continuous work organizations is "day off before leave". The complementary rule is that the day following leave must not be off.

LEAVE CONSTRAINT: The day preceding annual leave must be off (i.e. the weekly day off) and the day following leave must not be off.

This constraint requires that specific shifts must precede (in our example S_5) or follow leave (S_1 or S_2). Interpreted in weekly cycles, those weekly shifts with off on weekend must precede leave. The weeks with worked weekend can follow leave. In our example, S_5 being off at the end of W_2 and W_4, only the even-numbered codes can precede leave.

Figure 3 does not show complications when the duration of leave is a multiple of N in daily cycles, or when the leave is an even number of weeks in weekly cycles. More details on the problem and the relaxation methods used to solve it are proposed in Section 2.3.

Three copies of the weeks 2–16 would create a timetable over 45 weeks, but 9 weeks of leave exceeds the legal entitlement. In general, simple duplication of a work cycle would break some specific constraint described in the following paragraphs.

	1	2	3	4	5	6	7	8	9	10	11	12	13	14	15	16	17	18	19	20	21
R_1	Sp	A_1	A_1	A_1	A_1	A_1	A_1	A_1	A_1	A_1	A_1	S_1	S_2	S_3	S_4	S_5	S_1	S_2	S_3	S_4	S_5
R_2	S_5	S_1	S_2	S_3	S_4	S_5	S_1	S_2	S_3	S_4	S_5	A_1	A_1	A_1	A_1	A_1	A_1	A_1	A_1	A_1	A_1
R_3	S_4	S_5	S_1	S_2	S_3	S_4	S_5	S_1	S_2	S_3	S_4	S_5	S_1	S_2	S_3	S_4	S_5	S_1	S_2	S_3	S_4
R_4	S_3	S_4	S_5	S_1	S_2	S_3	S_4	S_5	S_1	S_2	S_3	S_4	S_5	S_1	S_2	S_3	S_4	S_5	S_1	S_2	S_3
R_5	S_2	S_3	S_4	S_5	S_1	S_2	S_3	S_4	S_5	S_1	S_2	S_3	S_4	S_5	S_1	S_2	S_3	S_4	S_5	S_1	S_2

(a)

	1	2	3	4	5	6	7	8	9	10	11	12	13	14	15	16
R_1	Sp	A_1	A_1	A_1	W_1	W_2	W_3	W_4	W_1	W_2	W_3	W_4	W_1	W_2	W_3	W_4
R_2	W_1	W_2	W_3	W_4	A_1	A_1	A_1	W_1	W_2	W_3	W_4	W_1	W_2	W_3	W_4	W_1
R_3	W_2	W_3	W_4	W_1	W_2	W_3	W_4	A_1	A_1	A_1	W_1	W_2	W_3	W_4	W_1	W_2
R_4	W_3	W_4	W_1	W_2	W_3	W_4	W_1	W_2	W_3	W_4	A_1	A_1	A_1	W_1	W_2	W_3
R_5	W_4	W_1	W_2	W_3	W_4	W_1	W_2	W_3	W_4	W_1	W_2	W_3	W_4	A_1	A_1	A_1

(b)

Fig. 3. (a) Application of 5 day cycle over 21 days with 10 day leave (A_L); (b) application of 4 week cycle over 16 weeks with 3 weeks leave

2.1.4 Daily Rest Constraint

Workers must have sufficient rest after each workday. For example, after a night shift finishing at 7 AM, workers must be given a rest day (in our example, S_4 must follow S_3).

DAILY REST CONSTRAINT: Some shifts must follow specific shifts to allow sufficient rest.

In weekly cycles, daily rest has been considered in the design of the work cycle, when it is applied *without* disruption. The corresponding constraint is that the last day of the week shift W_i must follow the first day of the following week shift W_{i+1}.

2.1.5 Constraints for Specific Periods

To avoid excess capacity, different cycles (or variants of a cycle) must be adopted during specific periods of the year. For example, when no resource is on leave over at least one week in a row, N day cycles are extended to $N + 1$ days, so that the extra time is redistributed equally to all workers (whatever the assignment may be). In groups where weekly cycles are used, one resource is assigned to "Spare" or Relief.

This means that work codes from the first cycle are followed by specific work codes from the second cycle such that daily rest is assured.

2.1.6 Spare Team Constraint

In some institutions, to design some flexibility into the work cycle, an additional team is added to the group. In addition to handling unforeseen situations, these workers may be used to reinforce the other teams. In the timetable, one team is cyclically assigned "Spare" in a row (typically as long as the leave itself) and typically precedes or follows leave.

When the Spare assignment is used before leave in a daily cycle, a "core period" defined, for example, by S_1–S_2–S_3, must be inserted between the Spare assignment and Leave, as shown in Figure 4(b).

This situation is captured by allowing specific work codes to precede (resp. follow) Spare and that Spare must precede (resp. follow) Leave.

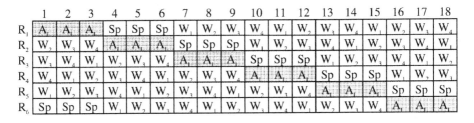

(a)

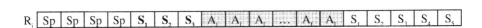

(b)

Fig. 4. (a) Example of Spare (Sp) team in a 4 week cycle; (b) example of "core period" (in bold) in a 5 day cycle

2.2   Other Constraints

We present some other hard constraints that must be respected before a work cycle can be used. They are solved at the design level and are invisible to the solver. This paper does not develop these constraints further.

Total annual leave. The real physical problem in organizing leave is to match the legal leave entitlement of each worker to the number of days and weeks in a year. In institutions where service is continuous all year round, public holidays and annual leave are compounded and accumulated to a total annual leave for each resource. Currently this is 51 days or 7 weeks (remaining days are taken on individual request). We have to ensure that everybody can take his/her leave entitlement.

The number of resources at most on leave at any moment (CA) is a key parameter. When $CA = 1$, N can be 5 (resp. 6), giving a total of 6 (resp. 7) teams. N cannot be 8 because 8 x 50 = 400 days of leave cannot be found in one year (unless two teams go on leave simultaneously). When $CA = 2$, N can be 10, 12 or 14.

Leave configurations. Historically, daily cycles have been used because of its flexibility since they can be locally extended or shortened in length. However, they give very few weekends off for the workers. Weekly cycles were created, based on a development of one daily cycle over N weeks, and adapted to supply the daily requirements while giving, for example, one weekend off per fortnight (see the sample 4 week cycle). In this situation, our experience shows that: N is forcibly even and we can use an abstraction whereby there are only two work codes Odd (denoted O) and Even (denoted E). Assuming that the even week codes have weekend off, it can precede leave and odd codes can follow leave.

Figure 5 presents an abstract timetable for an even number of weeks on leave. We assume that the leave is arranged in chronological order for each team (otherwise, the lines are re-arranged). It shows that when all resources are on leave continuously, on weeks preceding leave all resources are assigned to only even weeks, i.e. no odd weeks at all, and the requirements cannot be satisfied. By inserting weeks where no resource is on leave (*gaps*), equal numbers of odd and even work codes per week can be achieved. Each solution at the odd/even level of abstraction represents $N/2$ solutions where all work codes are enumerated.

2.3   Relaxation

In order to supply solutions even when the constraints described above are conflicting, we need to apply constraint relaxation in the following situations.

In *daily* cycles, so as to allow resources to take leave after a day off, the cyclical succession of codes may be relaxed locally giving cycles that are lengthened or shortened. In Figure 3(a), there was a problem on day 22 when R_2 normally gets S_1 on returning from leave. However, R_1 also gets S_1 due to the cycle. Figure 6 shows the solution by relaxing the cycle constraint, giving two local adaptations of the 5 day cycle, S_1–S_2–S_3–S_5 and S_1–S_2–S_3–S_4–S_4–S_5, used on days 15 and 16.

In *weekly* cycles, leave constraints give similar problems when the leave duration is an even number of weeks. It was not possible to relax cycle constraints (which are designed to allow one weekend off per fortnight, and this is contractually binding).

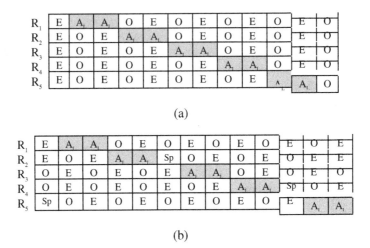

R_1	E	A_L	A_L	O	E	O	E	O	E	O	E	O
R_2	E	O	E	A_L	A_L	O	E	O	E	O	E	O
R_3	E	O	E	O	E	A_L	A_L	O	E	O	E	O
R_4	E	O	E	O	E	O	E	A_L	A_L	O	E	O
R_5	E	O	E	O	E	O	E	O	E	A_L	A_L	O

(a)

R_1	E	A_L	A_L	O	E	O	E	O	E	O	E	O	E
R_2	E	O	E	A_L	A_L	Sp	O	E	O	E	O	E	E
R_3	O	E	O	E	O	E	A_L	A_L	O	E	O	E	O
R_4	O	E	O	E	O	E	O	E	A_L	A_L	Sp	O	E
R_5	Sp	O	E	O	E	O	E	O	E	O	E	A_L	A_L

(b)

Fig. 5. Odd/even timetable representation for 2 week annual leave (AL): (a) insufficient numbers of odd and even codes per period; (b) timetable with gaps (Sp=Spare), giving 2 odd and 2 even codes

The solution is to advance or delay leave by one or two weeks. However, this is not allowed for summer leave.

	1	2	3	4	5	6	7	8	9	10	11	12	13	14	15	16	17	18	19	20	21	22
R_1	Sp	A_L	A_L	A_L	A_L	A_L	A_L	A_L	A_L	A_L	A_L	S_1	S_2	S_3	S_4	S_4	S_5	S_1	S_2	S_3	S_4	S_5
R_2	S_5	S_1	S_2	S_3	S_4	S_5	S_1	S_2	S_3	S_4	S_5	A_L	A_L	A_L	A_L	A_L	A_L	A_L	A_L	A_L	A_L	S_1
R_3	S_4	S_5	S_1	S_2	S_3	S_4	S_5	S_1	S_2	S_3	S_4	S_5	S_1	S_2	S_3	S_5	S_1	S_2	S_3	S_4	S_5	A_L
R_4	S_3	S_4	S_5	S_1	S_2	S_3	S_4	S_5	S_1	S_2	S_3	S_4	S_5	S_1	S_2	S_3	S_4	S_5	S_1	S_2	S_3	S_4
R_5	S_2	S_3	S_4	S_5	S_1	S_2	S_3	S_4	S_5	S_1	S_2	S_3	S_4	S_5	S_1	S_2	S_3	S_4	S_5	S_1	S_2	S_3
R_6	S_1	S_2	S_3	S_4	S_5	S_1	S_2	S_3	S_4	S_5	S_1	S_2	S_3	S_4	S_5	S_1	S_2	S_3	S_4	S_5	S_1	S_2

Fig. 6. Application of 5 day cycle over 22 days with 10 day leave

3 Solution Using Constraint Programming

3.1 Constraint Programming

CP is a technique arising from the Artificial Intelligence (AI) domain, whereby solutions are incrementally constructed by assigning possible values to variables. There are numerous AI problems that cannot be solved within polynomial time and timetabling is no exception. To obtain reasonable computation times, the CP approach is to reduce the search space by applying consistency checking methods such as forward checking and look-ahead (see [9]).

CHIP, created at ECRC (cf. [4]), was one of the earliest tools to extend the declarative aspects of logic programming with the efficiency of CP. Since 1991, the original CHIP team has created COSYTEC to commercialize and market CHIP V5. COSYTEC has introduced second-generation CP tools through the concept of *global constraints* [1]. In timetabling problems, to model and solve numerous work laws and regulations, the *sequence* and *among* constraints allow the building of suitable models for constraining patterns of values among consecutive variables. These constraints have been used to produce non-cyclical scheduling (see, for example, [2]).

We recall here the semantics of the *sequence* and *among* constraints applied on the variables $V_{i,j}$ which represent the shift to be assigned on period j to resource i, using the logic programming syntax of the CHIP language. Note that the period used in Section 3 corresponds to each work code and is not the period of the work cycle:

1) For each period j, R being the total number of resources,

```
among([N₁,..., N_N],[V_1,j,  ..., V_R,j],Zeros_R,[[S₁],...,[S_N]], all).
```

The variables V represent the assignments of the period j to the resources 1 to R. The constraint states that N_k ($k = 1,..., N$) is the exact number of occurrences of the value S_k among the variables V. $Zeros_R$ is a list of R zeros, indicating that we do not want any constant displacement to the value of variables V.

2) For each resource i, T being the total number of periods:

```
sequence([0,0,T],[V_i,1,  ..., V_i,T], Zeros_T, Pats, all).
```

The variables V represent the assignments of resource i during the periods 1 to T. The use of $Zeros_T$ is as above. Pats is the list of valid patterns on the variables V and the first argument expresses that there are exactly 0 instance of these patterns over the T variables. For example, in two successive variables of V, to constrain that the value X may be followed by any of the values in the list AllowedCodes, the pattern is the list of two constraints [[sum,1,#=,[X]],[sum,1,#\=,AllowedCodes]]. The first constraint states that the sum of 1 variable is equal to the value X and the second constraint that the sum of 1 following variable is *not* equal to the given codes.

No instance of inequality patterns is equivalent to equality patterns posted over successive variables. Examples are given in Section 3.2.

3.2   Model and Implementation

We first present the model unifying daily and weekly cycles, after which only weekly cycles will be treated. Then we show how to translate the various problem constraints into transition constraints, written as $X \rightarrow Y$. They state that the value Y can follow the value X in consecutive variables $V_{i,j}$ and $V_{i,j+1}$ corresponding to shift assignments for resource i on periods j and $j + 1$. Then we show how to implement these transitions using the sequence constraint. The weekly code W_i is encoded by the value $i + 1$. W_0 denotes leave whose value is 0 and the "Spare" assignment or W_{sp} has the value 1.

Construction of Weekly Cycles. Our system uses the same solver for both daily and weekly cycles. We outline here a method for *aggregating* daily into "weekly" cycles where the work codes do not always last 7 days.

Consider the 5 day cycle as presented in Figure 1. To build a 4 day weekly cycle, we use the shifts S_1, S_2, S_3 and S_4 for each day on row 1. If the assignment on the last day is S_k on row i, then on row $i+1$, the first assignment will be S_{k+1} (or S_1, if $k = N$). Figure 7 shows the resulting weekly cycle where the 5 day cycle is used exactly four times. The method can also generate other 5 week cycles of 3, 5 or 6 days, etc. For simplicity of explanation, we use only the 4 day week in this paper.

To use this aggregation method, we need to break up the given computation interval of N_{total} days into a number of periods of the correct size. Since the cycle construction method ignores leave during the week, all intervals (including leave) have to fall into a number of periods of different lengths. Taking N_p to be the number of periods of length P, the condition is

$$N_{total} = N_{N-1} \times (N-1) + N_{N+1} \times (N+1) + N_{N+2} \times (N+2).$$

	1	2	3	4
W_1	S_1	S_2	S_3	S_4
W_2	S_5	S_1	S_2	S_3
W_3	S_4	S_5	S_1	S_2
W_4	S_3	S_4	S_5	S_1
W_5	S_2	S_3	S_4	S_5

Fig. 7. Transforming the 5 day cycle into a weekly cycle of 4 days

Implementation of Cycle Constraints. Cycle constraints at the weekly level (based on weekly codes such as W_i) are created by verifying the daily succession for the last day of previous week and first day of the next week. The days within the cycle are assumed to respect daily succession constraints.

In the 4 day week of Figure 7, the cycle constraints are $W_1 \rightarrow W_2$, $W_2 \rightarrow W_3$, $W_3 \rightarrow W_4$, $W_4 \rightarrow W_5$ and $W_5 \rightarrow W_1$. The construction method allows the following patterns to be used in the sequence constraint.

Listing 1. Patterns for cycle constraint in a 5 period cycle:

```
Pats = [                                    % This is a comment
          [sum,1,#=,[2]],[sum,1,#\=,[3]],     % W₁ → W₂
          [sum,1,#=,[3]],[sum,1,#\=,[4]],     % W₂ → W₃
          [sum,1,#=,[4]],[sum,1,#\=,[5]],     % W₃ → W₄
          [sum,1,#=,[5]],[sum,1,#\=,[6]],     % W₄ → W₅
          [sum,1,#=,[6]],[sum,1,#\=,[2]]      % W₅ → W₁
                              ]
```

Implementation of Requirement Constraints. The among constraint is used directly for every period j for the resources 1 to R.

Listing 2. Requirement constraint:

```
among([N₁,…,Nₙ], [V₁,ⱼ, …,Vᵣ,ⱼ], Zerosᵣ, [[S₁],…,[Sₙ]], all)
```

Implementation of Leave Constraints. In the 4 day week of Figure 7, if S_5 is the day off then the daily transition $S_5 \rightarrow 0$ is translated into the weekly transition $W_5 \rightarrow W_0$ since S_5 occurs at the end of W_5. If $S_4 \rightarrow 0$ is allowed, the transition $W_1 \rightarrow W_0$ is also allowed.

The first day after leave cannot be off. This means that the transition $W_0 \rightarrow W_5$ is *not* allowed. Disallowed transitions are implicit since only allowed transitions are given. On the contrary, after leave certain daily shifts are highly favorable, such as S_1. We deduce $W_0 \rightarrow W_1$ since W_1 starts with S_1 in the example.

Listing 3. Patterns for cycle and leave constraints:

```
Pats = [
  [sum,1,#=,[0]],[sum,1,#\=,[0,2]],     % W₀ → W₀ and W₁
  [sum,1,#=,[2]],[sum,1,#\=,[3]],       % W₁ → W₂
  [sum,1,#=,[3]],[sum,1,#\=,[4]],       % W₂ → W₃
  [sum,1,#=,[4]],[sum,1,#\=,[5]],       % W₃ → W₄
  [sum,1,#=,[5]],[sum,1,#\=,[6]],       % W₄ → W₅
  [sum,1,#=,[6]],[sum,1,#\=,[0,2]]      % W₅ → W₀ and W₁
]
```

Daily Rest Constraint. In our example, S_3 (night) must be followed by S_4 (rest); i.e. $W_2 \rightarrow W_3$. We must not forget this transition in cases where additional cycles are used, such as that arising from constraints for specific periods: i.e. we get $W_8 \rightarrow W_9$, $W_2 \rightarrow W_9$ and $W_8 \rightarrow W_3$. See listing 4.

Constraints for Specific Periods. In the case of daily cycles, an N day cycle becomes an $(N + 1)$ day cycle during periods when no resource is on leave. By convention the extra shift is encoded $N + 1$, i.e. S_6 in our case. We compute the variants of this extended cycle for different "week" lengths, in the same fashion as an N day cycle.

In the example of Figure 9, the cyclic constraints are $W_6 \rightarrow W_7,…, W_{11} \rightarrow W_6$. To allow the use of N day cycle followed by the $(N + 1)$ day cycle in successive variables, we have to compare the ending codes of the first weekly cycle and the first codes of the second weekly cycle.

For example, from the transition $S_4 \rightarrow S_5$, to allow the N day cycle to precede the $(N + 1)$ day cycle, we deduce $W_1 \rightarrow W_8$, $W_2 \rightarrow W_9$, $W_3 \rightarrow W_{10}$, $W_4 \rightarrow W_{11}$, $W_5 \rightarrow W_7$ and $W_0 \rightarrow W_6$. Inversely, we have $W_6 \rightarrow W_1$, $W_7 \rightarrow W_2$, $W_8 \rightarrow W_3$, $W_9 \rightarrow W_4$, $W_{10} \rightarrow W_5$ and $W_{11} \rightarrow W_0$. The equivalent patterns are summarized in listing 4. This method can also be used to generate allowed transitions across different work cycles applicable in summer and winter.

	1	2	3	4	5
W_6	S_1	S_2	S_3	S_4	S_5
W_7	S_6	S_1	S_2	S_3	S_4
W_8	S_5	S_6	S_1	S_2	S_3
W_9	S_4	S_5	S_6	S_1	S_2
W_{10}	S_3	S_4	S_5	S_6	S_1
W_{11}	S_2	S_3	S_4	S_5	S_6

Fig. 8. Transforming a 6 day cycle into a 6 week (W_6–W_{11}) cycle of 5 days

In the case of weekly cycles, since all periods are one week in length, we do not have to create new cycles. We need to add the "Spare" value in the domain of the variables V and introduce extra patterns (since all weekly codes can follow Spare, and vice versa).

Listing 4. Patterns for Spare assignment and ($N + 1$) day cycle:

```
Pats =                                       %WEEKLY CYCLES
     [[sum,1,#=,[0]], [sum, 1, #\=,[0,1,6]],
      [Sum,1,#=,[1]], [sum, 1, #\=,[0,1,2, ...,6]],%SPARE
      ...                              %see Listing 3
      ]


Pats = [                                     %DAILY CYCLES
  [sum,1,#=,[0]], [sum, 1, #\=,[0,1,7]],   %W0 →W0,WSP,W6
  [sum,1,#=,[1]],[sum,1,#\=,[0,1,2, ..., 12]],%SPARE follows all
%N-day cycle
  [sum,1,#=,[2]], [sum,1,#\=,[1,3,9]],       %W1 → WSP,W2,W8
  [sum,1,#=,[3]], [sum,1,#\=,[1,4,10]],      %W2 → WSP,W3,W9
  [sum,1,#=,[4]], [sum,1,#\=,[1,5,11]],      %W3 → WSP,W4,W10
  [sum,1,#=,[5]], [sum,1,#\=,[1,6,12]],      %W4 → WSP,W5,W11
  [sum,1,#=,[6]], [sum,1,#\=,[0,1,2,8]]      %W5 → W0,WSP,W1,W7

% (N+1)-day cycle
  [sum,1,#=,[7]], [sum,1,#\=,[1,2,8]],       %W6 → WSP,W1,W7
  [sum,1,#=,[8]], [sum,1,#\=,[1,3,9]],       %W7 → WSP,W2,W8
  [sum,1,#=,[9]], [sum,1,#\=,[1,4,10]],      %W8 → WSP,W3,W9
  [sum,1,#=,[10]], [sum,1,#\=,[1,5,11]],     %W9 → WSP,W4,W10
  [sum,1,#=,[11]], [sum,1,#\=,[1,6,12]],     %W10 → WSP,W5,W11
  [sum,1,#=,[12]], [sum,1,#\=,[0,1,7]]       %W11 → W0,WSP,W1
]
```

Spare Team Constraint. In weekly cycles, "Spare" may be assigned before or after leave (according to specification). They are asserted without choice (i.e. not labeled).

In daily cycles with Spare assignment before leave, it was required that a specified part (3 day) of the daily cycle should appear, so that the resource achieves the core of the cycle. It was necessary to create a 3 day period before leave and use a weekly cycle of 3 days. In our example, the core part is S_1–S_2–S_3.

	1	2	3
W_{12}	S_1	S_2	S_3
W_{13}	S_4	S_5	S_1
W_{14}	S_2	S_3	S_4
W_{15}	S_5	S_1	S_2
W_{16}	S_3	S_4	S_5

Fig. 9. Weekly cycle over 3 days, deduced from 5 day cycle

These new shift values (W_{12} etc.) will have to be connected to all the other shifts according the allowed daily shifts: for example, $W_{12} \rightarrow W_0$, $W_1 \rightarrow W_{13}$ (since $S_4 \rightarrow S_5$), $W_2 \rightarrow W_{14}$, etc.

Relaxation of Daily Cycle Constraints. To make a slight deformation to the daily cycle, it suffices to add extra daily transition constraints from the same weekly shift, both to *shorten* and to *lengthen* the cycle. Let us consider the allowed daily transition $S_3 \rightarrow S_5$. At the weekly level, in addition to $W_2 \rightarrow W_3$, we allow $W_2 \rightarrow W_2$ and $W_2 \rightarrow W_8$. The system will backtrack over all allowed transitions to find the set of values that satisfy these constraints. The corresponding pattern in Pats is

$$[[\text{sum}, 1, \#=, [3]], [\text{sum}, 1, \#\backslash=, [1, 3, 4, 9]]].$$

In the same way, to lengthen the cycle, we may allow $S_4 \rightarrow S_4$, from which weekly assignments can be deduced.

Hence deformation of the cycle occurs only between two periods. The more periods there are, the more "flexible" the system gets, and vice versa. If a scheduling interval is split into periods of one day, there is the maximum flexibility, but the results may be very bad (i.e. the cycle is no longer recognizable). Similarly, allowing $S_i \rightarrow S_i$ over 2 variables may result in a long series of S_i. In such cases, we disallow $S_4 \rightarrow S_4 \rightarrow S_4$ by having no instance of the pattern $[[\text{sum}, 1, \#=, [5]], [\text{sum}, 1, \#=, [5]], [\text{sum}, 1, \#=, [5]]]$ defined over three consecutive variables V.

Relaxation of Leave Dates. In the case of weekly cycles, when a resource i is definitely assigned to leave on week j, variable $V_{i,\,j}$ is assigned to the value 0. When a leave can be assigned to week j, and possibly advanced/delayed by one period, then the domain of three weeks $j - 1, j, j + 1$ are allotted the value 0, in addition to the other possible shift values.

The following constraint is needed in order to assign the correct number of leave weeks, say $N_b = 1$: it states that the number of value 0 among the given variables is exactly N_b.

Listing 5. Total leave constraint:

```
among([N_b], [V_{i,j-1}, V_{i,j}, V_{i,j+1}], Zeros_3, [[0]], all)
```

When the leave is 3 weeks (on j, $j + 1$ and $j + 2$) and possibly advanced/delayed by one period, the period j is definitely on leave.

Unfortunately, in the case of daily cycles since the periods before leave may not be the same size as the periods after leave, this relaxation method cannot be used per se. It suffices to impose periods of the same size (e.g., 3 days).

3.3   Search Heuristic

In the CP paradigm, in addition to the constraint models, we need to define a search heuristic to arrive at coherent solutions rapidly.

Labeling. A very straightforward period-by-period labeling mechanism is sufficient to allow the global constraints to propagate deductions. For each period j, variables corresponding to all resources are labeled, to satisfy the requirements of that period for the various weekly or daily codes. The other constraints will be awakened when necessary and their coherence verified. During each period, variables are selected by the most-constrained heuristic, and values given in increasing order. Backtracking over periods would occur to satisfy the cyclic constraint. In the case of a complex cycle ($N = 12$, $CA = 2$) for 14 teams and where only two resource weeks are "Spare", this heuristic found solutions by propagation after enumerating the first 5 periods.

Conflicts. Like all applications, the system has to propose a solution even when the leave dates conflict with the application of work cycles. In the situation of Section 2.2 (Leave configurations) the requirement constraint is not posted and all resources are labeled "Spare". Such weeks *always* occur singly because there are no cycle constraints on the Spare assignment.

Resources are heuristically assigned in two steps. At the week level, during the Spare weeks, a special rule-based system is used to specify the days each resource must work, considering the previous and next weekend assignments. The idea is to try to maintain a weekend off every fortnight. If this is not possible, then assign more days off during the week.

At the day level, the system applies the working codes of the cycle for the day to resources in an increasing order. Remaining required codes are assigned to resources initially assigned off. This two-step procedure produces results similar to hand scheduling, where the cycle is adapted locally to one or two weeks that cannot be solved otherwise. In the rest of the year, the cycle can be applied without disruption.

4   Comparisons, Results, and Conclusion

4.1   Comparisons in Cyclical and Non-cyclical Timetables

Flexibility. In addition to the requirement constraint (Section 2.1.2), non-cyclical timetables respect constraints on vacations on succeeding days, such as the leave constraint (Section 2.1.3) or the daily rest constraint (Section 2.1.4). They do not have the cycle constraints (Sections 2.1.1 and 2.1.5) which cater for varying requirements in the cyclic context. Such timetables are very flexible and are capable of meeting both unforeseen events and individual wishes. However, they suffer from a lack of regularity. Consequently, workers have to consult the system for every day of the month and cannot plan their activities well ahead of time.

On the other hand, timetables produced using work cycles are largely invariant (especially weekly cycles). It can be argued that work cycles are too rigid to meet unforeseen requirements and are largely dependent on cycle designers' ability to incorporate flexibility into the cycle (e.g., by using relief workers). Cycles can be designed to offer an ideal balance between work and rest for each employee and encourage equal work for all members.

Cost Optimization. Optimization methods in staff scheduling [5] minimize the total cost of the schedule, given as the sum of individual costs of assigning each shift. The organization offers better pay for unpopular shifts (night hours and weekends). It must be realized that biorhythms are adversely affected and may lead to lower service quality.

The cyclical approach allows peers to share all activities organized in shifts and work stretches equally for all concerned. The organization stands to gain by not offering bonuses for unpopular work and encourages all workers to be equally competent in all its activities. We expect the workers to suffer less when the cycle is well balanced.

Other systems. Recent work [5], [6], [7], [8] has concentrated on designing work cycles. In particular, [8] reports that it remains more of an art than a science to produce workable rotating schedules, after 25 years of experience on the ground. Our work has highlighted the practical problems arising from the unraveling of work cycles over a yearly horizon. The essential feature is that the cycle has to be distorted locally to fit forecasted events such as annual leave, yet remaining sufficiently typical of the cycle. Users can specify acceptable "relaxations" by allowing transitions between successive daily or weekly codes. Based on a tree-search mechanism, the system provides solutions incorporating these relaxations.

4.2 Results

The concepts of cyclical staff scheduling using work cycles were built into an application, complete with software modules for the capture of various parameters as well as work cycles, for the solver, and finally for displaying the obtained schedule and statistics for each resource over the annual time horizon. The solver module is built using CHIP V5. The *among* and *sequence global constraints* were used to implement the requirement and weekly succession constraints and to introduce the necessary constraint relaxation methods.

The application has been in use in several regional sites in France since early 1999, producing annual timetables for some 150 agents per group. It was implemented with 14 daily and weekly cycles of different types, where each cycle can be used differently

(Spare teams, leave taken by whole teams or individually, etc.). The aggregation method discussed in Section 3.2 reduced the total number of periods to be handled.

Together with the constraints implementing a global reasoning, the system is able to build annual timetables for complex cycles. A typical example is a 12 week cycle applied over 14 agents (maximum of two on leave on any week). Search was limited only to the first five days and the rest of the problem was solved without cycle disruptions by constraint deduction in less than 20 s on a Pentium 233 MHz PC. In all cases, the application produces satisfactory results within one minute of computation.

Detailed results of our application are confidential, but private discussions with the first author are welcome. The system can achieve good results for weekly cycles, i.e. produce working timetables with very few cycle disruptions while respecting user constraints. Weekly cycles are very important since work and rest are generally well balanced; furthermore, they offer a weekend off every two weeks.

The method used can also handle special periods where staff requirements are different (see Section 2.1.5). More concretely, it is possible to apply two different cycles over different periods of the horizon, while respecting succession constraints at their junction.

In the case of daily cycles, the application is less successful, since cycle disruptions are introduced whenever local situations make it necessary. However, more work is deemed unjustified since daily cycles do not offer sufficient weekends off and are being phased out.

4.3   Conclusion and Future Work

We have shown that the application of given work cycles to a work force can become highly complex due to preprogrammed absences such as annual leave and professional training. It is necessary to relax various constraints, including those of the cycles themselves, to obtain feasible timetables. We propose a unified model with which it is possible to process both daily and weekly work cycles by translating daily succession constraints into weekly ones. This method also reduces the complexity of daily scheduling by increasing the time granularity. We have shown how to apply the method involving several different work cycles over different seasons to the same resources. We can thus modulate the supply of manpower during high and low seasons.

Further work would involve the creation of new weekly cycles to meet given workload, while respecting the various hard constraints discussed in Section 2.2. The system would then propose complete (unraveled) timetables using both current cycles and the new cycles.

Since the method has the inherent capacity to handle horizons of up to three years, we expect to be able to handle issues currently unexplored in human resource allocation: those of training, skill management and upgrading.

Work cycles represent a socially advantageous way of organizing work which encourages team spirit (by encouraging equal work for all, i.e. sharing unpopular shifts) and gives workers more control over their work life through regular work patterns.

References

1. Beldiceanu, N., Contejean, E.: Introducing Global Constraints in CHIP. J. Math. Comput. Modell. **20** (1994) 97–123
2. Chan, P., Kamel, H. Weil, G.: Nurse Scheduling with Global Constraints in CHIP: GYMNASTE. Practical Applications of Constraint Technology (1998)
3. Darwin, S., Sabah, U.R.: Nurse Scheduling Models: A State-of-the-Art Review. J. Soc. Health Syst. **2** (1990) 62–72
4. Dincbas, M., Van Hentenryck, P., Simonis, H., Aggoun, A., Graf, T., Berthier, F.: The Constraint Handling Language CHIP. Proc. Int. Conf. on Fifth Generation Computer Systems (FGC'88, Tokyo) (1988) 693–702
5. Hung, R.: Multiple-Shift Workforce Scheduling under the 3-4 Workweek with Different Weekday and Weekend Labor Requirements. Manage. Sci. **40** (1994) 280–284
6. Jarrah, A.I.Z., Jonathan, F.B., Anura, H.S.: Solving Large-scale Tour Scheduling Problems. Manage. Sci. **40** (1994) 1124–1144
7. Laporte, G., Norbert, Y., Biron, J.: Rotating Schedules. Eur. J. Oper. Res. **4** (1980) 24–30
8. Laporte, G.: The Art and Science of Designing Rotating Schedules. J. Oper. Res. Soc. **50** (1999) 1011–1017
9. Mackworth, A.K.: Consistency in Networks of Relations. Artif. Intell. **8** (1977) 99–118
10. Weil, G., Heus, K., Puget, F., Poujade, M.: Solving the Nurse Scheduling Problem Using Constraint Programming. IEEE Eng. Med. Biol. July–August (1995)

A Hyperheuristic Approach to Scheduling a Sales Summit

Peter Cowling, Graham Kendall, and Eric Soubeiga

Automated Scheduling, Optimisation and Planning (ASAP) Research Group,
School of Computer Science and Information Technology, The University of Nottingham,
Jubilee Campus, Wollaton Road, Nottingham NG8 1BB, UK
{pic|gxk|exs}@cs.nott.ac.uk

Abstract. The concept of a *hyperheuristic* is introduced as an approach that operates at a higher lever of abstraction than current metaheuristic approaches. The hyperheuristic manages the choice of which lower-level heuristic method should be applied at any given time, depending upon the characteristics of the region of the solution space currently under exploration. We analyse the behaviour of several different hyperheuristic approaches for a real-world personnel scheduling problem. Results obtained show the effectiveness of our approach for this problem and suggest wider applicability of hyperheuristic approaches to other problems of scheduling and combinatorial optimisation.

Keywords: hyperheuristics, metaheuristics, heuristics, personnel scheduling, local search, choice function

1   Introduction

Personnel scheduling involves the allocation of personnel to timeslots and possibly locations. The literature uses a variety of terms to describe the same or similar problems. For example, Meissels and Lusternik [11] used the term *employee timetabling* when utilising a constraint satisfaction problem (CSP) model to schedule employees. The term *rostering* can be found in Burke et al. [2], [3] where they employ a hybrid tabu search algorithm to schedule nurses in a Belgian hospital. Dodin et al. [5] use the term *(audit) scheduling* and employ tabu search to schedule audit staff. *Labour scheduling* is used by Easton et al. [7] where they utilise a distributed genetic algorithm technique to determine the number of employees and their work schedules based on predetermined work patterns.

Mason et al. [10] presented an integrated approach using heuristic descent, simulation, and integer programming techniques to schedule staff at the Auckland International Airport, New Zealand. They obtained results which triggered major changes in the attitude of the airport staff who became enthusiastic about the contribution of computer-based decision. Burke et al. [2] used a hybrid tabu search

E. Burke and W. Erben (Eds.): PATAT 2000, LNCS 2079, pp. 176–190, 2001.

algorithm to schedule nurses. Tabu search is hybridised with a memetic approach which combines a steepest descent heuristic within a genetic algorithm framework. The resultant search produces a solution which is better than either the memetic algorithm or the tabu search when run in isolation. The hybridised method was run using data supplied by a Belgian hospital and the results were much better than the manual techniques currently being used. Dowsland [6] uses tabu search combined with strategic oscillation to schedule nurses. Dowsland defined chain neighbourhoods as a combination of basic and simple neighbourhoods. Using these neighbourhoods, the search is allowed to make some moves into infeasible regions in the hope that it quickly reaches a good solution beyond the infeasible regions. The result is a robust and effective method which is capable of producing solutions which are of similar quality to those of a human expert.

However, the heuristic and metaheuristic approaches developed for particular personnel scheduling problems are not generally applicable to other problem domains (or even instances of the same or similar problems). Furthermore, heuristic and metaheuristic approaches tend to be knowledge rich, requiring substantial expertise in both the problem domain and appropriate heuristic techniques [1], and are thus expensive to implement. In this paper we propose a *hyperheuristic* approach, which operates at a level of abstraction above that of a metaheuristic. The hyperheuristic will have no domain knowledge, other than that embedded in a range of simple knowledge-poor heuristics. The resulting approach should be cheap and fast to implement, requiring far less expertise in either the problem domain or heuristic methods, and robust enough to effectively handle a wide range of problems and problem instances from a variety of domains.

Other researchers have investigated general-purpose heuristic-based methods for scheduling and optimisation problems. Hart et al. [9] used a genetic algorithm-based approach to select which of several simple heuristics to apply at each step of a real-world problem of chicken catching and transportation. Tsang and Voudouris [12] introduced the idea of having a fast local search (FLS) combined with a guided local search GLS) and applied it to a workforce scheduling problem. FLS is a fast *hill climbing* method which heuristically ignores moves used in the past without any improvement and GLS is a method which diversifies the search to other regions each time a local optimum is reached. Mladenovic and Hansen [8] introduced the idea of variable neighbourhood search (VNS) and applied it to many combinatorial optimisation problems including the travelling salesman problem and the p-median problem. VNS uses a range of higher-level neighbourhood operators for diversification. When a lower-level neighbourhood search operator reaches a local optimum, the search jumps to a random neighbour in the current high-level neighbourhood. When this diversification move proves ineffective, the next higher level neighbourhood is used.

Our hyperheuristic method does not use problem-specific information other than that provided by a range of simple, and hence easy and cheap to implement, knowledge-poor heuristics. A hyperheuristic is able to choose between low-level heuristics without the need to use domain knowledge, using performance indicators which are not specific to the problem each time a low-level heuristic is called, in order to decide which heuristic to use when at a particular point in the search space.

In order for our hyperheuristic approach to be applicable, we assume that implementing simple local search neighbourhoods and other heuristics (such as greedy constructive heuristics) for the problem in question is relatively easy. Our experience in real-world personnel and production scheduling problems suggests that this is often the case. Indeed, on first presenting a problem which is solved using manual or simple computer techniques, it is often easier for the manual scheduler to express the problem by discussing the *ways* in which the problem is solved currently, rather than the *constraints* of the problem. Usually these ways of manually solving a scheduling or optimisation problem correspond to simple, easy-to-implement heuristics. We may also implement very easily simple local search heuristics based upon swapping, adding and dropping events in the schedule. In addition we also require some method of numerically comparing solutions, i.e. one or more quantitative objective functions.

Each low-level heuristic communicates with the hyperheuristic using a common problem-independent interface architecture. The hyperheuristic can either choose to call a low-level heuristic in order to see what would happen if the low-level heuristic were used, or to allow the low-level heuristic to change the current solution. The hyperheuristic may also provide additional information to the low-level heuristic such as the amount of time which is to be allowed. When called, a low-level heuristic returns a range of parameters related to solution quality or other features (in the case we describe in this paper, a single objective function value is returned) and details of the CPU time used by the neighbourhood function, which allows us to monitor the expected improvement per time unit of each low-level heuristic. It is important to note that the hyperheuristic only knows whether each objective function is to be maximised or minimised (or kept within some range etc.) and has no direct information as to what the objective function represents. We illustrate this idea in Figure 1. All communication between the problem domain and the hyperheuristic is made through a barrier, through which domain knowledge is not allowed to cross.

The rest of the paper is organised as follows. In Section 2 we define a real-world sales summit scheduling problem that we use as a case study to test the effectiveness of our methods. In Section 3 we introduce our hyperheuristic approaches and in Section 4 we present the choice function which many of the approaches require. We then give the results of our experimentation in Section 5. Finally Section 6 presents conclusions and discusses the wider potential for application of hyperheuristic approaches.

2 The Sales Summit Scheduling Problem

The problem we are studying is encountered by a commercial company that organises regular sales summits which bring together two groups of company representatives. The first group, *suppliers*, represent companies who wish to sell some product or service and the second group, *delegates*, represent companies that are potentially interested in purchasing the products and services. Suppliers pay a fee to have a stand at the sales summit and provide a list of the delegates that they would like to meet, where each meeting requested by a supplier is classified as either a *priority* meeting which the supplier feels strongly may yield a sale, or a *nonpriority* meeting about which the supplier feels less strongly. Delegates do not pay a fee and have their travelling and hotel expenses paid by the organiser of the sales summit. In addition to

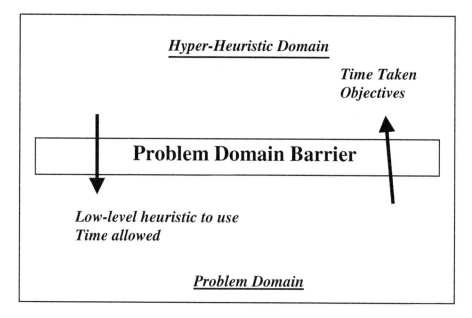

Fig. 1. The hyperheuristic approach and the problem domain barrier

meetings with suppliers, seminars are organised where delegates may meet other delegates. Each delegate supplies a list of the seminars which he will attend in advance of the sales summit, and is guaranteed attendance at all of the seminars requested. There are 24 meeting timeslots available for both seminars and meetings, where each seminar lasts as long as three supplier/delegate meetings. There are 43 suppliers, 99 potential delegates and 12 seminars. The problem is to schedule meetings consisting of (supplier, delegate, timeslot) triples subject to:

1. Each delegate must attend all seminars which they have requested;
2. Each delegate must have at most 12 meetings;
3. No delegate can be scheduled for more than one activity (meeting or seminar) within the same timeslot;
4. No supplier can be scheduled for more than one meeting within the same timeslot;
5. Each supplier should have at least 17 priority meetings;
6. Each supplier should have at least 20 priority and nonpriority meetings in total.

The objective is to minimise the number of delegates who actually attend the sales summit out of the 99 possible delegate attendees, and hence the variable cost of the sales summit, whilst ensuring that suppliers have sufficient delegate meetings. Several other commercial considerations are of secondary importance and are not considered in this paper.

Once delegates have been put into seminar groups, reducing the number of delegate timeslots available, a set of (supplier, delegate, timeslot) meeting triples must be found which minimises the number of attending delegates, whilst keeping all suppliers and the attending delegates happy. Analysis of the solutions produced using the method currently used by the company, a greedy heuristic still simpler than that which we use below to find an initial solution, suggests that in practice we may relax constraints 5 and 6, so long as no individual supplier has substantially fewer than 17 priority meetings, or 20 meetings in total. We have relaxed these constraints in the model given below.

We denote by S the set of suppliers, D the set of delegates and T the set of timeslots. Let P_{ij} be 1 if (supplier i, delegate j) is a Priority meeting and 0 otherwise ($i \in S$, $j \in D$). Our decision variables are denoted x_{ijk} ($i \in S$, $j \in D$, $k \in T$), where x_{ijk} is 1 if supplier i is to meet delegate j in timeslot k, otherwise x_{ijk} is 0. We can now formulate the problem as follows:

$$\text{minimise } E(x) = \sum_{i \in S} \left(\max \left\{ 0,17 - \sum_{j \in D} \sum_{k \in T} P_{ij} x_{ijk} \right\} \right)^2$$

$$+ 0.05 \sum_{i \in S} \left(\max \left\{ 0,20 - \sum_{j \in D} \sum_{k \in T} x_{ijk} \right\} \right)^2 + 8 \left(\sum_{j \in D} \min \left\{ 1, \sum_{i \in S} \sum_{k \in T} x_{ijk} \right\} - 72 \right)$$

subject to

$$\sum_{i \in S} \sum_{k \in T} x_{ijk} \le 12, \quad j \in D \tag{1}$$

$$\sum_{i \in S} x_{ijk} \le 1, \quad j \in D, k \in T \tag{2}$$

$$\sum_{j \in D} x_{ijk} \le 1, \quad i \in S, k \in T \tag{3}$$

$$x_{ijk} \in \{0,1\}, \quad i \in S, j \in D, k \in T. \tag{4}$$

The evaluation function $E(x) = B(x) + 0.05\ C(x) + 8\ H(x)$, where

$$B(x) = \sum_{i \in S} \left(\max \left\{ 0,17 - \sum_{j \in D} \sum_{k \in T} P_{ij} x_{ijk} \right\} \right)^2$$

$$C(x) = \sum_{i \in S} \left(\max \left\{ 0,20 - \sum_{j \in D} \sum_{k \in T} x_{ijk} \right\} \right)^2$$

$$H(x) = d(x) - 72 \quad \text{with} \quad d(x) = \sum_{j \in D} \min \left\{ 1, \sum_{i \in S} \sum_{k \in T} x_{ijk} \right\}$$

$B(x)$ represents the penalty associated with suppliers who have less than 17 priority meetings, where the quadratic nature of the penalty ensures that any suppliers with substantially less than 17 priority meetings result in a large penalty. $C(x)$ represents the penalty associated with suppliers who have less than 20 meetings in total, where again the quadratic nature of the penalty ensures that any suppliers with substantially less than 20 meetings are heavily penalised. However, these meetings are much less significant overall than priority meetings (and, for example, we would only want to include delegates with a large number of priority meetings) and $C(x)$ is multiplied by a factor of 0.05 to reflect this. $d(x)$ is the number of delegates who attend the sales summit in the meeting schedule. $H(x)$ represents the penalty associated with the cost of each delegate, and the factor of 8 reflects the fact that a delegate should be included only to satisfy a supplier who would otherwise have significantly less than 17 priority meetings, or eight suppliers who are missing a single meeting. Note that in a solution where each supplier had the required 20 meetings, there would be 43 x 20 = 860 meetings. Each delegate can attend at most 12 supplier meetings, so that $\lceil 860/12 \rceil$ = 72 delegates are required in this case. We penalise only delegates over 72, to avoid a large constant term in $H(x)$ dominating $B(x)$ and $C(x)$. This will be of particular importance for the roulette wheel approach which we will discuss later. Later, when the vector x to which we are referring is clear, we will simply refer to these quantities as E, B, C, d and H.

Currently, meetings are scheduled using a very simple heuristic which cycles through all suppliers and allocates the first (supplier, delegate, timeslot) triple available from an ordered list of delegates, where the order is simply the order in which the delegates were entered onto the database. The resulting solution has B = 226, C = 48.65, d = 99, H = 216, giving a total penalty of 444.43.

We find an initial schedule using a greedy approach INITIALGREEDY as follows:

INITIALGREEDY:

Do

   1. Let S_O be a list of suppliers ordered by increasing number of scheduled priority meetings (and increasing number of total meetings where two suppliers have the same number of priority meetings in the current schedule).

   2. Let D_O be a list of delegates who currently have less than 12 meetings scheduled, ordered by decreasing number of meetings scheduled.

   3. Find the first supplier $s \in S_O$ such that there is a delegate $d \in D_O$ where s and d both have a common free timeslot t, and (s, d, t) is a priority meeting.

   4. If no meeting triple was found in 3, then find the first supplier $s \in S_O$ such that there is a delegate $d \in D_O$ where s and d both have a common free timeslot t, and (s, d, t) is a non-priority meeting.

Until no meeting is found in either step 3 or step 4.

By considering the most priority-meeting dissatisfied supplier first at each iteration, we attempt to treat suppliers equitably. By attempting to choose the busiest possible delegate at each iteration, we try to minimise the number of delegates in the solution. The solution produced by the constructive heuristic is used as starting solution for all hyperheuristics that we consider below. It yields a solution with B = 52, C = 111, d = 93, H = 168, giving a total penalty of 225.55.

3   Hyperheuristic Approaches to the Sales Summit Scheduling Problem

Having introduced the general nature of hyperheuristic approaches in the introduction, we will now consider the specifics of our approach for the sales summit scheduling problem given above.

The low-level heuristics which we used may all be regarded as local search neighbourhoods which accept a current solution, perform a single local search move, and return a perturbed solution. We denote these neighbourhoods N_1, N_2,..., N_n. The neighbourhoods that we used are given in the appendix. It should be noted that all of our hyperheuristic approaches are independent of the nature or number of low-level heuristics N_1, N_2,..., N. Each neighbourhood can be requested to actually perform the best perturbation on the current solution, or investigate the effect upon the single objective function if the neighbourhood perturbation were performed. Each neighbourhood also returns the amount of CPU time which a call used.

We have considered three different categories of hyperheuristic approaches: random approaches, greedy approaches and choice-function-based approaches. Further, for each of the approaches, we investigate two varieties. In the first variety, denoted by the suffix OI (only improving), we will only accept moves which improve the current solution. In the second variety, denoted by the suffix AM (all moves), all moves are accepted. Each hyperheuristic will continue until a stopping criterion is met, which is a time limit in all cases.

We consider three random approaches. The first, SIMPLERANDOM, randomly chooses a low-level heuristic to apply at each iteration until the stopping criterion is met. The second, RANDOMDESCENT, again chooses a low-level heuristic at random, but this time, once a low-level heuristic has been chosen, it is applied repeatedly until a local optimum is reached where it does not result in any improvement in the objective value of the solution. The third, RANDOMPERMDESCENT, is similar to RANDOMDESCENT except that first we choose a random permutation of the low-level heuristics N_1, N_2,..., N, and when application of a low-level heuristic does not result in any improvement, we cycle round to the next heuristic in this permutation. Note that for the AM versions of RANDOMDESCENT and RANDOMPERMDESCENT, we will carry out one move which makes the current solution worse, before moving on to a new neighbourhood.

The GREEDY approach which we consider will evaluate, at each iteration, the change in objective function value caused by each low-level heuristic upon the current solution and apply the best low-level heuristic so long as this yields an improvement. The AM and OI versions of the GREEDY approach are then identical to each other.

In the third category of hyperheuristic approaches we introduce a *choice function* F, that the hyperheuristic will use to decide on the choice of low-level heuristic to be called next. For each low-level heuristic the choice function F aims to measure how likely that low-level heuristic is to be effective, based upon the current state of knowledge of the region of the solution space currently under exploration. We have implemented four different methods for using the choice function. The first three methods are independent of both the low-level heuristics used and the exact details of how the choice function is arrived at. The fourth method is also independent of the low-level heuristics used but decomposes the choice function into its component parts. We shall describe the fourth method later, once the definition of F is given. In the first (STRAIGHTCHOICE) method, we simply choose, at each iteration, the low-level heuristic which yields the best value of F. In the second (RANKEDCHOICE) method we rank the low-level heuristics according to F and evaluate the changes in objective function value caused by a fixed proportion of the highest ranking heuristics, applying the heuristic which yields the best solution. The third (ROULETTECHOICE) method assumes that for all low-level heuristics, F is always greater than zero. At each iteration a low-level heuristic N_i is chosen with probability which is proportional to $F(N_i)/\Sigma_i F(N_i)$. RANKEDCHOICE and ROULETTECHOICE are analogous to the rank-based selection and the roulette wheel selection from the genetic algorithms literature [4].

4   The Choice Function

The choice function is the key to capturing the nature of the region of the solution space currently under exploration and deciding which neighbourhood to call next, based on the historical performance of each neighbourhood. In our implementation we record, for each low-level heuristic, information concerning the recent effectiveness of the heuristic (f_1), information concerning the recent effectiveness of consecutive pairs of heuristics (f_2) and information concerning the amount of time since the heuristic was last called (f_3).

So for f_1 we have

$$f_1(N_j) = \sum_n \alpha^{n-1} \frac{I_n(N_j)}{T_n(N_j)}$$

where $I_n(N_j)$, respectively $T_n(N_j)$, is the change in the evaluation function (respectively the amount of time taken) the nth last time heuristic j was called, and α is a parameter between 0 and 1, which reflects the greater importance attached to recent performance. Then after calling heuristic N_j, the new value of $f_1(N_j)$ can be calculated from the old value using the formula

$$f_1(N_j) \leftarrow I_1(N_j)/T_1(N_j) + \alpha f_1(N_j).$$

f_1 expresses the idea that if a low-level heuristic recently improved well on the quality of the solution, this heuristic is likely to continue to be effective. Note that $I_n(N_j)$ is negative if there was an improvement and positive otherwise.

We consider that f_1 alone fails to capture much information concerning the synergy between low-level heuristics. Part of that synergy is measured by f_2 which may be expressed as

$$f_2(N_j, N_k) = \sum_n \beta^{n-1} \frac{I_n(N_j, N_k)}{T_n(N_j, N_k)}$$

where $I_n(N_j,N_k)$, resp. $T_n(N_j,N_k)$, is the change in the evaluation function (resp. amount of time taken) the nth last time heuristic k was called immediately after heuristic j and β is a parameter between 0 and 1, which again reflects the greater importance attached to recent performance. Then if we call heuristic N_k immediately after N_j, the new value of $f_2(N_j,N_k)$ can be calculated from the old value using the formula

$$f_2(N_j,N_k) \leftarrow I_1(N_j,N_k)/\, T_1(N_j,N_k) + \beta f_2(N_j,N_k).$$

f_2 expresses the idea that, if heuristic N_j immediately followed by heuristic N_k was recently effective and we have just used heuristic N_j, then N_k may well be effective again. Note that $I_n(N_j, N_k)$ is negative if there was an improvement and positive otherwise.

Both f_1 and f_2 are there for the purpose of intensifying the search. f_3 provides an element of diversification, by favouring those low-level heuristics that have not recently been used. Then we have

$$f_3(N_j) = \tau(N_j)$$

where $\tau(N_j)$ is the number of seconds of CPU time which have elapsed since heuristic N_j was last called.

For STRAIGHTCHOICE and RANKEDCHOICE hyperheuristics we will use the choice function F only to provide a ranking, and we will be indifferent as to the sign of F. However, for the ROULETTECHOICE hyperheuristic approach we want F to take only positive values, even for low-level heuristics which result in the objective function becoming much worse. Assume that the solution was perturbed most recently by low-level heuristic N_j. Recall that for our minimisation problem, large negative values of f_1 and f_2 are desirable. We define F as follows:

$$F(N_k) = \max \{ -\alpha\, f_1(N_k) - \beta\, f_2(N_j, N_k) + \delta\, f_3(N_k),$$
$$Q\rho^{\,\alpha f_1(N_k) + \beta f_2(N_j,N_k) - \delta f_3(N_k)} \}$$

Here δ is a parameter set at a value which leads to sufficient diversification,

$$Q = \frac{\sum_k \max \{0, -\alpha f_1(N_k) - \beta f_2(Nj, N_k) + \delta f_3(N_k) + \varepsilon\}}{10\eta}$$

where we have used $\varepsilon = 1$ and $\rho = 1.5$ to ensure that low-level heuristics which worsen the objective function value of the solution have a small, but non-zero probability of being chosen in the ROULETTECHOICE hyperheuristic, and that this probability falls rapidly for low-level heuristics which have exhibited very bad performance. The small term $\varepsilon/10\eta$ should enable every neighbourhood, no matter how bad, to be able to come around and diversify the solution after every other neighbourhood has been visited about 10 times.

The fourth (DECOMPCHOICE) method considers the individual components f_1, f_2 and f_3, of F. It tries the (up to four) low-level heuristics which yield the best values of

f_1, f_2, f_3, and F and performs the best move yielded by one of these low-level heuristics.

As we can see the information used by the hyperheuristic approaches to choose low-level heuristics is not specific to the summit scheduling problem (change in the evaluation function, time taken on the last call, time elapsed since last call for each heuristic).

5   Results

We used each of our hyperheuristics to solve the sales summit scheduling problem described in Section 2. The hyperheuristics were implemented in C++ and the experiments were conducted on a Pentium II 225MHz with 128MB RAM running under Windows NT version 4.0. In all experiments the stopping condition was 300 seconds of CPU time. $\eta = 10$ low-level heuristics were used, all of which are simple (and easy to implement). They are based either on the methods currently used for generating a schedule, or on simple moves such as swaps.

At this stage of development we determined values of α, β and δ experimentally. We chose $(\alpha, \beta, \delta) = (0.9, 0.1, 1.5)$ for all the AM cases and $(\alpha, \beta, \delta) = (0.2, 0.2, 0.8)$ for the OI ones. Experiments with a wide range of different values of (α, β, δ) showed that, while these values produced slightly better solutions on average, the hyperheuristic was insensitive to the parameter's values. It is undesirable that parameters need to be tuned in order for a general hyperheuristic approach to be effective, but the tuning process can be automated to preserve the problem-independence of the approach. Future work will investigate adaptively changing heuristic parameters during the solution process itself.

In RANKEDCHOICE the top r neighbourhoods (with respect to F) are tested and the best neighbourhood is retained. In our experiments we chose $r = \lceil 0.25\eta \rceil$.

In the roulette wheel approach we make the choice of the next neighbourhood randomly based upon a weighted probability function. The hyperheuristic chooses a random number v in the range $[0, A]$ where

$$A = \sum_{j=1}^{\eta} F(j).$$

Given

$$a_0 = 0$$

$$a_k = \sum_{j=1}^{k} F(j), k = 1,...,\eta$$

we choose neighbourhood k if $a_{k-1} \leq v < a_k$.

All choice-function-based hyperheuristics were started with the choice function values f_1, f_2, and f_3 set to 0 for each low-level heuristic. In order for the choice-function-based hyperheuristics to initialise the values of f_1, f_2, f_3, and F for each neighbourhood, we randomly call the neighbourhoods for an initial warm-up period. This warm-up period is included in the time allowed to the choice-function-based hyperheuristics. In our case the warm-up lasts 100 seconds of CPU time out of the total 300 seconds allowed. Apart from GREEDY which is entirely deterministic, all our hyperheuristic are averaged over 5 runs and in each run we changed the seed used to generate random values.

Results for all of the hyperheuristic approaches as well as the greedy heuristic, which is currently used for the sales summit problem, and the INITIALGREEDY heuristic which is used to generate an initial solution for each of our hyperheuristic approaches, are given in Table 1. For each algorithm, in addition to E, we give the values of B, C, H, d, defined in Section 2. We also give m and c where m is the number of meetings scheduled and c the total number of neighbourhood calls made.

We see that our INITIALGREEDY heuristic produced a much better solution than the algorithm currently used to schedule the sales summit (original greedy heuristic). All the hyperheuristics except SIMPLERANDOM and RANDOMDESCENT produced a better solution in the AM case than in the OI case. It appears that the OI version, which does not accept neighbour moves which yield a worse solution, has a greater tendency than the AM version to get stuck early in a local optimum, from which it never escapes. The SIMPLERANDOM and RANDOMDESCENT approaches use the different low-level heuristics in an erratic and unselective manner, and in their case accepting only improving moves limits the damage done by poor random choice of low-level heuristics. Note that the GREEDY approach will always produce identical results in AM and OI cases (since the only non-improving move ever accepted in the GREEDY-AM case is the final move).

We see that the choice-function-based approaches which accept non-improving moves are all significantly better than the other approaches. The large difference between AM and OI versions of these hyperheuristics is probably due to the diversification component of the choice function being rendered ineffective in the OI case, so that we become stuck in a local optimum too early. Encouragingly, all of the choice-function-based hyperheuristics produce good results. This would lend some support to the idea that each of these approaches is a general approach which could be used for a wide range of problem instances and a wide range of problems (so long as appropriate low-level heuristics were available). Overall DECOMPCHOICE hyperheuristic performed better than all the others. It also appears that the controlled randomness of the ROULETTECHOICE yields improvement over the STRAIGHTCHOICE and RANKEDCHOICE hyperheuristics. All of these simple choice-function-based approaches appear worthy of further investigation.

Table 1. Experiment results

Algorithm	B	C	d	H	E	m	c
Original Greedy Heuristic	226.00	48.65	99.00	216.00	**444.43**	823.00	-
INITIALGREEDY	52.00	111.00	93.00	168.00	**225.55**	811.00	-
SIMPLERANDOM-AM	27.00	83.20	89.80	142.40	**173.56**	828.00	1102.60
SIMPLERANDOM-OI	57.80	47.20	80.80	70.40	**130.56**	838.00	786.60
RANDOMDESCENT-AM	53.80	32.60	86.80	118.40	**173.83**	847.80	825.60
RANDOMDESCENT-OI	56.40	35.40	85.00	104.00	**162.17**	844.60	789.40
RANDOMPERMDESCENT-AM	57.60	28.20	85.20	105.60	**164.61**	849.20	850.40
RANDOMPERMDESCENT-OI	52.40	23.80	87.80	126.40	**179.99**	852.60	866.00
GREEDY-AM	56.00	27.00	86.00	112.00	**169.35**	851.00	847.00
GREEDY-OI	56.00	27.00	86.00	112.00	**169.35**	851.00	836.00
STRAIGHTCHOICE-AM	60.00	118.00	76.20	33.60	**99.50**	811.80	774.40
STRAIGHTCHOICE-OI	47.80	53.20	83.20	89.60	**140.06**	841.40	908.00
RANKEDCHOICE –AM	44.40	84.40	78.80	54.40	**103.02**	824.80	880.80
RANKEDCHOICE –OI	49.40	56.60	83.20	89.60	**141.83**	838.40	1007.00
ROULETTECHOICE –AM	59.20	132.60	76.00	32.00	**97.83**	809.40	765.20
ROULETTECHOICE –OI	53.80	43.60	83.60	92.80	**148.78**	842.60	937.20
DECOMPCHOICE –AM	38.80	74.80	78.40	51.20	**93.74**	826.00	782.40
DECOMPCHOICE –OI	47.20	61.80	83.00	88.00	**138.29**	837.80	1014.00

6   Conclusion

We have presented the idea of a hyperheuristic, that allows us to use knowledge-poor low-level heuristics, which generally lead to poor local optima when considered in isolation, in a framework which yields results which may, in some cases, be as good or better than those provided by knowledge-rich metaheuristic approaches. We have applied a range of hyperheuristics to a real-world sales summit scheduling problem. The results obtained are far superior to those provided by the system currently used to generate schedules. We believe that this approach is promising for a wide range of scheduling and optimisation problems.

Hyperheuristics have three important advantages over knowledge-rich approaches for practical scheduling and optimisation problems. The first is that, for many practical problems, modelling the problem using simple heuristics which describe the way that the system is currently solved (often by hand) is an easy way for problem owners to consider their problem. The second advantage is that simple heuristics based upon current user practice, simple local search neighbourhoods and greedy methods are quick to implement on a computer. Since this is all that is required in order to apply a hyperheuristic method, this should yield a method for fast prototyping of decision support systems for practical scheduling and optimisation problems. Indeed, we might simply keep adding low-level heuristics until we are satisfied (following experimentation) that we have a sufficient set of heuristics to provide good results. We then use cheap computer time to find out how best to manage the low-level heuristics, rather than expensive expertise. The third advantage is that the approach should generalise readily to small changes in the model (and indeed to large changes in the problem through the addition and modification of low-level heuristics, if necessary) yielding an approach which is robust enough to effectively handle a very wide range of problems and problem instances.

We do not regard hyperheuristics as a panacea to solve all problems in scheduling and optimisation (and so long as no fast algorithm is found for an NP-hard problem this is unlikely even to be possible). It is simply that for a very wide range of real-world problems where a reasonable solution is required in an acceptable amount of time, hyperheuristics should prove to be a useful tool.

In this paper we have introduced a range of simple choice-function-based hyperheuristic approaches, which are effective in spite of their simplicity, for the real-world problem which we have considered. While the details of the choice function are relatively complex, the user is shielded from this, simply supplying the objective function and the low-level heuristics. At this stage of development our hyperheuristic has several parameters, which may be tuned automatically to preserve domain-knowledge independence of the approach.

Several issues will be dealt with in future work. Parameter values will be set adaptively by the hyperheuristic in order for it to be a genuine problem-independent method applicable to a wide range of problems and instances, using different sets and types of low-level heuristics. We shall apply our approaches not only to other instances (with different but realistic objective functions) of this real-world problem but also to other personnel scheduling problems. We shall also consider how we may embed a range of more sophisticated methods into our hyperheuristic. In particular, we will consider the development of hyperheuristics which use metaheuristic techniques to decide which low-level heuristic to use, including population-based choice functions, tabu search and simulated annealing. We also intend to consider a genetic-programming approach to choice function evolution.

Appendix: The Low-Level Sales Summit Scheduling Heuristics

We used ten low-level heuristics:

1. *Remove one delegate:* This heuristic removes one delegate who has at least one meeting scheduled. It chooses the delegate with the least number of priority meetings, and the least number of meetings in total where there is a tie.
2. *Increase priority of one meeting:* This heuristic replaces one non-priority meeting with a priority meeting involving the same supplier, by changing the assigned delegate, if possible, without adding any new delegates.
3. *Add one delegate:* This heuristic adds one delegate (the delegate with the largest number of potential priority meetings) who currently has no meetings and greedily schedules as many meetings involving the new delegate as possible.

4. *Add meetings to dissatisfied supplier – version 1:* This heuristic adds as many meetings as possible to one dissatisfied supplier until the supplier is satisfied (if possible), without adding new delegates. This may only involve the deletion and rearrangement of meetings already arranged between delegates and other supplier, but only for "saturated" delegates who already have 12 meetings.

5. *Add meetings to dissatisfied or priority-dissatisfied supplier:* Same as the previous heuristic except that here the heuristic considers priority-dissatisfied suppliers (who may already have enough meetings, but not of sufficient priority) as well as dissatisfied ones.

6. *Add meetings to dissatisfied supplier – version 2:* Same as in heuristic 4, except that here the heuristic may move meetings of nonsaturated delegates who have less than 12 meetings as well as saturated ones.

7. *Cut surplus supplier meetings:* This heuristic takes each supplier who has more than 20 meetings scheduled and removes all the extra meetings (in increasing order of priority).

8. *Add meetings to priority-dissatisfied supplier:* This heuristic takes a supplier who has too few priority meetings and adds as many priority-meetings as possible to him, without adding delegates or violating the limitation on the maximum number of meetings per delegate .

9. *Add meetings to dissatisfied supplier:* Same as 8 but considers only suppliers who have enough priority meetings but too few meetings in total, and adds non-priority meetings, without adding delegates or violating the limitation on the maximum number of meetings per delegate.

10. *Add delegates and meetings to priority-dissatisfied supplier:* Same as 8 except we allow the addition of new delegates (those who do not currently have any meetings).

References

1.   Aickelin, U., Dowsland, K.: Exploiting Problem Structure in a Genetic Algorithm Approach to a Nurse Rostering Problem. J. Scheduling **3** (2000) 139–153

2.   Burke, E.K.: Cowling, P., De Causmaecker, P., Vanden Berghe, G.A.: Memetic Approach to the Nurse Rostering Problem. Int. J. Appl. Intell. to appear

3.   Burke, E., De Causmaecker, P., Vanden Berghe, G.A.: Hybrid Tabu Search Algorithm for the Nurse Rostering Problem. Selected Papers of the 2nd Asia–Pacific Conference on Simulated Evolution and Learning (SEAL '98). Lecture Notes in Artificial Intelligence, Vol. 1585: Springer, Berlin Heidelberg New York (1998) 186–194

4.   Back, T., Fogel, D.B., Michalewicz, Z. (eds.): Handbook of Evolutionary Computation. IOP Publishing and Oxford University Press (1997)

5.   Dodin, B., Elimam, A.A., Rolland, E.: Tabu Search in Audit Scheduling. Eur. J. Oper. Res. **106** (1998) 373–392

6.   Dowsland, K.A.: Nurse Scheduling with Tabu Search and Strategic Oscillation. Eur. J. Oper. Res. **106** (1998) 393–407

7.   Easton, F.F., Mansour, N.A.: Distributed Genetic Algorithm for Deterministic and Stochastic Labor Scheduling Problems. Eur. J. Oper. Res. **118** (1999) 505–523

8.   Mladenovic, N., Hansen, P.: Variable Neighborhood Search. Comput. Oper. Res. **24** (1997) 1097–1100

9.   Hart, E., Ross, P., Nelson, J.: Solving a Real-World Problem Using an Evolving Heuristically Driven Schedule. Evol. Comput. 6 (1998) 61–80
10. Mason, A.J., Ryan, D.M., Panton. D.M.: Integrated Simulation, Heuristic and Optimisation Approaches to Staff Scheduling. Oper. Res. **46** (1998) 161–175
11. Meisels, A., Lusternik, N.: Experiments on Networks of Employee Timetabling Problems. Practice And Theory of Automated Timetabling II: Selected papers. Lecture Notes in Computer Science, Vol. 1408. Springer, Berlin Heidelberg New York (1997) 130–155
12. Tsang, E., Voudouris, C.: Fast Local Search and Guided Local Search and their Application to British Telecom's Workforce Scheduling Problem. Oper. Res. Lett. **20** (1997) 119–127

Solving Rostering Tasks as Constraint Optimization

Harald Meyer auf'm Hofe

GWI-SIEDA GmbH
Richard-Wagner-Str. 91, D-67655 Kaiserslautern
Harald.Meyer@gwi-ag.com

Abstract. Based on experiences with the ORBIS Dienstplan system [12] – a nurse rostering system that is currently used in about 60 German hospitals – this paper describes how to use constraint processing for automatic rostering. In practice, nurse rostering problems have many varying parameters: working time accounts, demands on crew attendance, set of used shifts, working time models, etc. Hence, rostering requires a flexible formalism for representing the variants of the problem as well as a robust search procedure that is able to cope with all problem instances. The described approach differs in mainly two points from other constraint-based approaches [1], [22] to rostering.

On the one hand, the used constraint formalism allows the integration of fine-grained optimization tasks by fuzzy constraints, which a roster may partially satisfy and partially violate. Such constraints have been used to optimize the amount of working time and the presence on the ward. In contrast, traditional frameworks for constraint processing consider only crisp constraints which are either completely violated or satisfied. On the other hand, the described system uses an any-time algorithm to search for good rosters. The traditional constraint-based approach for solving optimization tasks is to use extensions of the branch&bound. Unfortunately, performance of tree search algorithms is very sensitive to even minor changes in the problem representation. Therefore, ORBIS Dienstplan integrates the branch&bound into local search. The branch&bound is used to enable the optimization of more than one variable assignment within one improvement step. This search algorithm converges quickly on good rosters and, additionally, enables a more natural integration of user interaction.

Keywords: employee timetabling, artificial intelligence, commercial packages, local search, soft computing, constraint based methods.

1 Introduction

The objective of rostering tasks is to label employees with a working (or idle) shift for each day of a certain period of time. Hence, these problems may be viewed as constraint satisfaction problems (CSPs), that concern the assignment of values out of a known domain to a finite number of variables [19]. This paper describes a system for solving nurse rostering problems of various kinds by constraint-based reasoning.

The nurse rostering system ORBIS Dienstplan (currently sold under different trademarks) is the result of collaboration between the GWI-SIEDA GmbH Kaiserslautern and the German Research Center for Artificial Intelligence (DFKI) and is currently operational in about 60 hospitals and in fire departments [12], [14]. This paper describes

E. Burke and W. Erben (Eds.): PATAT 2000, LNCS 2079, pp. 191–212, 2001.
© Springer-Verlag Berlin Heidelberg 2001

representation and search in a prototype on constraint-based rostering that reflects the lessons learned from the commercial system. From the perspective of research on constraint reasoning, the system exemplifies an integration of branch&bound search into iterative search algorithms. Furthermore, this system demonstrates how to use soft and even fuzzy or non-crisp constraints from a constraint library that is inspired by common practice in constraint logic programming (CLP) [13].

Nurse rostering is a comparably hard rostering task since shift assignments are required to comply with many different constraints concerning rest time, preferred sequences of shifts, working time accounts, compensation of working shifts on weekends, the expected expenditure of work and, last but not least, employees' preferences. Several authors have stressed the relevance of constraint-based methods to solve nurse rostering problems [1], [5], [22]. As Section 5 describes in detail, the work presented here differs in two major points from these approaches:

1. In several realistic situations, constraints on a roster cannot be satisfied completely. Consider, for instance, the case that some overtime work is necessary to guarantee a crew of appropriate size on the ward. In these cases, nurse rostering is rather a problem of constraint optimization than a problem of constraint satisfaction. Consequently, special techniques had to be invented to optimize compliance with *soft* and partly *fuzzy constraints.*
2. All approaches from the literature apply several heuristics which exploit characteristics of the current application. Such heuristics restrict the applicability of the resulting system to some special instances of nurse rostering where, for instance, only three shifts (early-morning shift, late shift, night shift) have to be assigned to the nurses. In contrast, ORBIS Dienstplan is applicable to arbitrary sets of shifts including longer day-turns and on-call duties. Additional parameters of the problem comprise working time accounts and a specification of a minimal and a preferred size of crew attendance *that may differ from day to day.* Obviously, such flexibility requires much more generic search procedures.

Additionally, constraint-based representations of nurse rostering problems are quite large. Since one needs a variable for each nurse on each day, rostering problems typically comprise 150 to 1200 variables. The structure of the constraints between these variables characterize nurse rostering as a problem of high computational complexity.

This paper is organized as follows. The first technical section (Section 2) presents a representation of nurse rostering as constraint problem, starting with the description of a special formalization of constraint problems that covers optimization aspects as well as use of constraint libraries. Based on this formalism, a representation of rostering problems is given. Section 3 briefly describes the standard algorithms for searching constraint optimization problems: iterative improvement and branch&bound search enhanced by constraint propagation. This section concludes with an integration of both search paradigms that may be considered as an iterative search allowing complex improvement steps. However, this integration of complex improvements raises the question of how to distinguish promising from useless improvement steps. Section 4 presents a heuristic that turned out to be successful on this task. Finally, Section 5 provides a comparison with related work: (1) the proposed formalization of soft constraints relates, for instance, to the valued constraint satisfaction (VCSP) formalism [3], [16] and (2) literature on compelling systems for constraint-based nurse rostering is used to point out the differences of this approach to the state of the art in this field.

2   Rostering Problems as Constraint Optimization

The basic framework for the proposed representation of rostering problems originates from the CSP whose basic ingredients are variables, a set of values called domain, and constraints that impose restrictions on how to assign values from the domain to variables. This section develops a formal representation of rostering problems in two steps:

1. The traditional notion of constraint problems is extended to deal with optimization. Additionally, constraints are derived from a *constraint library*, i.e. the problem representation refers explicitly to a number of *constraint types* out of a library that provides elaborate methods of reasoning on constraints.
2. In a second part, this section presents the required constraint types and the way they are used to represent rostering problems. Rosters are represented by an assignment of shifts to variables. Constraints represent demands on appropriate rosters.

2.1   Hierarchical Constraint Optimization Problems

Problems of constraint satisfaction or optimization concern the task of finding assignments of values from a certain domain D to variables from a set X. These assignments have to respect constraints from a set C. Specializations of this general task differ with respect to the representation of domains (numeric or finite set of symbols), constraints (binary or out of a constraint library), and the way that constraints assess assignments (hard or soft constraints). Rostering problems require a very general and therefore quite complex framework for the representation of constraint problems.

Assignments to the variables in X represent rosters. In order to distinguish better from worse rosters, the constraints of a rostering problem define a preference ordering $\succ$ among assignments (rosters) in such a way that $A_1 \succ A_2$ holds true for two assignments A_1 and A_2 iff A_2 complies better with the constraints than A_1. Each constraint $c \in C$ states a valuation of the assignments to some of the variables in X by use of a propagation method that is provided by a *constraint library*. The effect of such propagation methods can be described by a function φ that maps assignments to the local variables to a *degree of constraint violation*. A constraint violation of 0 says that the assignment satisfies the constraint whereas a degree of 1 indicates a total violation. Values in between these extremes represent a partial constraint violation. Thus, constraints of this framework are *non-crisp* or *fuzzy* in the sense that there is something in between complete violation and complete satisfaction.

Specifications of constraints comprise some more attributes in order to enhance applicability of propagation methods since the implementation of these methods often requires a large effort. A parameter p enables the user of the constraint library to adapt the used propagation method φ to the current optimization problem. Two additional attributes of constraints specify which constraints are more important than others. The hierarchy level h of a constraint is a positive integer that represents the constraint's categorical importance: each constraint of a more important hierarchy level is more important than all constraints in the levels of lower importance together. Traditionally, hierarchy level 0 is the most important hierarchy level, comprising the mandatory constraints [4]. Additionally, each constraint exhibits a weight ω in order to state its gradual importance. This means that importance of constraints *of the same hierarchy level* grows with their weight. In contrast to hierarchy levels, satisfying more constraints of smaller weight may be preferred to the satisfaction of a single constraint of larger weight. These commitments result in the following definition of a constraint formalism.

A hierarchical constraint optimization problem (HCOP) is a tuple $\mathcal{P} = \langle X, D, C \rangle$ where X is a set of variables, D is a finite set of values, and C is a set of constraints.

A constraint $\langle h, \omega, \boldsymbol{x}, p, \varphi \rangle \in C$ is composed of the *hierarchy level* $h \in \mathbb{N}_0$, the constraint's weight $\omega \in \mathbb{R}^+$, the vector of n local variables $\boldsymbol{x} \in X^n$, the *parameter* $p \in P$, and the *degree of constraint violation* $\varphi : P \times D^n \to [0; 1]$.

Let A be an assignment of values from D to all variables in X and $A \downarrow \boldsymbol{x}$ result in a vector $\boldsymbol{d} \in D^n$ where $\boldsymbol{d}[i]$ is the value that is assigned to $\boldsymbol{x}[i]$ by A.

The weight of an assignment A in hierarchy level i is defined as

$$\Omega_h(A) = \sum_{\langle h, \omega \, \boldsymbol{x}, p, \varphi \rangle \in C} \omega \cdot \varphi(p, A \downarrow \boldsymbol{x}).$$

Assume A_1 and A_2 to denote two assignments of values from D to the variables in X, then the preference ordering $\succ$, which distinguishes less from more preferred assignments, is defined as follows: if A_1 violates some mandatory constraints (hierarchy level 0) to a degree larger than 0, equivalent to $\Omega_0(A_1) > 0$, and A_2 satisfies all mandatory constraints, equivalent to $\Omega_0(A_2) = 0$, then A_2 is better than A_1 or $A_1 \succ A_2$. Additionally, $A_1 \succ A_2$ holds true, iff both assignments satisfy the mandatory constraints and there is a hierarchy level i with $\Omega_i(A_1) > \Omega_i(A_2)$ and for all levels j with $1 \le j < i$: $\Omega_j(A_1) = \Omega_j(A_2)$.

According to this definition, an assignment that satisfies all mandatory constraints is preferred to an assignment that violates some mandatory constraints. Two assignments that both violate mandatory constraints are considered to be equally bad. Preference, referring to the soft constraints, is determined according to the most important hierarchy level where one assignment violates less important constraints than the other assignment. Of course, HCOP is related to other generalizations of the standard CSP, namely lexicographic VCSP [3,16]. See Section 5 for further notes on this issue.

2.2   Rostering Problems as HCOP

The first task in representing real-world problems as a constraint problem is to identify the variables and the domains. In our representation, a constraint variable is generated for each nurse on each day in the roster. The planning period is typically a month, i.e. 30 or 31 days. In the following, x_{ij} denotes the constraint variable of nurse i on day j. Variable x_{ij} is labeled with the shift that nurse i has to serve on day j. The descriptor of a shift is typically composed of a letter representing the shift type that is followed by a number. Examples: **F1** represents early-morning shift variant 1, **S2** is the second variant

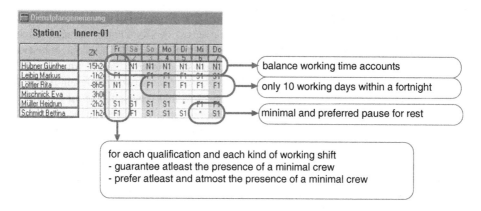

Fig. 1. The constraints on a roster

of a late shift, N1 is a night shift. Often, longer day-turns like T3 are used in addition. The descriptors "∗" and "−" denote idle shifts whereas holidays are indicated by UL. Shifts have three properties: start time s, end time e and working time w, where the latter represents the influence of the shift on working time accounts. The set of working shifts used often varies from hospital to hospital. Thus, the problem representation has to be able to deal with arbitrary sets of shifts. Figure 1 shows a portion of a schedule to illustrate this representation.

Requirements on a roster are given as constraints which partly have a large number of local variables. Figure 1 indicates instances of the main classes of constraints mapping local variables to textual descriptions of the constraints' extensions. The full representation of realistic problems comprises from 300 to more than 2000 constraints. However, these constraints are weighted instances of a rather small set of constraint types. The following paragraphs describe these types more formally, also illustrating their role in the representation of rostering problems.

Rest times. Two consecutive shifts of the same employee have to allow a minimal and a preferred time of rest. This demand can be ensured by two binary constraints between all consecutive shifts that may be implemented easily as so-called *extensional* constraints by use of propagation ext_n. This propagation simply receives the set of all consistent combinations of values as a parameter. The degree of constraint violation is defined as follows:

$$ext_n : 2^{D^n} \times D^n \longrightarrow 0; 1 : ext_n(p, d) = \begin{cases} 0 : d \in p \\ 1 : \text{otherwise.} \end{cases}$$

Hence, a tuple d complies with the constraint if it is element of the extension p of the constraint.

The required constraints can be generated from this propagation defining parameter p, i.e. the extension. Usually, the minimal time of rest between shifts is 11 hours. Hence, on these constraints p comprises all pairs of shifts where the second shift starts more than 11 hours after the first shift ends. Analogously, a preferred rest time of 16 hours can be specified.

Working time models. So-called working time models are preferred sequences of shifts that are stored in a database and range typically over two weeks. On the one hand, each nurse is preferred to work according to a certain working time model. The management of the ward assigns one working time model to each nurse in order to control which nurses are preferred to serve night shifts and which nurses alternate between different types of shifts. On the other hand, all working time models together represent a set of generally preferred shift sequences that nurses are accustomed to serve. As a consequence, the shift assignments to each nurse are also preferred to be consistent with an *arbitrary* working time model. Working time models differ strongly from hospital to hospital.

It is quite easy to state both demands by extensional constraints. For each nurse, consistency with the assigned working time model is represented by unary extensional constraints for each day whose extension comprises only the shift that the assigned working time model schedules on that day. Additionally, extensional constraints of 14 local variables have all specified working time models as extension.

Undesired sequences of shifts. There are also some sequences of shifts that shall be generally avoided. Examples are: too long chains of night shifts and sequences of the form working shift, idle shift, working shift. Details of undesired sequences of shifts also vary from hospital to hospital.

The representation of such demands requires a constraint type whose parameter describes forbidden instead of allowed tuples of values. Let μ be a function mapping k-ary tuples to $\{0; 1\}$ with $0 < k \leq n$ where $\mu(d) = 1$ iff tuple d should be avoided. Then the following degree of constraint violation states a penalty on any sequence of tuples that μ maps to 1:

$$avoid_{n,k} : avoid_{n,k}(\mu, d) = \begin{cases} 1 : \exists i : \mu(\langle d[i], \ldots, d[i+k-1]\rangle) = 1 \\ 0 : \text{otherwise.} \end{cases}$$

Different μ can now be used to specify certain sequences of shifts that have to be avoided.

Ensure minimal crew. One of the fundamental demands on rosters is to guarantee a minimal crew on the ward. Example: On day 5 the roster has to schedule at least one early-morning shift **F1** or **F2** to nurses 2, 4, 7, or 8 since these nurses have a special qualification that is required at that time on the ward. The rostering system enables the management to formulate arbitrary demands of this form. Hence, a constraint type is needed that counts the occurrences of certain values in the assignment to variables and compares this to a goal sum g. Let μ be a function that maps shifts from the domain D to real numbers. Then, g and μ are parameters of a propagation method $\varphi = atleast_n$ implementing the following relation:

$$atleast_n(\langle g, \mu \rangle, d) = \begin{cases} 0 : \sum_{i=1}^{n} \mu(d[i]) \geq g \\ 1 : \text{otherwise.} \end{cases}$$

Thus, a constraint of this type with the local variables $x = \langle x_{2,5}, x_{4,5}, x_{7,5}, x_{8,5} \rangle$ and parameter $p = \langle 1, \mu_{(F1,F2)} \rangle$ represents the demand of the example, where $\mu_{F1,F2}(d)$ is one iff d is either **F1** or **F2** and 0 otherwise.

Compensate work on weekends. Working shifts on Saturdays or Sundays need to be compensated within a fortnight. This demand is translated into a slightly tighter constraint that allows only ten days of work within a fortnight. The type of this constraint is called $atmost_n$ and is defined analogously to $atleast_n$.

Prefer standard crew. The second demand on crew attendance is to prefer a standard crew that is often larger then the minimal crew. Example: On day 5 the roster should schedule preferably 3 early-morning shifts **F1** or **F2** to nurses 2, 4, 7, or 8. This means that also scheduling 2 early-morning shifts is preferred to scheduling only 1, etc. Propagation method $approx_n$ is used to represent such demands. Again, parameters are a function μ and a goal sum g. Additionally, this type receives an exponent e as parameter that will be used later on. In the meantime assume $e = 1$. The following definition of $approx_n$ shows that this propagation implements a fuzzy relation between assignments to the local variables:

$$approx_n\left(\langle g, \mu, e\rangle, \boldsymbol{d}\right) = \frac{\left| g^e - \left(\sum_{i=1}^{n} \mu(\boldsymbol{d}[i])\right)^e \right|}{n \cdot \max\{\mu(d)^e \mid d \in D\}}.$$

Hence, the degree of constraint violation grows with the difference between goal sum g and sum of the results of μ.

A constraint of the local variables $\boldsymbol{x} = \langle x_{2,5}, x_{4,5}, x_{7,5}, x_{8,5}\rangle$ can be used to represent the demand from the example. The parameter of the constraint is set to $p = \langle 3, \mu_{F1,F2}, 1\rangle$. In contrast to constraints of the previously described types *ext*, *avoid*, *atmost* and *atleast*, constraints of type *approx* are fuzzy or non-crisp constraints: most assignments to the local variables are neither completely satisfied nor fully violated. A roster assigning exactly three of the requested shifts to the affected nurses satisfies this constraint perfectly. Assigning only two of the requested shifts leads to a constraint violation of degree $\frac{1}{4}$. The minimal crew (where only one of the requested shifts gets assigned) causes a larger constraint violation of degree $\frac{1}{2}$, and so on.

Keep working time accounts in balance. Depending on time credits and working contracts, each nurse has to serve a certain amount of working time. The scheduled shifts have to approximate the available working time as well as possible. The propagation $approx_n$ is also appropriate to this task. In contrast to the management of crew assignment, on this demand the mapping w of shifts to working time is used as parameter μ. For each nurse i a constraint is required with $\boldsymbol{x} = \langle x_{i,1}, \ldots, x_{i,31}\rangle$ as local variables. Parameter g of the constraint type reflects the amount of working time that nurse i has to serve in that month. Parameter e is set to 2 in order to distribute possibly necessary overtime work on as many shoulders as possible.

2.3 Use of Constraint Weights in Semi-automatic Rostering

The previous section described local variables, constraints, and their parameters that are used to represent rostering problems. Hierarchy levels and weights of the constraints also need to be defined.

Moreover, complex rostering tasks cannot be solved in an off-line manner since some demands on rosters will always be tacit (i.e. in the mind of responsible persons). Hence, rostering systems have to enable this responsible person to intervene into the rostering process. This section shows that hierarchy levels and weights of the constraints relate closely to the organization of this dialog.

The system performs as follows. Once started, the system searches for a roster that complies well with the constraints as described in the previous section according to shifts, working time accounts, holidays, and working time models from a data base. Provided with information on the quality of the best yet found roster, the user of the system is always able to stop the search for better rosters and state new demands as unary, extensional constraints: for example, defining some shifts of a certain nurse. Then, the rostering system shall be able to improve the best yet found roster respecting the newly introduced constraints until it is stopped again or the roster is perfect. This procedure enables the operator of the system to support the search for good rosters and to make tacit demands on the roster explicit. The portion of a roster in Figure 1 is drawn from the presentation of an imperfect roster to the operator of the system. The shaded rows indicate days where the system yet failed to achieve a standard crew on the ward. Hence, the operator can easily recognize deficiencies of the best available roster.

As a consequence of user dialog, hierarchy level 0 of mandatory constraints only comprises the constraints that all rosters have to satisfy when presented to the operator. These are constraints on minimal resting times and approved holidays.

Hierarchy level 1 comprises the constraints that have been generated on previous interactions with the user. Hence, satisfying one interactively formulated constraint is more important than satisfying the complete initial problem representation – the operator of the system may override any non-mandatory part of the initial problem.

Hierarchy level 2 contains all demands on a legal roster: minimal crew, at maximum 10 working days within a fortnight, and some mandatory bounds on the working time accounts.

Hierarchy level 3 is about preferred resting times. Hierarchy level 4 usually prefers the standard crew on the ward, whereas level 5 holds the constraints that try to keep working time accounts in balance. Sometimes, the levels 3 to 5 are put into another order if, for instance, too many time credits respectively time debts are known in advance. In this situation the constraints on working time accounts are more important than others.

Level 6 holds the constraints on working time models. The system enables nurses to state personal preferences in advance. These preferences are then translated into unary extensional constraints and put into hierarchy level 7. However, this feature of the rostering system is currently more important for selling the system than for running the system, i.e. many hospitals want to have the opportunity to deal with personal preferences of the employees but currently most of the hospitals do not use this feature. Within the hierarchy levels, the weights of the constraints are usually set to 1: i.e. the solver tries to satisfy as many demands as possible.

This section shows that representing rostering as a constraint problem leads to a formalization that allows user interaction and supports problem reformulation on adding new constraints. In particular, hierarchy levels proved to be useful in adopting constraint models on rostering to the current situation. Although the formal definition of HCOPs is rather complex, users of the system gain, surprisingly fast, a sufficient understanding of the semantics of hierarchy levels and constraint weights.

2.4   Comparison with Standard CLP

At first glance, the crisp constraint types *ext*, *avoid*, *atmost*, and *atleast* seem to be quite similar to constraints which are provided by programming languages like CHIP [6] or ECl^{i}PS^e [20] from the field of CLP. However, Section 3.2 shows that constraints of the proposed kind allows the combination of branch&bound search *á la* partial constraint satisfaction [9] with propagation procedures from CLP. So, the algorithms for constraint propagation are quite similar but, in contrast to CLP, constraint propagation does not only prune the domains of unlabeled variables.

Moreover, non-crisp constraints like the ones of type *approx* are unknown to CLP. Similar dependences are represented by higher-order predicate minimize(G(X), f(X)). This predicate allows only substitutions to the variables in X that are minimal referring to the result of function f(X) among the substitutions that satisfy the goals G(X). Fages presented a declarative semantics of this predicate [8] and some examples of its non-intuitive meaning. The implementation of this constraint runs a branch&bound search on the part of the problem as described by G(X) and f(X) without taking any advantage from constraint propagation as described in Section 3.4 of this paper.

3   Mixed-Paradigm Search

Several search algorithms have been proposed to solve constraint optimization problems, two being iterative improvement and branch&bound search. The iteration of user dialogue and solving process as described in Section 2.3 suggests the use of an iterative repair algorithm. However, as pointed out below, standard algorithms in this field exhibit some significant drawbacks that are relevant to rostering problems. Hence, some work on solving constraint problems in general has been necessary to implement the described nurse rostering system. This system uses a hybridization of both iterative improvement and branch&bound.

3.1   Iterative Improvement

Figure 2 shows the structure of algorithms on iterative improvement. These algorithms start by computing an initial assignment to all variables. Then, a loop of improvement steps follows. The standard algorithm in this field is the MinConWalk [21] which chooses in line 3 a single variable at random. The operation in line 5 depends on a so-called *walk probability p*. With probability $1 - p$, line 5 assigns an optimal value to the variable in X'. In order to escape from local minimal, line 5 assigns with probability of p a randomly chosen value. Although these algorithms have been fairly successful on several kinds of constraint problems, they perform not so well on problems implying constraints of many local variables that are hard to satisfy.

Consider Figure 3 as an example. The upper box (situation 1) presents a portion of a roster comprising shifts of the nurses A and B. The area within the dotted box complies with working time models concerning early-morning shifts. However, there is a problem with the minimal resting time for nurse A between the days 0 and 1: the morning shift F1 starts too soon after late shift S1. The lower box (situation 2)

1: compute an initial assignment A to all variables in X.
2: **loop**
3:     set $X' \subseteq X$ to hold a region where A possibly is suboptimal.
4:     **if** $X' = \emptyset$ **then break, end if**
5:     change assignments of A to the variables in X'.
6: **end loop**
7: **return** A

Fig. 2. Structure of algorithms on iterative improvement

presents an improvement to this roster: it complies with the same working time models: working time and crew attendance is the same as in the roster above. Nevertheless, the previously violated constraint on times of rest is now satisfied. This improvement has been achieved by changing assignments to 8 variables at once, but any change of less than 8 assignments makes the roster worse. Note that compliance with working time models, although the corresponding constraints form a deeper part of the hierarchy, is very important since it provides the only *sufficient* conditions on acceptable sequences of shifts (whereas constraints on number of working days and times of rest provide only *necessary* conditions). Hence, the rostering system is required to satisfy working time models well.

Standard algorithms on iterative improvement can, in theory, deal with this situation, but the probability of finding the described improvement is very low. Typical extensions of iterative search like tabu lists [18] fail to increase this probability significantly. Such situations obviously require the availability of more complex improvement steps in which the assignments to more than one variable get changed. The branch&bound algorithm is appropriate to conduct such complex improvement steps.

3.2   Extending Branch&Bound Search by Constraint Propagation

Propagation of soft constraints provides an optimistic estimate on the quality of a solution that can be found in the current branch of the search tree [9]. For each labeling $x \leftarrow d$ of a variable $x \in X$ with a value $d \in D$, tree search algorithms with propagation of soft constraints maintain a data structure $conflicts[x][d]$ to collect a measure for the importance of the constraints that are violated by all leaves in the current branch of the search tree that label x with d. The term *constraint propagation* denotes algorithms that detect new conflicts to be stored in $conflicts[x][d]$ referring to the properties of the propagated constraint and previously detected conflicts. Estimate $conflicts[x][d]$ can then be used by the A*-heuristic [15] to guide depth-first search. Additionally, adoptions of the well-known minimum remaining values (MRV) heuristic to constraint hierarchies are known to improve the order of the variables in the search tree [11], [13].

Here we present a new integration of constraint propagation into branch&bound search with two major characteristics. (1) The search algorithm is able to integrate special algorithms for the efficient propagation of non-binary constraints which may have non-crisp extensions as exemplified by constraint type *approx*. (2) The algorithm uses fuzzy sets [23] on the set of constraints C to represent estimate $conflicts[x][d]$.

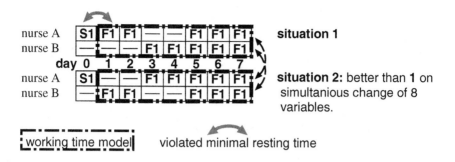

Fig. 3. Example for large bad regions caused by working time models

These sets map a certain degree of membership to each constraint c. During search, this degree of membership corresponds with the minimal degree of violating constraint c in the current branch of the search tree assigning d to x.

More formally, a fuzzy set on values from the crisp set S is a tuple $\tilde{S} = \langle \mu, S \rangle$ where μ is a mapping from S to numbers from the interval $[0; 1]$ that is called *membership function*. The membership function assigns a membership degree to each value $s \in S$, where $\mu(s) = 1$ designates full members and $\mu(s) = 0$ characterizes non-members of the fuzzy set $\tilde{S}$. Thus, a crisp set S can be treated as a fuzzy set assigning membership 1 to all the elements and membership 0 to all non-elements. The empty set $\emptyset$ is equivalent to a fuzzy set assigning only membership 0. Intersection $\sqcap$ and union $\sqcup$ of fuzzy sets are defined as follows: $\langle \mu_1, S_1 \rangle \sqcap \langle \mu_2, S_2 \rangle = \langle \mu_3, S_1 \cap S_2 \rangle$ where $\mu_3(c) = \min\{1, \mu_1(s), \mu_2(s)\}$ and $\langle \mu_1, S_1 \rangle \sqcup \langle \mu_2, S_2 \rangle = \langle \mu_4, S_1 \cup S_2 \rangle$ with $\mu_4(c) = \max\{0, \mu_1(s), \mu_2(s)\}$.

Figure 4 presents a branch&bound algorithm that uses fuzzy sets on the constraints C to record conflicts. Assume assignment A to label each variable in X with a value from D, and fuzzy set $\overline{C}(A)$ is intended to provide a measure for the conflicts of assignment A. For each constraint $c = \langle h, \omega, \boldsymbol{x}, p, \varphi \rangle$, the $\varphi(p, A \downarrow \boldsymbol{x})$ is A's degree of violating c. Thus, $\varphi(p, A \downarrow \boldsymbol{x})$ provides a measure for A's conflict with constraint c that can serve as a membership in a fuzzy set of constraints reflecting A's conflicts with all constraints: $\overline{C}(A) = \langle \mu, C \rangle$ with $\mu(\langle h, \omega, \boldsymbol{x}, p, \varphi \rangle) = \varphi(p, A \downarrow \boldsymbol{x})$.

Measuring conflicts in terms of fuzzy sets on constraints requires a notion of preference that complies with the preference among assignments. Thus, the preference ordering $\succ$ of Section 2.1 needs to be lifted to fuzzy sets of constraints in such a way that for all assignments A_1 and A_2 of domain values to variables, $A_1 \succ A_2$ is equivalent to $\overline{C}(A_1) \succ \overline{C}(A_2)$. Again, a Ω_i is defined to sum up the weights in hierarchy level i: $\Omega_i(\langle \mu, C' \rangle) = \sum_{c \in C'} \omega \cdot \mu(c)$ with $c = \langle h, \omega, \boldsymbol{x}, p, \varphi \rangle$. This sum of weights within a hierarchy level is used to lift the preference ordering on two fuzzy sets of constraints $\tilde{C}_1$ and $\tilde{C}_2$ as follows:

$$\tilde{C}_1 \succ \tilde{C}_2 :\Longleftrightarrow \left(\Omega_0(\tilde{C}_1) > 0 \wedge \Omega_0(\tilde{C}_2) = 0 \right) \vee$$

$$\left((\Omega_0(\tilde{C}_1) = \Omega_0(\tilde{C}_2) = 0) \wedge \right.$$

$$\left(\exists i > 0 : \Omega_i(\tilde{C}_1) > \Omega_i(\tilde{C}_2) \right) \wedge$$

BB-Start(X, D, C)
1: Store problem variables X, domain D, and constraints C as global variables.
2: Generate a 2-dimensional array $conflicts[X][D]$ and initialize each entry with the empty set.
3: **return** BB-Search$(\emptyset, C, \emptyset, \emptyset, conflicts, X)$.

BB-Search$(S, \tilde{\beta}, A, \tilde{\delta}, conflicts, X')$
1: **if** $\tilde{\beta} \not\succ \tilde{\delta}$ **then return** $\langle S, \tilde{\beta} \rangle$ **end if.**
2: **if** $X' = \emptyset$ **then return** $\langle A, \tilde{\delta} \rangle$ **end if.**
3: **for all** $x \in X'$ and $d \in D$ **do** $conflicts[x][d] \leftarrow conflicts[x][d] \sqcup \tilde{\delta}$. **end for**
4: Choose a variable $x \in X'$ according to a specialized MRV heuristic.
5: Copy the $conflicts[x]$ to the new array $conflicts'$.
6: **for all** $d \in D$ with $conflicts[x][d] \not\succ \tilde{\beta}$ choosing d's with smaller conflicts first **do**
7:     **for all** $d' \in D$ with $d' \neq d$ **do** $conflicts[x][d] \leftarrow C$. **end for**
8:     **for all** $c = \langle h, \omega, \boldsymbol{x}, p, \varphi \rangle \in C$ where x is a local variable **do**
9:         Update $conflicts$ according to propagation of c with an algorithm corresponding with φ
        with the arguments $(c, \boldsymbol{x}, \tilde{\beta}, p, conflicts')$.
10:     **end for**
11:     **if** for all variables x' there is a $d' \in D$ with $\tilde{\beta} \succ conflicts[x'][d']$ **then**
12:         $\langle S, \tilde{\beta} \rangle \leftarrow$ BB-SEARCH$(S, \tilde{\beta}, A \cup \{x \leftarrow d\}, conflicts[x][d], conflicts, X' \setminus \{x\})$.
13:         **if** $\tilde{\beta} = \emptyset$ **then return** $\langle S, \emptyset \rangle$, **end if.**
14:     **end if**
15:     $conflicts[x] \leftarrow conflicts'$.
16: **end for**
17: **return** $\langle S, \tilde{\beta} \rangle$.

Fig. 4. Branch&bound for solving HCOPs

$$\left(\forall j : 1 \leq j < i \Rightarrow \Omega_j(\tilde{C}_1) = \Omega_j(\tilde{C}_2) \right).$$

The proposed version of the branch&bound as presented in Figure 4 exploits three advantages of fuzzy sets of this kind as representation of detected conflicts:

1. Assume that the two fuzzy sets on constraints $\tilde{\delta}_1$ and $\tilde{\delta}_2$ hold conflicts of a certain assignment with the constraints. Then, $\tilde{\delta}_1 \sqcup \tilde{\delta}_2$ holds also valid conflicts of this assignment, since fuzzy set union simply collects the largest degrees of constraint violation that have been proven previously and stored either in $\tilde{\delta}_1$ or in $\tilde{\delta}_2$.

2. Additionally, fuzzy set union is an idempotent operation, i.e. $\tilde{\delta} \sqcup \tilde{\delta} = \tilde{\delta}$ for any fuzzy set of conflicts $\tilde{\delta}$. This means that conflicts are counted only once even if they have been added twice or more.

3. Assume that assignment S is the best currently known solution to the problem and that fuzzy set $\tilde{\delta}$ comprises conflicts of the currently explored partial assignment. Then, $\tilde{\beta} = \overline{C}(S)$ may serve as a bound, i.e. the algorithm can backtrack without losing solutions better than S as soon as $\tilde{\delta} \succ \tilde{\beta}$ holds true.

Algorithm BB-SEARCH conducts a depth-first tree search of the branch in the search tree that is described by partial assignment A, where distance $\tilde{\delta}$ is a fuzzy set of constraints comprising the previously detected conflicts of A, and $X' \subseteq X$ is the set of variables to be labeled by the algorithm. S is the best yet found solution and bound $\tilde{\beta}$ contains the conflicts of this solution. The algorithm results in the best assignment S to the variables in X' according to preference ordering $\succ$ or, equivalently, the least important conflicts $\overline{C}(S)$.

Function BB-START simply starts BB-SEARCH with initial arguments: S, A, and distance $\tilde{\delta}$ are empty. Bound $\tilde{\beta}$ has a maximal value. All variables are to be labeled. Array *conflicts* holds the empty set for each labeling since constraint propagation has not yet proven any conflict. So, let us return to BB-SEARCH.

Line 1 returns S if the distance $\tilde{\delta}$ is not better than bound $\tilde{\beta}$, i.e. the currently explored branch of the search tree does not contain any assignment that is better than S. Line 2 says that otherwise the currently explored assignment A is an improvement of S if all variables are labeled by A. Thus, A is a new solution. Line 3 performs an update of the *conflicts*. This is necessary because $\tilde{\delta}$ and the entries of *conflicts* both hold valid conflicts and from here on *conflicts*$[x][d]$ is required to hold all known conflicts that result from adding assignment $x \leftarrow d$ to A. These conflicts are typically needed by the heuristics for dynamic variable reordering in line 4 that determine which variable to label next. Heuristic MRV [2] (sometimes called *first fail*) is one of the most successful heuristics that is applicable here. It chooses the variable where the number of values which are not proven to be inconsistent with mandatory constraints is as small as possible (see [11], [13] for extensions of this heuristic that consider the complete constraint hierarchy). In this implementation this is the variable x where a minimal number of values d do not have a mandatory constraint in *conflicts*$[x][d]$ with a membership larger 0.

Branching is done by the loop in lines 6 to 15. Note that *conflicts*$[x][d]$ now holds all conflicts which are known to result from assigning d to x. Solutions to the problem are required to be better than bound $\tilde{\beta}$. Only values with smaller conflicts are *admissible*, i.e. need to be considered for branching. Additionally, *conflicts*$[x][d]$ allows us to follow the A*-heuristic [15]: try the assignments with the least important conflicts first.

Line 7 simulates assigning value d to variable x. All other values are marked with a maximal set of conflicts. After selecting value d for variable x, BB-SEARCH conducts an *extended forward checking* of soft constraints calling propagation methods. This procedure is related to *forward checking* on soft constraints [9], but this algorithm calls propagation without regarding the number of variables that have not yet been labeled.

Propagation of constraint $c = \langle h, \omega, \boldsymbol{x}, p, \varphi \rangle$ adds c with a certain membership to the *conflicts* of the local variables. The membership of constraint c in *conflicts*$[x][d]$ is intended to be an estimate of the effect of adding $x \leftarrow d$ to the currently explored partial assignment A. The rostering system uses a propagation rule for inferring this membership that allows the application of algorithms which are similar to the ones that are used in CLP (unfortunately, the propagation rule of the max–min algorithm [17] does not comply with algorithms for constraint propagation from CLP [13]). The concrete algorithm is chosen according to the constraint type φ, but all these procedures obey the same specification. Propagation of c looks for the smallest degree of violating c that is caused by assignments to the local variables of constraint c that exclusively consider *admissible* labelings. More formally, assume that propagation of c has to determine c's new membership to *conflicts*$[x][d]$ where $x = \boldsymbol{x}[i]$. D^n is the set of vectors representing assignments to the n local variables of c. Now, filter out any vector d with $d[i] \neq d$. These vectors represent assignments to the local variables that are not relevant to labeling $x \leftarrow d$. Furthermore, prune all vectors d concerning a non-admissible labeling: there is a j with $1 \leq j \leq n$ and $conflicts[x[j]][d[j]] \succ \tilde{\beta}$. These assignments consider labelings that cause too many conflicts in order to be part of any solution. Let D denote the remainder. Then, $\mu = \min\{\varphi(p, d) \mid d \in D\}$ is the minimal degree of violating c that is a consequence of assigning d to x. So, c is added with this membership to $conflicts[x][d]$.

Constraint propagation adds stronger memberships to the $conflicts$. This may lead to the case that for some labelings $x' \leftarrow d'$, $conflicts[x'][d']$ becomes larger than bound $\tilde{\beta}$. As a consequence, search in deeper branches of the search tree will ignore this labeling since it is not admissible. Line 11 tests for the case that a variable x' does not have any admissible labelings. In this case, there is no need to explore the current branch of the search tree because it does not comprise improvements to S.

Otherwise, branching is done in line 12 by a recursive call of BB-SEARCH where A is extended by the new labeling, $\tilde{\delta}$ is replaced by the conflicts caused by the new labeling and x is removed from the variables to be labeled. Note that the presentation of Figure 4 assumes a call by value that copies the arguments. Thus, the recursive call of BB-SEARCH returns the best labeling that has been found in the deeper branches of the search tree without touching the current data on conflicts. This can be considered as an implementation of the backmarking strategy [9], [10]. Finally, line 15 undoes the selection of d in x.

3.3 A Very Small Example

Reconsider the small example of Figure 3. This problem concerns two nurses over 8 days and, consequently, 16 variables. The representation as described in Section 2.2 requires more than 80 constraints. To cut a long story short, this section only refers to the constraints on minimal pause of rest, some constraints on working time models, and 7 constraints $atleast_1$ to $atleast_7$ enforcing at least one morning shift F1 on each of the days 1 to 7. Let $x_{A,3}$ denote the constraint variable for the shift of nurse A on day 3. For each pair of variables $x_{N,i}, x_{N,i+1}$ concerning consecutive shifts of the same nurse N, a constraint $pause_{N,i}$ allows the assignment of (F1, F1), (F1, S1), (S1, S1), or all combinations with the idle shift $-$. Two constraints $wmod_A$ and $wmod_B$ affecting the variables $x_{A,1}, \ldots, x_{A,7}$ and $x_{B,1}, \ldots, x_{B,7}$ respectively both allow the following two sequences of shifts: (F1, F1, $-$, $-$, F1, F1, F1) and ($-$, $-$, F1, F1, F1, F1, F1). Furthermore, assume that $x_{A,0}$ and $x_{B,0}$ hold the carry-over from the previous plan. So, $x_{A,0} = S1$ and $x_{B,0} = -$ is fixed by mandatory hard constraints.

Figure 5 sketches the search tree of this problem. After adding labeling 1 assigning S1 to $x_{A,0}$, propagation of $pause_{A,0}$ adds $pause_{A,0}$ as a full member to $conflicts[x_{A,1}][F1]$. The entries $conflicts[x_{A,1}][S1]$ and $conflicts[x_{A,1}][-]$ remain empty. Labeling 2 adds $atleast_1$ as a full member to $conflicts[x_{B,1}][-]$ and $conflicts[x_{B,1}][S1]$, since assigning F1 to $x_{B,1}$ is the only admissible opportunity to satisfy this constraint. Additionally, constraint $wmod_A$ is added as full member to the conflicts of all labelings concerning shifts of nurse A that do not comply with sequence ($-$, $-$, F1, F1, F1, F1, F1). Variable $x_{B,1}$ has the smallest domain, now. Labeling 3 is the best alternative here. Propagation of $wmod_B$ adds a conflict to all labelings concerning shifts of nurse B that do not comply

Fig. 5. A possible search tree for the example of Figure 3

with sequence $(F1, F1, -, -, F1, F1, F1)$. At this point, the labelings with empty entries in array *conflicts* characterize exactly "situation 2" in Figure 3. So, the search proceeds quickly to this solution that does not exhibit any conflict. A solution without conflicts is always optimal, so line 13 of algorithm BB-SEARCH terminates the search immediately. However, proving optimality of a known solution is often harder than finding it. In order to investigate the benefits of backmarking in proving optimality, assume that line 13 has been missed in algorithm BB-SEARCH.

Finding a solution lowers bound $\tilde{\beta}$ and turns all labelings with larger conflicts into non-admissible ones. Consequently, all variables except $x_{A,1}$ have only one admissible label: the one that is concerned with the solution. Recall that the backmarking strategy records all inferred conflicts. Thus, backtracking is done without any constraint propagation until changing the label of $x_{A,1}$. Here, labeling 4 is the only admissible alternative. However, propagation of $wmod_A$ reveals that this constraint cannot be satisfied since S1 is not concerned by any working time model. As a consequence, propagation of $wmod_A$ adds this constraint to any entry in *conflicts* referring to a labeling of one of $wmod_A$'s local variables turning these labelings into non-admissible ones. This is a reason for an immediate backtracking: the whole search tree has been traversed.

3.4   How to Propagate Soft Constraints

The above-mentioned extension of branch&bound search by constraint propagation presented a specification of the propagation rule but neglected the issue of implementation. Most of the literature on constraint processing is on binary constraint that can be propagated efficiently enumerating all assignments to the local variables. However, rostering applications exhibit constraints that affect a large number of local variables. These constraints require, thus, special purpose algorithms for propagation. This section presents two examples.

The $atleast_n$ constraint has been invented in section 2.2 to formalize least requirements on the presence of the crew. It is similar to the one-dimensional cardinality constraints which are provided by many CLP languages. In addition to the important role of this constraint in the nurse rostering domain, it is cited here to demonstrate how to use propagation methods from CLP to derive soft conflicts.

Recall that $\mu(d)$ defines a weight for each value $d \in D$. Within each complying assignment to the local variables of $atleast_n$-constraint c, the sum of these weights has to be larger than or equal to the goal sum g. Propagating $atleast_n$-constraints means: Look for values d where $\mu(d)$ is too small in order to allow satisfaction of the constraint. Figure 6 presents an implementation of a corresponding propagation method. Lines 1–8

ATLEAST $(c, \boldsymbol{x}, \tilde{\beta}, \langle g, \mu \rangle, conflict)$

  1: Initialize each entry of $\hat{\mu}[1 \ldots n]$ with 0 where n is the number of entries in $\boldsymbol{x}$.

  2: $\hat{\mu}_\Sigma \leftarrow 0$.

  3: **for all** $i \in \{1 \ldots n\}$ **do**

  4:    **for all** $d \in D$ with $\tilde{\beta} \succ conflicts[\boldsymbol{x}[i]][d]$ **do**

  5:       **if** $\mu(d) > \hat{\mu}[i]$ **then** $\hat{\mu}[i] \leftarrow \mu(d)$ **end if**

  6:       $\hat{\mu}_\Sigma \leftarrow \hat{\mu}_\Sigma + \hat{\mu}[i]$.

  7:    **end for**

  8: **end for**

  9: **for all** $i \in \{1 \ldots n\}$ **do**

10:    **for all** $d \in D$ with $\tilde{\beta} \succ conflicts[\boldsymbol{x}[i]][d]$ **do**

11:       **if** $\mu(d) + \hat{\mu}_\Sigma - \hat{\mu}[i] < g$ **then**

12:          $conflicts[\boldsymbol{x}[i]][d] \leftarrow conflicts[\boldsymbol{x}[i]][d] \sqcup \langle 1, \{c\} \rangle$.

13:       **end if**

14:    **end for**

15: **end for**

16: **return** $conflicts$.

Fig. 6. Propagating $atleast_n$ constraints

determine for each local variable $\boldsymbol{x}[i]$ of the constraint the largest $\mu(\cdot)$ of admissible values. Additionally, the sum of these largest weights is stored. This information is used in line 11 to determine whether constraint c can be satisfied by assigning a certain value to a certain local variable. This constraint is crisp since it is either satisfied or violated. The next example is on the propagation of fuzzy constraints.

Figure 7 shows the propagation of $approx_n$-constraints. As Section 2.2 mentioned, a tuple $\boldsymbol{d}$ violates such constraints with a degree growing with $|g - \sum_{i=1}^{n} [\mu(\boldsymbol{d}[i])]|$. Additionally, an exponent e may specify that the geometric distance is used instead of the arithmetic distance. Propagation of these constraints starts with finding for each variable $\boldsymbol{x}[i]$ the maximal weight $\hat{\mu}[i]$ and the minimal weight $\check{\mu}[i]$ of an admissible value. Additionally, the sum of the maximal and minimal weights gets stored. This information is used in lines 10–18 to decide for each admissible assignment to a local variable, whether goal g can still be reached. If the sum of weights of assigned values is too small, then the degree of constraint violation is determined due to $\hat{\mu}$ (line 13). Otherwise, the minimal weights $\check{\mu}$ are used to determine the degree of constraint violation (line 15).

The effort for propagating both constraints grows only linearly with the number of local variables n. Hence, these propagation methods are also appropriate to constraints affecting many local variables as they appear in rostering problems.

3.5   Branch&Bound as Improvement Step

The previous section stressed that both iterative search and branch&bound have their advantages and drawbacks. An integration of both algorithms according to Figure 8 can inherit the advantages of both search paradigms. The basic idea of this algorithm is to follow the main procedure in iterative algorithms but enable more complex improvement

APPROX $(c, \boldsymbol{x}, \tilde{\beta}, \langle g, \mu, e\rangle, conflict)$
1: Initialize each entry of $\hat{\mu}[1\ldots n]$ with 0 and $\breve{\mu}[1\ldots n]$ with -1.
2: $\hat{\mu}_{\Sigma} \leftarrow 0.\breve{\mu}_{\Sigma} \leftarrow 0.$
3: **for all** $i \in \{1\ldots n\}$ **do**
4:      **for all** $d \in D$ with $\tilde{\beta} \succ conflicts[\boldsymbol{x}[i]][d]$ **do**
5:          **if** $\mu(d) > \hat{\mu}[i]$ **then**$\hat{\mu}[i] \leftarrow \mu(d)$ **end if**
6:          **if** $\breve{\mu}[i] = -1 \vee \breve{\mu}[i] > \mu(d)$ **then**$\breve{\mu}[i] \leftarrow \mu(d)$ **end if**
7:          $\hat{\mu}_{\Sigma} \leftarrow \hat{\mu}_{\Sigma} + \hat{\mu}[i].\breve{\mu}_{\Sigma} \leftarrow \breve{\mu}_{\Sigma} + \breve{\mu}[i].$
8:      **end for**
9: **end for**
10: **for all** $i \in \{1\ldots n\}$ **do**
11:     **for all** $d \in D$ with $\tilde{\beta} \succ conflicts[\boldsymbol{x}[i]][d]$ **do**
12:         **if** $\mu(d) + \hat{\mu}_{\Sigma} - \hat{\mu}[i] < g$ **then**
13:             $conflicts[\boldsymbol{x}[i]][d] \leftarrow conflicts[\boldsymbol{x}[i]][d] \sqcup \left\langle \frac{(\mu(d)+\hat{\mu}_{\Sigma}-\hat{\mu}[i])^e - g^e}{(\max\{\mu(d)^e \mid d\in D\}}, \{c\} \right\rangle .$
14:         **else if** $\mu(d) + \breve{\mu}_{\Sigma} - \breve{\mu}[i] > g$ **then**
15:             $conflicts[\boldsymbol{x}[i]][d] \leftarrow conflicts[\boldsymbol{x}[i]][d] \sqcup \left\langle \frac{g^e - (\mu(d)+\hat{\mu}_{\Sigma}-\hat{\mu}[i])^e}{\max\{\mu(d)^e \mid d\in D\}}, \{c\} \right\rangle .$
16:         **end if**
17:     **end for**
18: **end for**
19: **return** $conflicts$.

Fig. 7. Propagating $approx_n$ constraints

ITERATIVESEARCH(X, D, C)
1: $\tilde{\beta} = C.\ S \leftarrow \emptyset.$
2: Set initial solution S according to the working time models that have been assigned to each employee by the management.
3: Generate $conflicts[X][D]$ with $conflicts[x][d] = \emptyset$ forall $x \in X$ and $d \in D$.
4: **loop**
5:      $X' \leftarrow$ CHOOSEBADREGION$(X, D, C, S, \tilde{\beta})$.
6:      **if** $X' = \emptyset$ **then break, end if**
7:      $\tilde{\delta} \leftarrow \tilde{\beta}$. Load A with all labelings in S that do not affect variables in X'.
8:      **for all** $x \in X', d \in D$ **do** set $conflicts[x][d]$ to the empty set if S assigns d to x and to C otherwise. **end for**.
9:      **for all** $c \in C$ with c has a local variable in X' **do**
10:         $\tilde{\delta} \leftarrow \tilde{\delta} \sqcap (C \setminus \{c\})$. Set $conflicts$ due to propagation of c.
11:     **end for**
12:     $\langle S, \tilde{\beta}\rangle \leftarrow$ BB-SEARCH$(S, \tilde{\beta}, A, \tilde{\delta}, conflicts, X')$.
13: **end loop**
14: **return** $\langle S, \tilde{\beta}\rangle$

Fig. 8. Iterative repair using branch&bound and constraint propagation

steps by use of the branch&bound. Hence, Figure 8 details the algorithm of Figure 2: the initial assignment to the variables is generated in line 2 according to the working time models that have been assigned to the employees (see Section 2.2).

The loop of improvement steps ranges over lines 4–13. A function CHOOSEBAD-REGION is assumed to point at a region $X' \subseteq X$ in the current solution S where changes in the current assignments are likely to result in an improved solution. The algorithm terminates in line 6 if this procedure cannot find any region in the solution that is suboptimal. Otherwise, the branch&bound is called to find an optimal assignment to the variables in X'. Therefore, the distance is initialized to comprise all conflicts of the persistent assignments in S to the variables in $X \setminus X'$. Additionally, *conflicts* is initialized to hold all conflicts between variables in X' with the persistent assignments to the variables in $X \setminus X'$. After these preparations, the call of BB-SEARCH in line 12 is able to find improvements to the variables in X' if this is possible.

This procedure may be stopped after each improvement step by user input. Hence, this algorithm supports user interaction as described in Section 2.3. Moreover, improvement steps are, in contrast to standard algorithms, not restricted to affecting only the assignment to one variable – assignments to many variables may get changed depending on the output of function CHOOSEBADREGION. However, the implementation of this function is the main problem of this approach. As the next section shows, the ORBIS Dienstplan system uses heuristics for this purpose which are based upon some general assumptions on characteristics of rostering problems.

4   How to Find Bad Regions in a Roster

The currently available ORBIS Dienstplan system uses heuristics to detect the shift assignments in a roster that are responsible for deficiencies. The implementation of these heuristics is provided by the function CHOOSEBADREGION that is used in algorithm ITERATIVESEARCH. This section provides a description of the heuristic to achieve the minimal and preferred crew on the ward in order to illustrate this procedure. This heuristic is called whenever a constraint on crew attendance is more or less violated.

Figure 9 presents (intermediate) results of the *ORBIS Dienstplan* system to a small sample problem that considers six nurses (186 variables and about 600 constraints). The upper roster represents the initial assignment of shifts before any step of repair. The lower roster is an acceptable solution which has been computed by conducting several repair steps. Light shading represents suboptimal but acceptable deficiencies in crew attendance. Dark shading represents unacceptable deficiencies in crew attendance.

First of all, consider the problem on Sunday the 17th where an early-morning shift (F1) is missing and two late shifts (S1) have been scheduled but we prefer to have only one on that day. For such easy problems, changing only a single assignment of a shift suffices to achieve an improvement: assigning an early-morning shift instead of a late shift to Bettina Schmidt on that day satisfies both previously violated constraints.

On Sunday the 10th, the schedule exhibits a worse problem: again an early-morning shift is missing. In such situations, at least two assignments of shifts have to be changed: a currently idle person has to serve an early-morning shift on Sunday the 10th and this additional working time has to be compensated for on another day because of

Dienstplangenerierung

Station: Innere-01   von: 01.11.1996   bis: 30.11.1996

	ZK	Fr 1	Sa 2	So 3	Mo 4	Di 5	Mi 6	Do 7	Fr 8	Sa 9	So 10	Mo 11	Di 12	Mi 13	Do 14	Fr 15	Sa 16	So 17	M
Hübner Günther	-15h24	N1	N1	N1	N1	N1	N1	N1	N1	-	*	x	-	-	-	-	N1	N1	N
Leibig Markus	-1h24	F1	-	F1	F1	S1	S1	S1	S1	S1	S1	S1	*	F1	F1	F1	*	-	S
Löffler Rita	-8h54	S1	-	-	F1	F1	F1	F1	F1	F1	UL	UL	UL	UL	UL	UL	-	-	F
Mischnick Eva	3h06	N1	-	-	-	-	-	-	N1	N1	N1	N1	N1	N1	N1	N1	*	-	
Müller Heidrun	-2h24	S1	S1	S1	S1	*	F1	F1	F1	-	-	S1	S1	S1	S1	S1	S1	S1	S
Schmidt Bettina	-1h24	F1	F1	S1	S1	S1	*	S1	S1	-	-	F1	F1	F1	F1	F1	F1	S1	S

	ZK	Fr 1	Sa 2	So 3	Mo 4	Di 5	Mi 6	Do 7	Fr 8	Sa 9	So 10	Mo 11	Di 12	Mi 13	Do 14	Fr 15	Sa 16	So 17	M
Hübner Günther	-15h24	-	N1	N1	N1	N1	N1	N1	*	-	F1	F1	F1	*	-	-	N1	N1	N
Leibig Markus	-1h24	F1	-	F1	F1	F1	S1	S1	S1	S1	S1	S1	*	F1	F1	F1	x	-	
Löffler Rita	-8h54	N1	-	F1	F1	F1	F1	F1	F1	F1	UL	UL	UL	UL	UL	UL	*	-	
Mischnick Eva	3h06	-	-	-	-	-	-	-	N1	N1	N1	N1	N1	N1	N1	N1	*	-	
Müller Heidrun	-2h24	S1	S1	S1	S1	*	F1	F1	F1	-	-	S1	S1	S1	S1	S1	S1	S1	
Schmidt Bettina	-1h24	F1	F1	S1	S1	S1	*	S1	S1	-	-	F1	F1	F1	F1	F1	F1	F1	

Fig. 9. Initial (above) and improved (below) roster

working time restrictions. Procedure CHOOSEBADREGION recognizes that one of the idle persons Hübner, Müller or Schmidt is required to serve an additional shift. To achieve this, CHOOSEBADREGION looks for days when one of these persons is surplus and stores the corresponding variables in a pool – this pool may also be filled by some other heuristics. Finally, CHOOSEBADREGION makes random choices from this pool and returns the resulting set of variables as a description of the next step of repair. In this example, taking the variable referring to Günther Hübner on Friday the 1st enables the system to compensate with an additional early-morning shift of Günther Hübner on Sunday the 17th. Often, only a very small number of opportunities is available to cope with such situations.

The lower roster in Figure 9 shows the result of such repair steps that is available in about one minute. On larger problems, finding first solutions that satisfy the constraints on minimal crew attendance without exceeding the working time may take about five minutes. Running times of more than half an hour have been reported for problems with inappropriate assignments of working time models.

5   Conclusion

Based on experiences with the ORBIS Dienstplan system [12] – a nurse rostering system that is already used in many German hospitals – this paper describes how constraint processing can be used to implement automatic rostering systems that can be flexibly adapted to different situations at different hospitals. In practice, the current formalization of the rostering problem is still not complex enough to enable fully automatic planning procedures. Thus, complexity of the problem representation can be expected to grow in the future. Additionally, nurse rostering problems have many varying parameters: working time accounts, demands on crew attendance, set of used shifts, working time models. Hence, rostering requires both a flexible formalism for representing the currently known and future variants of the problem and a robust search procedure that is able to cope with quite different problem instances and that allows the responsible managers to intervene in the planning process. This paper shows that constraint reasoning meets all these demands if the state of the art is extended in several directions.

The paper introduces a new notion of fuzzy or non-crisp constraints – constraints that may be partially violated and partially satisfied – in HCOPs to improve problem representation and problem solving. In contrast to standard formalisms of fuzzy constraints [17], [7] that sum up degrees of constraint violation by taking the minimum or maximum, the formalism introduced in this paper uses more flexible operations to combine different constraints. As a consequence, the well known branch&bound has been extended to detect conflicts with fuzzy constraints by constraint propagation. Of course, the proposed framework relates to other formalizations of soft constraints like, for instance, VCSP [3], [16], especially lexicographic VCSP. However, the framework of Section 2.1 uses fuzzy sets of constraints as a valuation structure that is idempotent but not totally ordered, and this framework is able to borrow algorithms for the propagation of non-binary constraints from CLP. VCSP neglects the advantages of specialized algorithms for constraint propagation [13]. Lexicographic VCSP can be translated into HCOPs with binary constraints of type *ext* that all have the same weight.

In combination with new extensions of generic search algorithms, this framework forms the basis of a nurse rostering system that differs in two main points from other constraint-based prototypes on this application [1], [22]:

- Fuzzy constraints allow the integration of very fine-grained optimization tasks. Such constraints have been used to optimize the amount of working time and the presence on the ward. In contrast, traditional frameworks for constraint processing in nurse rostering consider only crisp constraints which are either completely violated or satisfied.
- The traditional constraint-based approach for solving optimization tasks is to use an extension of the branch&bound. Unfortunately, performance of tree search algorithms is very sensitive to even minor changes in the problem representation. For instance, INTERDIP [1] tries to overcome this problem by the use of highly developed heuristics on problem decomposition. Unfortunately, these heuristics rely on properties of the available working shifts which are, in practice, parameters of the problem that vary from hospital to hospital. In contrast, *ORBIS Dienstplan* integrates the branch&bound into search by iterative improvements. The branch&bound is used to enable the optimization of more than one variable assignment within one improvement step. The resulting search algorithm converges quickly on good rosters and, additionally, enables a more natural integration of user interactions.

Unfortunately, the ability to deal with complex improvement steps affecting more than one variable increases the number of applicable improvement steps compared to standard local search where only a single value assignment gets changed within each step of repair. Therefore, Section 4 illustrates a heuristic that is very successful in finding the reason for deficiencies of the best yet found roster. Such heuristics are able to guide the proposed local search effectively.

References

1. Abdennadher, S., Schlenker, H.: INTERDIP – an Interactive Constraint Based Nurse Scheduler. In: Proc. Int. Conf. on Practical Applications of Constraint Technology and Logic Programming. Practical Applications Expo, London (1999)
2. Bacchus, F., van Run, P.: Dynamic Variable Ordering in CSPs. In: Montanari, U., Rossi, F. (eds.): Proc. 1st Int. Conf. on Principles and Practice of Constraint Programming. Springer-Verlag, Berlin Heidelberg New York (1995) 258–275
3. Bistarelli, S., Fargier, H., Montanari, U., Rossi, F., Schiex, T., Verfaillie, G.: Semiring-Based CSPs and Valued CSPs: Basic Properties and Comparisons. In: Jampel, M. (ed.): Over-Constrained Systems. Springer-Verlag, Berlin Heidelberg New York (1996)
4. Borning, A., Freeman-Benson, B., Wilson, M.: Constraint Hierarchies. Lisp and Symbolic Comput. 5 (1992) 233–270
5. Darmoni, S.J., Fajner, A., Leforestier, A., Mahè, N.: Horoplan: Computer-Assisted Nurse Scheduling Using Constraint-Based Programming. J. Soc. for Health Systems 5 (1995) 41–54
6. Dincbas, M., Van Hentenryck, P., Simonis, H., Aggoun, A., Graf, T., Berthier, F.: The Constraint Logic Programming Language CHIP. In: Proc. Int. Conf. on Fifth Generation Computer Systems (1988)
7. Dubois, D., Fargier, H., Prade, H.: The Calculus of Fuzzy Restrictions as a Basis for Flexible Constraint Satisfaction. In: Proc. 2nd IEEE Int. Conf. on Fuzzy Systems (San Francisco, CA) (1993) 1131–1136
8. Fages, F., Fowler, J., Sola, T.: Handling Preferences in Constraint Logic Programming with Relational Optimization. In: Jampel, M. (ed.): Over-Constrained Systems. Springer-Verlag, Berlin Heidelberg New York (1996)
9. Freuder, E.C., Wallace, R.J.: Partial Constraint Satisfaction. Artif. Intell. 58 (1992) 21–70
10. Gaschnig, J.: A General Backtrack Algorithm that Eliminates Most Redundant Checks. In: Proc. IJCAI-77 (Cambridge, MA) (1977)
11. Meyer auf'm Hofe, H.: Partial Satisfaction of Constraint Hierarchies in Reactive and Interactive Configuration. In: Hower, W. Ruttkay, Z. (eds.): ECAI-96 Workshop on Non-Standard Constraint Processing (1996) 61–72
12. Meyer auf'm Hofe, H.: ConPlan/SIEDAplan: Personnel Assignment as a Problem of Hierarchical Constraint Satisfaction. In: PACT-97: Proc. 3rd Int. Conf. on the Practical Application of Constraint Technology (Practical Application Expo, London) (1997) 257–272
13. Meyer auf'm Hofe, H.: Kombinatorische Optimierung mit Constraintverfahren – Problemlösung ohne anwendungsspezifische Suchstrategien. DISKI – Dissertationen zur Künstlichen Intelligenz, Vol. 242. Infix, Akademische Verlagsgesellschaft (2000) ISBN 3-89838-242-7
14. Meyer auf'm Hofe, H., Abecker, A.: ConPlan: Solving Scheduling Problems Represented by Soft Constraints. In: ISIAC-98: Proc. Int. Symp. on Intelligent Automation and Control (Anchorage, AK) (1998) 245–250
15. Nilsson, N.: Principles of Artificial Intelligence. In: Symbolic Computation. Springer-Verlag, Berlin Heidelberg New York (1982) chapter 3.2
16. Schiex, T., Fargier, H., Verfaillie, G.: Valued Constraint Satisfaction Problems: Hard and Easy Problems. In: Mellish, C. (ed.): IJCAI-95: Proc. 14th Int. Joint Conf. on Artif. Intell. Kaufmann, San Francisco, CA (1995) 631–637
17. Snow, P., Freuder, E.C.: Improved Relaxation and Search Methods for Approximate Constraint Satisfaction with a Maximin Criterion. In: Proc. 8th Biennial Conf. of the Can. Soc for Comput. Studies of Intelligence (1990) 227–230

18. Steinmann, O., Strohmeier, A., Stützle, T.: Tabu Search vs. Random Walk. In: Brewka, G., Habel, C., Nebel, B. (eds.): KI-97: Advances in Artificial Intelligence. LNAI, Vol. 1303. Springer-Verlag, Berlin Heidelberg New York (1997) 337–348

19. Tsang, E.: Foundations of Constraint Satisfaction. Computation in Cognitive Science. Academic, London (1993) ISBN 0-12-701610-4

20. Wallace, M., Novello, S., Schimpf, J.: Ecl^{i}ps^e – a Platform for Constraint Logic Programming. ICL Systems J. **12** (1997) 159–200

21. Wallace, R.J.: Analysis of Heuristic Methods for Partial Constraint Satisfaction Problems. In: CP–96: Proc. 2nd Int. Conf. on Principles and Practice of Constraint Processing. Springer-Verlag, Berlin Heidelberg New York (1996) 482–496

22. Weil, G., Heus, K.: Eliminating Interchangeable Values in the Nurse Scheduling Problem Formulated as a Constraint Satisfaction Problem. In: CONSTRAINT-95: The FLAIRS-95 Int. Workshop on Constraint-Based Reasoning (Melbourne Beach, FL) (1995)

23. Zadeh, L.A.: Fuzzy Sets. Information and Control **8** (1965) 338–353

Assigning Resources to Constrained Activities

Amnon Meisels and Ella Ovadia

Department of Computer Science
Ben-Gurion University of the Negev
Beer-Sheva, 84-105, Israel
am@cs.bgu.ac.il

Abstract. Resource allocation problems (RAPs)are naturally represented as constraint networks (CNs), with constraints of inequality among activities that compete for the same resources at the same time [5]. A large variety of timetabling problems can be formulated as CNs with inequality constraints (representing time conflicts among classes of the same teacher, for example). A new algorithm for solving networks of RAPs is described and its detailed behavior is presented on a small example. The proposed algorithm delays assignments of resources to selected activities and processes the network during the assignment procedure, to select delays and values.

The proposed algorithm is an enhancement to standard intelligent backtracking algorithms [13], is complete and performs better than two former approaches to solving CNs of resource allocation [15], [3], [4]. Results of comparing the proposed resource assignment algorithm to other ordering heuristics, on randomly generated networks, are reported.

Keywords: constraint-based methods, timetabling networks

1 Introduction

Let us imagine a common problem of school timetabling, in which pairs of class-teacher have to be assigned a time slot in order to construct a weekly schedule [2]. This formulation of a timetabling problem (TTP) assigns resources (i.e. time slots) to activities (class–teacher pairs), thus generating an instance of the resource allocation problem (RAP) [5], [10]. A RAP can be naturally represented by a discrete constraint satisfaction problem (CSP) [18], [3] in which variables are activities, the domains of values of variables are the resources that can be assigned to the activities and there are constraints of inequality among pairs of variables that compete for the same resource. These inequality constraints are related to the fact that resources can only be assigned *one at a time to activities that overlap in time*. This type of binary constraint has been termed *mutual exclusion* constraints in the scheduling community [4,8], and *constraints of difference* in the constraint processing community [15].

In Section 2 the constraint network (CN) of RAPs is defined and an algorithm for solving them is described and demonstrated. The proposed resource

E. Burke and W. Erben (Eds.): PATAT 2000, LNCS 2079, pp. 213–223, 2001.

assignment algorithm (RAA) is based on standard intelligent backtracking methods and orders variables dynamically at every assignment step. The dynamic ordering of variables takes the form of *a series of delays of assignments* that generates additional delays and finally assigns values to the delayed variables in the reverse order of delays. The RAA is traced in detail on a small example in Section 3 and compared with former heuristics for decomposing and solving RAP networks [4], [3].

An approach to *preprocessing* of constraints of difference has been presented by Regin [15]. Addressing the non-binary constraints of difference that are part of every RAP, Regin proposes an efficient consistency enforcing algorithm for these *all-diff* constraints [15]. The behavior of all three procedures is traced on the example in section 3 and illuminates the differences among these approaches and the advantages of RAA.

An interesting test for the proposed heuristics and resulting RAA is to solve random instances of RAPs. A realistic method for generating random networks of inequality constraints is to use the structure of real-world timetabling problems [10,16]. In Section 4 real-world employee timetabling problems (ETPs) are used as a template for random constraint networks and the proposed method is used to solve them. The experiments in Section 4 solve the randomly generated CNs by three different algorithms – forward checking (FC) with static variable ordering; FC with dynamic ordering; and the proposed RAA. The performance of the RAA is better than that of the competing algorithms.

2   The Resource Assignment Algorithm

RAPs are naturally represented by CNs in which variables are activities and values are resources. The major binary constraint is a constraint of difference, expressing the intuitive fact that each resource can only be assigned to activities that are non-overlapping in time. A network that consists solely of constraints of difference, defines a *list coloring* problem with resources as lists of colors for each variable [4]. List coloring is the problem of coloring an interval graph [6], [7] by using a different *list* of colors for different variables (nodes) and has been shown to be NP-complete [1].

The general RAP has activities which can be grouped into sets that are *overlapping in time*. These sets form *complete subnetworks* with respect to the binary constraint of mutual exclusion. Any value (resource) in the domain of any variable in the network of mutually exclusive variables can be assigned *to only one of these variables* (i.e. at one time).

The problem of resource assignment to activities, which forms the heart of all scheduling problems, has been addressed by Choueiry and Faltings [5], [4]. The main maxim of Choueiry and Faltings is that the assignment of resources can be delayed differently for activities that differ in their degree of dependency on resources of different contentions. A different approach to problems that involve CNs with constraints of inequality (or constraints of difference, to use a different terminology) has been taken by Regin [15]. Regin proposes a special procedure for preprocessing the CN, to impose consistency on $n-ry$ constraints of difference (which he calls *all-diff*). These approaches will be compared with the present proposed algorithm on an example network in Section 3.

The RAA which is proposed in this paper is a backtracking algorithm that proceeds to select the next variable, select the next value for the selected variable, check the consistency of the proposed assignment and then assign the selected value to the selected variable. The basic search algorithm can be any intelligent BT algorithm, such as forward checking (FC), backjumping (BJ) or any combination of them (see FC-CBJ [13]). Before each step of the search procedure (starting before the first step), the RAA attempts to select variables whose assignment can be safely delayed to the end of the procedure. This is equivalent to a partial ordering of variables in the search procedure. Variables that are delayed are added to the end of the queue of variables for assignment of values. In case of backtracking, delayed variables are returned to the general pool, if the backtracking unassigns a variable that was assigned a value before they were delayed.

The selection procedure for delaying variables scans the network and selects unassigned variables that can be assigned values that *do not affect other assignments*. There are two criteria for selecting such variables:

1. variables whose domain size is greater than the number of constraining variables,
2. Variables which have in their domain a value that is unique to their domain (i.e. a value that does not appear in the domain of any constraining variable),

Having selected these variables, the algorithm delays their assignment to the end and *eliminates* from the CN all the constraints that connect the delayed variables to other variables. The resulting CN has a reduced number of constraints and variables and the RAA is called again to scan the variables for candidates to be delayed. This series of steps is performed until there are no more variables which satisfy one of the above criteria, i.e. no variables can be delayed. Next, an assignment step is performed (i.e. a variable that was not delayed is selected and a value is selected for it and then assigned), followed by another step of delaying variables, if any. In other words, the operation of finding variables that can be delayed is performed before every assignment action and in particular before the beginning of the whole search process.

The RAA builds a variable ordering through a series of iterations. It keeps two lists of unassigned variables at all times: one list holds unordered variables (which have not been delayed yet), the other list holds ordered variables in the reverse order to their delays. At each iteration more variables are delayed and the ordered list grows. The main point of the algorithm is that the delayed variables have the following two properties.

Property 1. For each consistent assignment of variables from the "undelayed list", there exists a legal assignment for all "delayed" variables. If no legal assignment exists for the subproblem that consists of the undelayed variables, than the whole CN has no consistent assignment.

Property 2. If a solution exists for the subproblem that consists of the undelayed variables, then a consistent assignment of the delayed variables can be found *without backtracking.*

Thus, RAA splits the original problem into two smaller ones, where backtracking occurs only in one of them. The heuristic for selecting the variable for the next assignment can be any effective ordering heuristic. In our experiments we use the *fail-first* principle, that selects the variable with the smallest domain that has the maximal number of constraints. After every assignment step the new subproblem is scanned for possible delays of variables.

Since the RAA is essentially a special type of a variable ordering heuristic, it can be naturally implemented on top of any backtracking search algorithm. Take, for example, the FC algorithm (see [18], [13]). Following every assignment of a value to some variable, FC propagates the constraints to all unassigned variables that are influenced by the current assignment. The result of this propagation takes the form of reduced domains of values for future (unassigned) variables in the search procedure [13]. The FC search procedure can be easily combined with any type of variable ordering. Think of the variable ordering heuristic as a step of selecting the next variable after each assignment. FC propagates the current assignment to all future variables in the search. A variable ordering heuristic such as *fail-first* [18] selects as the next variable for assignment the variable with the smallest domain of values. These two steps suit one another perfectly. For the proposed RAA method the same idea applies. After each step of assignment, forward check the assignment against all future variables and then scan them for possible delays. The undelayed variables can now be ordered by domain size, exactly as the *ffc* ordering heuristic. This version of the RAA is described recursively below and is compared experimentally to the domain-size ordering heuristic, on random constraint networks, in Section 4.

A recursive description of FC-RAA:

1. Call the procedure **Delay(problem)** with the problem as its argument, to split the problem into two parts: delayed variables and undelayed variables. Call the set of undelayed variables - **subproblem**.
2. **solve_by_BT(subproblem)**
   a) If **subproblem is empty**, then return solution.
   b) Select a variable and remove it from **subproblem** (call the remainder **subsubproblem**).
   c) Assign the selected variable a value and erase the value from all constraining variables in sub^2**problem**.
   d) sub^3**problem** ← **Delay**(sub^2**problem**).
   e) **solve_by_BT**(sub^3**problem**)
3. If no solution was found for **subproblem**, then return **fail**.
4. Assign delayed variables in the reverse order to their delays.
5. Return the complete solution.

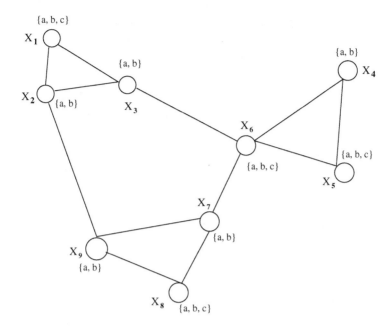

Fig. 1. An example RA constraint network

3 A Simple Example

Consider the example CN in Figure 1. There are nine variables $X_1, \ldots, X_9$ and their domains have values from the set $\{a, b, c\}$. Three of the variables in the network have more values in their domains than constraints connecting them to other variables. These are X_1 with three values and two arcs, X_8 with three values and two arcs, and the same condition for X_5. Delaying these three variables and eliminating their connecting arcs creates new variables that can be delayed: X_4 with two values and one constraint, X_6 with three values and two constraints, X_3, X_7, X_2, and finally X_9 with two values and one arc. The end result in the example of Figure 1 is that all variables are delayed.

The assignment of values to the delayed variables (which in the current example is the whole network) is performed in the reverse order of delays, in the RAA algorithm. The first variable to be assigned a value is X_9 and since all variables have been delayed, the network is solvable with no backtracks. Assigning the value a to X_9 and propagating the assignment (by erasing the assigned value from the domains of all constraining variables, we end up with a solution that is composed of the following list of assignments: $< X_2, b >$, $< X_7, b >$, $< X_3, a >$, $< X_6, c >$, $< X_4, a >$, $< X_5, b >$, $< X_8, c >$, $< X_1, c >$.

A heuristic of value assignment delays was proposed for CNs of RAPs by Choueiry and Faltings [4]. The approach of [4] performs a preprocessing proce-

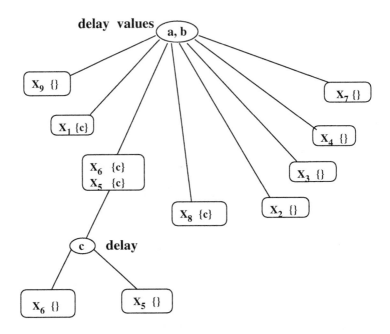

Fig. 2. Value assignment delay heuristic applied to the RA constraint network

dure that delays the assignment of certain values and creates A clustering of variables that compete for these values. This preprocessing heuristic can help in decomposing the RAP. The value-assignment delay (VAD) heuristic of [4] looks for the most solicited resources (i.e. those that are in the domains of the largest cliques in the network) and delays the assignment of these values by taking them off the domains of the relevant variables. Applying the VAD heuristic to the example problem of Figure 1 produces the result in Figure 2. The most solicited values are a and b, that appear in cliques of size 3, each. Both have to be delayed in order to split the cliques of the network and the result of delaying a, b is a set of variables with empty domains X_9, X_7, X_4, X_3, X_2, and three variables with single-valued domains X_6, X_5, X_8 (with the value c). Delaying the last value c from the domains of the variables in the clusters of X_5, X_6 and X_8, creates empty domain variables that are completely separated (Figure 2).

The goal of a series of delays in the VAD heuristic is to arrive at clusters that can be assigned the remaining values (see [4]). As can be easily seen in Figure 2, the assignment of any of the delayed values to any of the generated clusters does not necessarily lead to a solution. For example, c can be assigned to either X_5 or X_6: one leads to a solution and the other does not.

Yet another approach that can be used to simplify a network of RAPS is to preprocess the network by arc consistency algorithms. Moreover, since there are cliques of inequality constraints in the network of Figure 1, one proposed approach is to enforce arc consistency on the non binary (*all-diff*) constraints (see Regin [15]). Performing Regin's arc consistency algorithm on the non-binary constraints in our example eliminates the values a, b from the domains of the variables X_1, X_8. The domains of X_1, X_8 are now single-valued and can be assigned their only value c. However, the rest of the assignments are dependent on the order of assigning the cliques of the network. Assigning the X_6, X_4, X_5 clique first can generate backtracking. For example, assigning the value b to X_6 does not lead to a consistent global solution, even though it is consistent with assigning the values a, c to the members of the clique (i.e. to X_4, X_5 respectively).

It is easy to see from the above analysis and comparison that only RAA solves the example without backtracking and it does so by delaying variables. The other two methods attempt to reduce the problem, but do not succeed well on this simple example. An important comparison for testing the performance of the RAA is to test it against variable ordering heuristics. Moreover, since the RAA can be simply combined with FC and any variable ordering heuristic, such as the *ffc* or domain-size, it is interesting to test experimentally the performance of RAA against standard domain size ordering. To make these experiments meaningful, the random CNs must have different domains for different variables and their inequality constraints must have a non-uniform structure, to resemble realistic RAPs. In the next section we will present experimental tests of the proposed RAA on randomly generated problems that resemble real-world ETPs. The random networks have a large number of variables, different domains for different variables and a tunable density of inequality constraints.

4   Testing RAA on Random Networks

A typical real-world problem of resource allocation is that of assigning employees to scheduled tasks. Tasks have domains of values, composed of those employees that can be assigned to perform the task (i.e. those who are *able* to perform the task and are also *available* to do it). Sets of tasks that overlap in time, or are consecutive, with durations larger than those allowed for employees to work, form complete subnetworks with inequality constraints. This model for ETPs and their CNs has been discussed in [9], [10] and forms the basis of a recent model for realistic ETPs [16], [11].

TTPs are characterized by variables of differing domains (different time slots assignable to different activities) and non-uniform constraint density. The latter feature is related to the fact that conflicting activities are grouped, like shifts of the same day that cannot be assigned the same employee. In a former study we have generated random ETPs that are characterized by similar parameters to real-world problems [10]. The random CNs of ETPs have several additional parameters to standard random CSPs that are used in many studies [14], [17]. The non-uniformity of the networks is determined by the number of groups g and the density of constraints between variables of different groups. We denote this density by p_{1out} [10].

The other important parameter of CNs of timetabling is the distribution of values in domains of variables. The domains of variables are generated by assigning values (employees) randomly to variables (tasks), in analogy to real-life ETPs. This is equivalent to randomly selecting the subsets of employees that can be assigned to each and every task of the problem. The resulting domains are characterized by their sizes and the intersection of their domains (which is relevant for RAA).

The random networks that are used in the experiments are characterized by the number of variables n, the total number of values k, the number of groups of (complete networks) of conflicting variables g, the density of (intergroups) inequality constraints p_{1out} and by their effective average domain size avg_size. In the experiments we take $n = 30$ variables (tasks), $k = 10$ values (total of 10 employees) and $g = 5$ (5 groups for 5 weekly working days). The remaining parameters are two: density of constraints p_{1out}, and average domain size avg_size. In the experiments we varied the density of constraints and performed the experiments for randomly generated CNs with different values of average domain size.

Two different search algorithms were compared with the RAA. All three algorithms, including the RAA, use the FC as their underlying search algorithm. One reference algorithm has a static variable order, by domain size. The other reference algorithm uses dynamic variable ordering by domain size. This is a natural comparison with the RAA for two reasons:

- domain size ordering is used also by the RAA *for non-delayed* variables,
- ordering variables by domain size is known to be one of the best ordering heuristics [18].

Figure 3 depicts the fraction of solved random problems for values of $0.1 \leq p_{1out} \leq 0.6$ and for two values of average domain size $avg\_size = 5, 6$. All runs of all algorithms were limited to 3 CPU min. It is clear that RAA has the largest fraction of solved problems within the given time limit.

One measure of search effort for CNs which is very common to the CSP community is the number of constraint checks [18], [13], [17]. Comparing the average number of constraint checks for solving the random networks by dynamically ordered FC and the RAA produces the graph in Figure 4. All problems were generated 100 times and results are averaged over all runs. As can be seen in Figure 4, the RAA performs a little better than dynamically ordered FC (domain size variable ordering). Both algorithms used the same value ordering heuristic.

Figure 5 compares run times of the two algorithms, instead of constraints checks. Here again the RAA is a little better than FC with dynamic variable ordering. It is important to note that FC with dynamic variable ordering (*ffc*) and value ordering has been found by many experimental studies on random CNs to be the best performing algorithm [13]. From the experiments of the present study it seems that RAA improves on FC without losing any important feature. Versions of RAA that incorporate hybrid search algorithms, such as FC-CBJ and others (see [13] for hybrid algorithms), can be easily constructed [12].

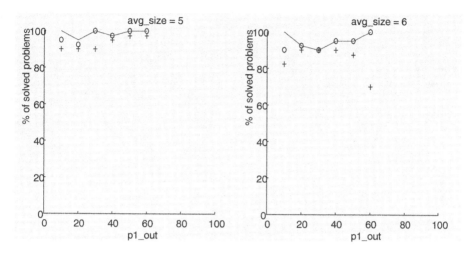

Fig. 3. Percentage of solved problems for the three solving algorithms: (+) FC with static variable ordering; (o) dynamically ordered FC; lines for RAA

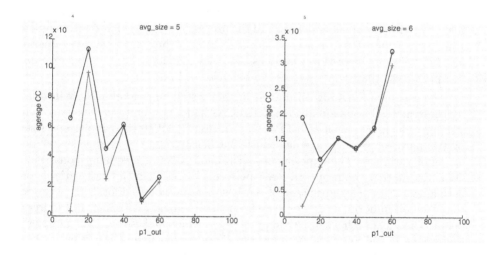

Fig. 4. Number of constraints checks for the two best solving algorithms: (o) dynamically ordered FC; (+) RAA

5   Conclusions

A new assignment algorithm for RAPs is proposed. The new RAA processes the CN by analyzing domains of values and delaying assignments to selected activities. The delay of selected activities dynamically changes the topology of

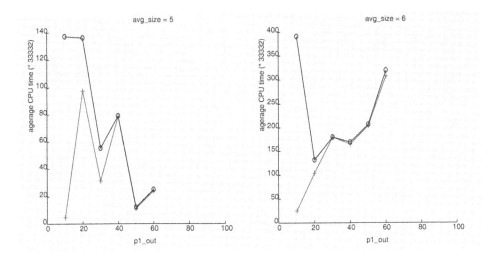

Fig. 5. CPU time for the solving algorithms: (o) dynamically ordered FC; (+) RAA

the CN, giving rise to more delays, and the process is integrated into the complete search procedure. RAA is a complete search algorithm and is always superior to a backtracking algorithm that does not delay assignments. A simple example was presented, that demonstrates the behavior of RAA.

The proposed RAA algorithm was further tested on random CNs of inequality constraints. The randomly generated CNs had a non-uniform density of constraints, which is typical of TTPs. Our random RAP testbed is parametrized by two parameters – the average constraint density and the average size of domains of variables. The experiments on random CNs of inequality constraints demonstrate that the proposed RAA performs better than the best variable ordering heuristics. The advantage of RAA is more pronounced for TTP networks that are non-uniform, as can be seen from the left-hand side panels of Figures 4 and 5. These panels have lower values of p_{1out}, which makes the CN "clumpier", the groups being complete subnetworks. The bottom line of the experimental study seems to be that, for realistic inequality CNs, FC-RAA is the best policy, especially in view of the fact that the procedure of delaying variables is naturally combined with the best search strategy (FC) and any variable and value ordering heuristics.

Going over to real-world RAPs is complex. One central problem is that additional constraints are needed. To give just a simple example with ETPs, one needs to limit the number of assignments per schedule of every employee (i.e. finite capacity). The addition of such non-binary constraints transforms the RAA into a non-complete algorithm and therefore a heuristic. We have studied TTPs in the past which contain a combination of RA and non-binary constraints of finite capacity [10]. The extension of RAA (as proposed in this paper) for the combination of constraints that are typical for RAPs in real-world problems is now being studied.

References

1. Arkin, E., Silverberg, E.: Scheduling Jobs with Fixed Start and End Times. Discrete Appl. Math. **18** (1987) 1–8
2. Carter, M., Laporte, G.: Recent Developments in Practical Examination Timetabling. In: Lecture Notes in Computer Science, Vol. 1153. Springer-Verlag, Berlin Heidelberg New York (1996)
3. Choueiry, B.Y.: Abstraction Methods for Resource Allocation Ph.D. Thesis, EPFL, Lausanne (1994)
4. Choueiry, B.Y., Faltings, B.: A Decomposition Heuristic for Resource Allocation. In: Proc. 11th Eur. Conf. on Artificial Intelligence (Amsterdam, 1994) 585–9
5. Choueiry, B.Y., Faltings, B.: Interactive Resource Allocation Problem Decomposition and Temporal Abstractions. In: Backstrom and Sandwell (eds.): Current Trends in AI Planning. IOS Press (1994) 87–104
6. Golumbic, M.C.: Algorithmic Graph Theory and Perfect Graphs. Academic, New York London (1980)
7. Golumbic, M.C.: Algorithmic Aspects of Perfect Graphs. Ann. Discrete Math. **21** (1984) 301–323
8. Meisels, A., El-Saana, J., Gudes, E.: Decomposing and Solving Timetabling Constraint Networks. Comput. Intell. (1997) **13** (1997) 486–505
9. Meisels, A., Gudes, E., Solotorevsky, G.: Combining Rules and Constraints for Employee Timetabling. Int. J. Intell. Syst. **12** (1997) 419–439
10. Meisels, A., Liusternik, N.: Experiments on Networks of Employee Timetabling Problems. In: Proc. PATAT'97 (Toronto, Canada). Lecture Notes in Computer Science, Vol. 1408. Springer-Verlag, Berlin Heidelberg New York (1997) 130–141
11. Meisels, A., Schaerf, A.: Modelling and Solving Employee Timetabling Problems. Appl. Intell. submitted October, 1999
12. Ovadia, E.: Delay and Order Variables in Timetabling Problems. M. Sc. Thesis, Ben-Gurion University (1999)
13. Prosser, P.: Hybrid Algorithms for the Constraint Satisfaction Problem. Comput. Intell. **9** (1993) 268–299
14. Prosser, P.: Binary Constraint Satisfaction Problems: Some Are Harder than Others. In: Proc. 11th Eur. Conf. on Artificial Intelligence (Amsterdam, 1994) 95–99
15. Regini, J.C.: A Filtering Algorithm for Constraints of Difference in CSPs. In: AAAI-94 (Seattle, WA, 1994) 362–367
16. Schaerf, A., Meisels, A.: Solving Employee Timetabling Problems by Generalized Local Search. In: Proc. Italian AI Assoc. (May, 1999) 493–502
17. Smith, B.M.: Phase Transition and the Mushy Region in CSP. In: Proc. 11th Eur. Conf. on Artificial Intelligence (Amsterdam, 1994) 100–104
18. Tsang, E.: Foundations of Constraint Satisfaction. Academic, New York (1993)

Other Timetabling and
Related Problems

Fleet Scheduling Optimization: A Simulated Annealing Approach

Danuta Sosnowska and José Rolim

CUI, University of Geneva,
24 rue Général Dufour, 1211 Geneva 4, Switzerland,
Danuta.Sosnowska@obs.unige.ch, Jose.Rolim@cui.unige.ch

Abstract. The fleet assignment problem belongs to large-scale combinatorial optimization problems. It consists in assigning aircraft types to a fixed schedule in order to obtain the best possible financial gain with the best fit of aircrafts capacities to estimated demand. The next step after the assignment of aircraft types is the aircraft routing, i.e. the assignment of flight sequences to particular aircrafts with respect to scheduled maintenances. In this paper we present an optimization method for the whole fleet scheduling, i.e. the fleet assignment together with the aircraft routing. Our method is based on the simulated annealing heuristic. We have successfully optimized the fleet scheduling for the period of one month and obtained results some 10–12% better than that of the airline company. All tests were performed on a real set of data.

1 Introduction

In order to be operationally ready for the season, an airline company has to execute several steps. It starts with the schedule planning, followed by fleet scheduling divided into fleet assignment, i.e. the assignment of aircraft types to flights, and aircraft routing, i.e. the construction of sequences of flights assigned to particular aircrafts, called rotations, with respect to scheduled maintenances. The next step is crew assignment. In this paper, we are mostly concerned with the fleet scheduling problem.

The fleet assignment and the aircraft routing problems have been well studied during the last 20 years. Most of the proposed approaches deal with a period of one day ([1], [2], [3], [4], [5], [6]), assuming that the same schedule repeats every weekday and that changes for the weekend are minor. In these approaches, the problem is modelled as a time–space network and formulated in terms of integer programming. The solving method used is usually a branch-and-bound algorithm with linear or Lagrangian relaxation applied in the nodes of the branch-and-bound tree. In [7] and [5] the swapping operation is considered only in terms of the daily fleet assignment.

The aircraft routing problem with respect to maintenances was investigated in [8], [9], [10]. All the approaches solve the problem for periods of up to four days. The solutions are based on graph operations and integer programming.

E. Burke and W. Erben (Eds.): PATAT 2000, LNCS 2079, pp. 227–241, 2001.

An airline company of medium size has about 300 flights per day and about 60 aircrafts to serve all these flights. Since the problem of finding the best fleet assignment is NP-complete, a simple exhaustive search is impossible. For periods longer than one day the deterministic methods of operational research are computationally too weak, thus imposing the use of heuristics.

In this paper we deal with the fleet assignment and aircraft rotation problems simultaneously. The optimization is done on real data for a period of one month.

In Section 2 we present the formalism for the fleet scheduling with maintenances. Section 3 presents the operation of swap, which is the basis of the simulated annealing approach. In Sections 4 and 5 we present the simulated annealing algorithm, its application to the fleet scheduling problem, some tests on the real data for one month and the results obtained.

2 Formalism

In the literature many different formalisms for the fleet assignment ([1], [2], [6]) and aircraft routing ([8], [9], [10]) problems have already been presented. However, since our approach differs considerably from other methods, we introduce our own formalization of the problem.

2.1 Definitions

Let

- FL be the set of flight legs,
- MT be the set of maintenances,
- AC be the set of aircrafts,
- AP be the set of airports,
- $\{rc^k\}_{j=0,...,MAX}$ be the rotation cycle of the kth aircraft, where MAX is the maximum number of flight legs in one rotation,
- rc_i^k be the ith flight leg or maintenance in the rotation rc^k,
- $|rc^k| \in \mathbb{N}$ be the length of rc^k,
- c_{ik} be the cost of the ith flight leg flown by the kth aircraft, the maintenance cost always being equal to zero.

2.2 Cost Function

The problem consists in minimizing the following cost function:

$$C = \sum_{k \in AC} \sum_{i \in rc^k} c_{ik}.$$

The cost function is the sum of the costs of every assigned flight leg. The cost depends only on the type of aircraft assigned, not on the assignment of the connection flights, which are two consecutive flights assigned to the same aircraft. It is calculated as the difference between the total operating cost (TOC) and the total revenue (TR) ([11]). TOC consists of the block hour cost (BHC) multiplied by the flight duration, the so-called block time (BT). The BHC includes the pilot and cabin crew cost, the cost of fuel and oil, the insurance and maintenance costs, the landing cost and the passenger service cost. The administrative and commission costs are omitted as non-significant to the prototype implementation. The TR consists of the average ticket price multiplied by the total number of passengers on the flight. In order to have better performance of the optimization algorithm, we penalize also the empty seats and the overbooked passengers. To simplify the cost calculation we have only taken into consideration two major factors: BHC to determine TOC and estimated passengers demand for the calculation of TR. From the computational point of view this assumption is sufficient to perform the optimization. Finally, the cost for a single flight leg is calculated as follows:

$$
\begin{aligned}
c_{ik} = {} & \mathrm{BHC} \cdot \mathrm{BT} \\
& + \text{empty seats cost} \\
& + \text{overbooked passengers cost} \\
& - \text{demand} \cdot \text{standard ticket price.}
\end{aligned}
$$

The *empty seats cost* is equal to the number of empty seats multiplied by the ticket price. It is equal to zero, if all seats are sold. The *overbooked passengers cost* is equal to the estimated loss due to impossibility of transportation of passengers, who will in this case go to another airline company. It is equal to zero, if demand is lower than the aircraft capacity.

2.3   Auxiliary Functions

To define the constraints for the fleet scheduling problem (FSP) we need auxiliary functions describing departure and arrival time and airport for each flight leg and the notion of ground time - the amount of time necessary to perform the ground service for each aircraft in different airports.

Given $k \in AC$, $i \in FL \cup MT$, $a \in AP$, $j \in MT$:

- $DT(i) \in \mathbb{N}$ determines the departure time of the flight leg i or the beginning of the maintenance i,
- $AT(i) \in \mathbb{N}$ determines the arrival time of the flight leg i or the end of the maintenance i,
- $DA(i) \in AP$ determines the departure airport of the flight leg i or the maintenance station,
- $AA(i) \in AP$ determines the arrival airport of the flight leg i or the maintenance station,
- $GT(a, k) \in \mathbb{N}$ determines the ground time needed by aircraft k in the airport a,
- $MA(j) \in AC$ indicates the aircraft to which the maintenance is assigned.

2.4   Constraints

The optimization of the FSP consists of assigning flight legs to the available aircrafts in the best possible way, in order to obtain the biggest economic profit from the point of view of the airline company. The feasible solution has to respect a number of operational constraints demanded by the assignment problem or specific to the airline company, depending on its law, country laws, geographical environment, etc. In our approach, only constraints independent of the local situation of the airline company were taken into consideration. The maintenances are fixed in time and assigned to aircrafts before the process of the fleet scheduling. Every feasible solution has to respect imposed maintenances, i.e. each aircraft has to be at the maintenance station at the beginning of the maintenance and can depart only after the maintenance is finished.

The following four equations describe the imposed constraints:

- Every flight or maintenance can be assigned only once:

$$(rc_{i_1}^{k_1} = rc_{i_2}^{k_2}) \Rightarrow (k_1 = k_2 \quad \text{and} \quad i_1 = i_2) \tag{1}$$
$$\forall k_1, k_2 \in AC \quad \forall\, 0 \le i_1 < |rc^{k_1}| \,\forall\, 0 \le i_2 < |rc^{k_2}|$$

- Every aircraft must depart after its arrival plus a margin necessary for ground service:

$$DT(rc_i^k) > AT(rc_{i-1}^k) + GT(AA(rc_{i-1}^k), k) \tag{2}$$
$$\forall k \in AC \,\forall\, 0 < i < |rc^k|$$

- Every aircraft has to depart from the airport it has landed:

$$AA(rc_{i-1}^k) = DA(rc_i^k) \tag{3}$$
$$\forall k \in AC \quad \forall\, 0 < i < |rc^k|$$

- Every maintenance is assigned always to the same aircraft

$$(rc_i^k = j) => (MA(j) = k) \tag{4}$$
$$\forall k \in AC \quad \forall\, j \in MT$$

2.5   Representation

The choice of representation is determined by the fact that we always operate on a solution which is the set of aircraft rotations. One rotation is assigned to each aircraft. A rotation is the sequence of flight legs together with maintenances, assigned in a specific order. We can see a small example of two very short rotations in Figure 1. This example contains two aircrafts and four flight legs. It illustrates one of four feasible solutions conserving all constraints. Both rotations are sequences of two consecutive flight legs.

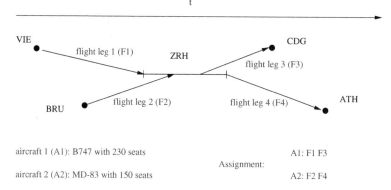

Fig. 1. A simple example of one feasible solution. Four flight legs are assigned to two aircraft. The first (A1) will fly from Vienna (VIE) to Zurich (ZRH) and then to Paris (CDG). The second will start in Brussels (BRU), go to Zurich (ZRH) and then to Athens (ATH). In this solution, all constraints are satisfied

3   Initial Solution

One of the non-negligible problems in our method for solving the fleet scheduling problem is to find a feasible solution to start with. The simplest first-in/first-out (FIFO) method works very well when the maintenances do not have to be taken into consideration. The FIFO method consists in assigning flights sorted by the departure times, one by one, to the first aircraft available at the departure airport. However, this method cannot guarantee that the aircraft will be at the maintenance airport on time. This is the reason why we used a variation of the FIFO method to incorporate maintenances in the first feasible solution.

The modification to the simple FIFO method consists in treating all the maintenances as regular flights, with the departure and arrival times equal to the start and the end of the maintenance. The departure and arrival airports must therefore be identical and equal to the maintenance airport.

Such extended flights are sorted by the departure time and assigned to available aircrafts by the FIFO method. As we have mentioned, this will not guarantee the correct maintenance assignment. We then use the swap operation (described in the next section) in order to place maintenances on the aircrafts they are assigned to.

4   The Swap Operation

The simulated annealing approach for solving the fleet scheduling problem is based on the operation of swapping parts of rotation cycles. This operation is used for creating a new feasible solution starting from the existing one. The basic idea for this operation is to find so-called *break points* in two chosen rotation cycles and perform the swap of the parts of these cycles. The place of the break is chosen in such a way that the rotation remains feasible after the swap (Figure 2): i.e. the first aircraft can continue its rotation with flight legs previously assigned to the second one and vice versa.

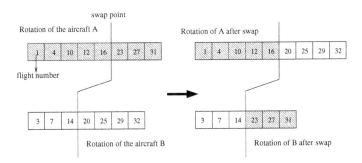

Fig. 2. Swap of two rotation cycles assigned to two aircrafts. After the swap the first aircraft obtains the second part of the second assignment and the second aircraft the second part of the first. The choice of the swap point ensures the feasibility after the swap

The simple swap operation described above is in fact insufficient to take the maintenances into account. Indeed, they are assigned to an aircraft only once and for all and cannot be moved from one rotation to another. The only parts of a rotation that can be swapped are those without any maintenance inside. To ensure this constraint for maintenances, only the parts between two swap points are allowed to be exchanged, as shown in Figure 3.

4.1 Conservation of Constraints

The goal of the swapping operation is to pass from one feasible solution to another. It requires the conservation of all imposed constraints. Constraint (1) is conserved automatically, as two cycles are swapped without adding or deleting any flight leg. To conserve constraints (2) and (3) the swapping point has to be chosen carefully. The first aircraft has to be operationally ready after the last flight before the swapping point and on time to take off for the first flight after the swapping point in the second rotation cycle. Also, the landing airport of the first flight should be equal to the departure airport of the second flight, and vice versa. The maintenances are not moved during the swap operation (Figure 3), so the constraint (4) is also preserved.

There are many ways of choosing the swapping point. In this implementation we initially find all possible swap points and subsequently all pairs of swap points without maintenance in between. Finally, one of these pairs is chosen randomly. This way of choosing swap points was taken because it ensures breadth of the neighbour search space and is not very expensive in time.

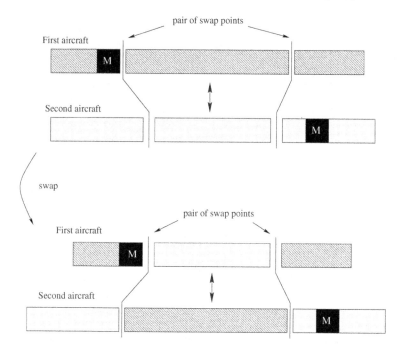

Fig. 3. Swap points are chosen in order to conserve all constraints. Only the parts without maintenance are allowed to be exchanged. The maintenances remain assigned to the same aircraft

5   Solution by Simulated Annealing (SA)

5.1   SA Algorithm

The general scheme of the SA algorithm is presented in Figure 4. The probability of accepting a worse solution depends on the cost difference between two subsequent solutions (ΔC) and the so-called cooling schedule (T), which is a positive function converging slowly to zero.

5.2   Fleet Scheduling Optimized with SA

To apply the simulated annealing method to the FSP we need a method for generating the initial and the neighbouring feasible solutions and a function for comparing two subsequent solutions. The initial feasible solution is constructed with a variation of the FIFO method, as described in the previous section. To construct the neighbour feasible solution starting from a given one, we use the swap operation. The method of applying the swap guarantees the preservation of constraints. In order to compare two solutions, the cost function is used.

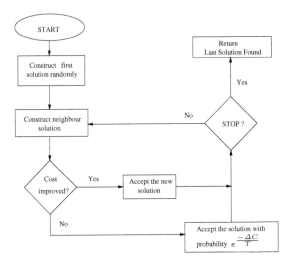

Fig. 4. Scheme of the SA algorithm

For the optimization we have tested several commonly used cooling schedules and adjusted the initial temperature in an empirical way, by observing the optimization process for different values on a small number of iterations.

6   Results

Our testing example contains about 2600 flights flown by 20 aircraft during a period of one month. Data were obtained from a real airline company. One optimization process running time is about 5.5 h, which is acceptable from the point of view of an airline company, which usually uses up to one night of computation time per one month scheduling.

In SA algorithm tests we have used several cooling schedules and we have tried to find the best initial temperature (T_0) for each schedule. If the temperatures decreases too quickly the algorithm stays in a local optimum. If, on the other hand, it decreases very slowly, the time to reach the global or near-global optimum is very long. The quality of the solution depends also on the value of T_0. It should be chosen in the range of the biggest difference between the cost values of two neighbouring solutions. This choice is empirically made by testing different values on a small number of iterations.

6.1   Parameter Tuning

In the work of Hajek [12] the proof of the optimality of the logarithmic cooling schedule is presented. In practical applications, the logarithmic cooling process is too slow and the algorithm takes a very long time to get to the global or near-global optimum. In the real-world optimization process the processing time is limited. Therefore, even if theoretically leading to optimality, this cooling schedule proved to be the worst among the tested ones.

We have tested five cooling schedules:

- $T = T_0/t$
- $T = T_0 \cdot 0.99^t$
- $T = T_0/e^t$
- $T = T_0/\sqrt{t}$
- $T = T_0/\log t$

where t is the number of iterations.

The first two schedules behave in a similar way, which is why we present only the first one in Figure 5. The simple $T = T_0/t$ seems to be slightly better. The third one cools down too quickly (Figure 8) and its result is equivalent to a randomized hill-climbing method. The schedule $\sqrt{t}$ (Figure 6) is slightly worse than the first two and the last one (Figure 7) is much worse in tests with a reasonable number of iterations. All four graphics are results of ten tests. The upper trace is the worst case (max), the middle is the average and the lower is the best (min) of all ten runs.

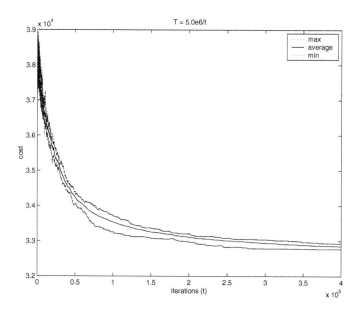

Fig. 5. Results of ten tests of the cooling schedule $T = 5.0 \cdot 10^6/t$

The quality of the final solution depends not only on the cooling schedule but also on the initial temperature, the constant T_0. For each cooling schedule

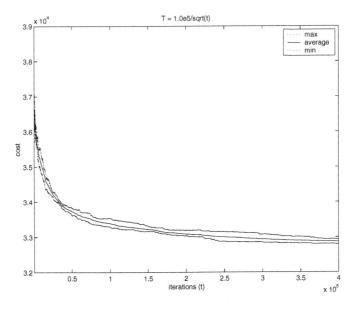

Fig. 6. Results of ten tests of the cooling schedule $T = 1.0 \cdot 10^5/\sqrt{t}$

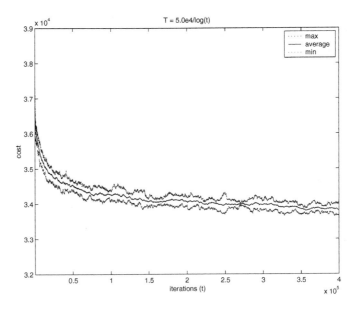

Fig. 7. Results of ten tests of the cooling schedule $T = 5.0 \cdot 10^4/\log(t)$

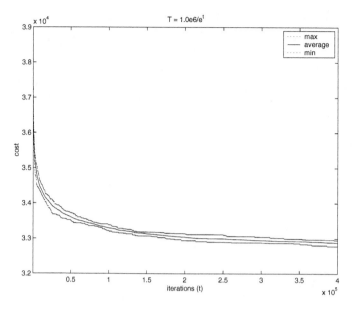

Fig. 8. Results of ten tests of the cooling schedule $T = 1.0 \cdot 10^6 / e^t$

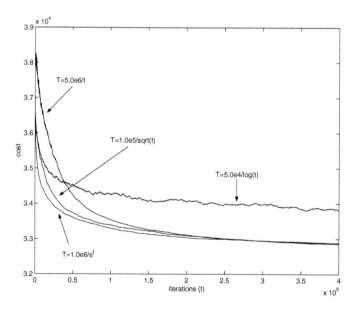

Fig. 9. Comparison between the average of ten runs of the four different cooling schedules

we have tested several different constants. The T_0 should be large enough to allow the algorithm to pass easily between two neighbouring solutions, but not so large as to limit the acceptance of solutions with much worse cost function value.

In Figure 10 we can see the difference between the optimization process with the same cooling schedule but with three different initial temperatures. For small T_0 the process falls quickly, but it could be blocked in a local minimum. For high T_0 the process easily escapes from the local minima, but it takes much more time to reach the globally optimal or near-optimal solution.

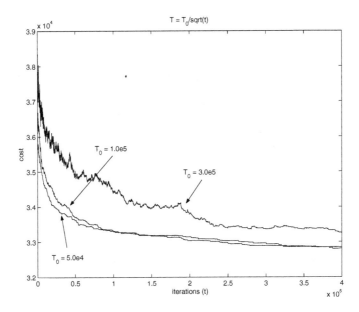

Fig. 10. Comparison between three different values of T_0 for the cooling schedule $T = T_0/\sqrt{t}$

6.2   Comparison with Other Solutions

In order to evaluate our results we solved our data samples in a different way by using the commercial software developed by ILOG called CPLEX.[1] Another comparison was made by calculating the cost of the airline company assignment.

To solve the FSP by CPLEX, we used the formulation presented by Abara in [1]. This formulation is very efficient for the fleet assignment problem, because it only treats the aircraft types, which reduces considerably the number of variables and constraints. Unfortunately, this is not possible for the aircraft routing, where

---

[1] http://www.cplex.com and http://www.ilog.com/products/cplex/

each aircraft has to be treated separately. In this case the solution cannot be found in a reasonable amount of time (our calculations were stopped after one week without giving any solution).

The only method to obtain any reference solution for the completely formulated problem was to solve it day by day with constraints for aircraft departures taken from the solution of the previous day. Obviously, the sum of the optimal solutions for each day is not the optimal solution for the whole period, but it can give a reference point for other optimization methods. Effectively, the SA gives slightly better results, as can be seen in Figure 11 (lower panel). The first three schedules give a ratio a little smaller than 1.0. The last one has a worse performance.

The calculation of the air company results is straightforward. As we know their assignment for all flight legs and the cost of each flight leg assigned to the aircraft type, the overall cost will be the sum of the costs of all assigned flight legs minus the profit made by the effective demand. In Figure 11 (upper panel) the cost of SA with different cooling schedules is compared with the airline company cost. For three of the schedules the ratio is about 0.9: i.e. the cost of our solution is about 10% better than that of the airline company.

7   Conclusions

The heuristic approach does not guarantee the optimal solution. However, it provides a much better method than the current airline company solution and is even slightly better than the solution composed of daily optimal solutions.

The complexity of the algorithm is polynomial in terms of the length of the optimized period and number of aircraft in the fleet, though it grows slowly with the size of data. This reasonable time complexity is due to the random choice of the rotations to swap and random choice of the swapping points pair. In these conditions we could even consider optimization of the whole season: i.e. a six month period.

Another argument for attempting a longer period optimization is the fact that large problems have a higher number of longer rotations, which increases the space of swap operations and thus the number of paths leading from one solution to another. Therefore, larger problems should give better algorithm performance.

The FSP presented in this paper relates to operations performed by the airline company many months before the schedule is realized. It is off-line fleet scheduling. However, the same problem can occur during the execution of the schedule or shortly before: for example, when an aircraft has a technical problem or is late. In this case it is called on-line fleet scheduling. Here the data set is less important than for off-line scheduling, but the computing time restrictions are very strong. In future research we plan to adapt the presented algorithm to its on-line version.

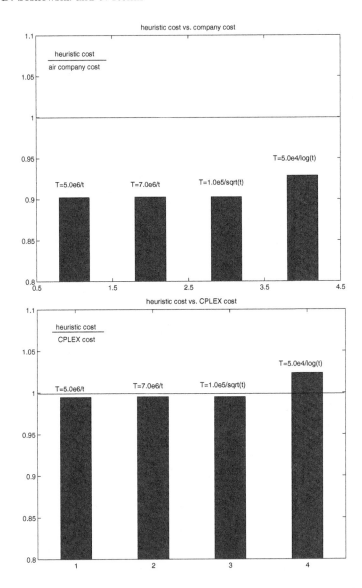

Fig. 11. Comparison between different SA solutions, the airline company solution and the CPLEX solution. Above: ratio between SA solution and the company solution. Below: ratio between SA solution and the CPLEX solution

References

1. Abara, J.: Applying Integer Linear Programming to the Fleet Assignment Problem. Interfaces **19** (1989) 20–28

2. Hane, C.A., Barnhardt, C., Johnson, E.L., Marsten, R.E., Nemhauser, G.L., Sigismondi, G.: The Fleet Assignment Problem: Solving a Large-Scale Integer Problem. Math. Program. **70** (1995) 211–232

3. Dillon, J., Gopalan, R., Petrofsky, R., Ramachandran, S., Rushmeier, R., Talluri, K.: Fleet Assignment. Presentation inside USAir, March 1993

4. R. Subramanian et al.: Coldstart: Fleet Assignment at Delta Airlines. Interfaces **24** (1994) 104–120

5. Berge, M.E., Hopperstad, C.A.: Demand Driven Dispatch: A Method for Dynamic Aircraft Capacity Assignment, Models and Algorithms. Oper. Res. **41** (1993) 153–168

6. Barnhart, C., Boland, N.L., Clarke, L.W., Johnson, E.L., Nemhauser, G.L., Shenoi, R.G.: Flight String Models for Aircraft Fleeting and Routing. In: Proc. AGIFORS Symp. (1997)

7. Talluri, K.T.: Swapping Applications in a Daily Airline Fleet Assignment. Transport. Sci. **30** (1996) 237–248

8. Gopalan, R., Talluri, K.T.: The Aircraft Maintenance Routing Problem. Oper. Res. **46** (1998) 260–271

9. Talluri, K.T.: The Four-Day Aircraft Maintenance Routing Problem. Transport. Sci. **32** (1998) 43–53

10. Clarke, L.W., Hane, C.A., Johnson, E.L., Nemhauser, G.L.: Maintenance and Crew Considerations in Fleet Assignment. Transport. Sci. **30** (1995) 921–934

11. Al-Haimy, A.A.: Development of a Mathematical Model for Airline Fleet Planning. In: Proc. 24th AGIFORS Symp. (1984)

12. Hajek, B.: Cooling Schedules for Optimal Annealing. Math. Oper. Res. **13** (1998) 311–329

A Schedule-Then-Break Approach to Sports Timetabling

Michael A. Trick

GSIA, Carnegie Mellon, Pittsburgh PA 15213, USA
trick@cmu.edu
http://mat.gsia.cmu.edu

Abstract. Sports timetabling algorithms need to be able to handle a wide variety of requirements. We present a two-phase method in which the first phase is to schedule the teams ignoring any home and away requirements and the second phase is to assign the home and away teams. This approach is appropriate when there are requirements on the schedule that do not involve the home and away patterns. Examples of this include fixed game assignments and restrictions on the effect carry-overs can have. These requirements can be met early in the process and the best home and away patterns can then be found for the resulting schedules.

1 Introduction

Creating a sports timetable is challenging due to the variety of different needs that must be addressed. This has led to a plethora of alternative approaches. While many of the approaches have a common thread due to their division of the problem into a number of common subproblems, these methods differ in the order in which subproblems are solved and the solution method(s) chosen.

We present results on an alternative approach for solving a standard sports timetabling problem. In this problem, teams must play a round-robin tournament among themselves, with each game played at the home field of one of the competing teams. We assume that the number of teams is even and that during every time period, or slot, every team plays exactly one game. This sort of timetable occurs very often in competitions where each team represents a city or other region and the competition occurs over an extended period of time. This contrasts with the situation where many teams share a single field, or where all teams are brought to a single location for the competition.

This round-robin scheduling, or variants thereof, has appeared often in the literature. Examples include minor league baseball (Russell and Leung [17]), college basketball (Ball and Webster [2]; Nemhauser and Trick [12]; Walser [22]; Henz [9]), Australian basketball (de Werra et al. [25]), and Dutch professional football (Schreuder [20]). It has also been extensively explored independent of a particular sport (de Werra [23], [24]; Schreuder [19]; Henz [8]; Schaerf [18]).

In this paper, we look at an alternative approach to solving these types of problems. In this approach, we divide the problem into two phases. In the first

E. Burke and W. Erben (Eds.): PATAT 2000, LNCS 2079, pp. 242–253, 2001.

phase (the scheduling phase), we assign games to be played without concern for the home/away decision. Our second phase (the break phase) then assigns home and away teams in such a way as to minimize the occurrence of bad structures (here defined to be consecutive home games or consecutive away games for a team, or breaks). This approach is particularly useful when there are requirements on the schedule that do not involve home/away patterns. An example is to create a schedule where a number of games are fixed *a priori*.

This approach uses constraint and integer programming to solve the two phases. The scheduling problem is a natural combinatorial design problem. This can easily be extended to allow fixed games and to implement the carry-over restrictions of Russell [16].

The break problem has been studied by Régin in [13,14]. In this paper, we present an integer programming formulation that is able to solve larger instances than the constraint programming method of Régin. Our success is due in part to applying Régin's discoveries in constraint programming to our integer programming formulation.

Section 2 outlines the standard multiphase approach to sports timetabling and gives our general approach to this problem. Section 3 then develops our scheduling phase and Section 4 gives our break phase. We conclude with some open problems.

2   Multiphase Approach to Sports Timetabling

Sports timetabling problems come in two broad types: temporally relaxed and temporally constrained. In a temporally relaxed schedule, the number of days on which games can be played is larger than the minimum number needed. Examples of such timetables include that of (American) National Basketball (Bean and Birge [3]) and National Hockey League schedules (Ferland and Fleurent [7]). In these leagues, teams may play on any day of the week, but would typically only play on two or three of them in any week. These schedules may look difficult, due to the additional decision of choosing a day on which to play, but many effective heuristic techniques have been developed. Other examples of competitions with temporally relaxed schedules include the scheduling of cricket matches (Armstrong and Willis [1]; Willis and Terrill [26]; Wright [27]).

In a temporally constrained schedule, the length of the schedule is chosen so that there is just enough time to play all of the games. While there is no longer a decision on when to play (each team will play as much as possible), there is also no longer the extra flexibility needed to apply the heuristic techniques listed above. Determining a reasonable local search neighborhood seems a difficult question in a temporally constrained problem, though Walser [22] reports success using his general local search methods for integer programs.

A number of researchers have attacked this problem in various ways (Ball and Webster [2]; Henz [8,9]; Nemhauser and Trick [12]; Russell and Leung [17]; Schaerf [18]; Schreuder [19,20]; de Werra [23,24]; de Werra, Jacot-Descombes, and Masson [25]). All of this work can be seen as addressing a number of sub-problems. These subproblems include the following:

1) Finding home–away patterns (HAPs). Find a set of n strings of H (home) and A (away) of length $n - 1$ that corresponds to the home and away sequence of a team. For six teams, a possible HAP would be

```
1: HAHAH
2: AHAHA
3: HHAAH
4: HAHHA
5: AAHHA
6: AHAAH
```

2) Assign games consistent with the HAP (so if i plays j in a slot, then either i is home and j is away, or the reverse). For the above, we might get the following timetable (+ denotes at home; − is away), to get a basic match schedule (BMS):

```
1: +2 -3 +6 -4 +5
2: -1 +4 -5 +6 -3
3: +6 +1 -4 -5 +2
4: +5 -2 +3 +1 -6
5: -4 -6 +2 +3 -1
6: -3 +5 -1 -2 +4
```

3) Assign teams to the BMS to complete the timetable. If the teams were A, B, C, D, E, F, then we might assign them to 3, 6, 5, 4, 2, 1 respectively to get the final timetable:

```
F: +E -A +B -D +C
E: -F +D -C +B -A
A: +B +F -D -C +E
D: +C -E +A +F -B
C: -D -B +E +A -F
B: -A +C -F -E +D
```

Alternative approaches to sports timetabling differ in the order in which they solve the subproblems and the method employed to solve each of the subproblems. For instance, Nemhauser and Trick [12] solved the subproblems in the order given above and used integer programming for problems 1 and 2, and complete enumeration for problem 3. Henz [9] then improved on this by using constraint programming for all three steps, where the speedup on problem 3 was enormous. Walser [22] used a heuristic integer programming technique to avoid breaking the problem into three subproblems.

The decision on which order to do the subproblems depends on the importance of the decision being made. In general, the more critical an aspect of a schedule, the more important it is to fix that decision early in the process. The above ordering makes the form of the HAPs most important. Leagues with other requirements may need to consider alternative orderings.

We will consider the case where the game assignment issue in step 2 is critical, even ignoring the home and away aspects. For instance, there might be a number of pre-assigned games that must be played in particular slots. Examples where this occurs include required television matchups and required "traditional matchup" requirements. Both of these occurred in the creation of the Atlantic Coast Conference basketball schedule of Nemhauser and Trick [12]. While the requirements in the year given in that paper were not extensive, later years added many more required match opponents, which made the original multiple-phase approach presented less appealing.

Another reason to consider step 2 (even before home/away assignment) is the work of Russell [16]. Russell looked at the issue of carry-over effects: if A plays B in round 1, then plays C in round 2, then C gets a carry-over effect from B. If B is a very difficult team, then C may gain an advantage. Russell showed that if the number of teams is a power of 2, then the advantages can be spread out evenly among the teams, and he gave a construction in this case. For other sizes, he gave a heuristic solution to minimize the variance of the carry-over effect. These give schedules to which home and away decisions would then be applied.

Our approach is therefore to first schedule the teams and then find the home team for each game. This is the following two-phase approach:

1. Find a schedule of the teams (ignoring home and away) that is consistent with the required match opponents or other requirements on the schedule (schedule phase).
2. Assign the home team to each game to give good HAPs (break phase).

The definition of good HAPs is very league dependent. Wallis [21] looked at this problem in the context of creating a schedule so that the teams are paired with one being at home and one away during each period. Schreuder [19] and de Werra [23], [24] addressed this in the context of minimizing the breaks in a timetable. In these models, the ideal pattern is alternating homes and aways (this is a very reasonable requirement for many real leagues). Any deviation from this, in the form of two consecutive aways or two consecutive homes, is a break. Schreuder and de Werra developed a number of structural conditions and bounds on the number of breaks. This work combined the schedule and the break phase to give schedules with the minimum number of breaks.

Régin [13], [14] looked at the break phase for arbitrary schedules and gave a constraint programming approach to the problem of minimizing the number of breaks. We will continue this direction by adopting a definition of "good HAPs" to mean the set of patterns that minimizes the number of breaks.

The next two sections look at each phase of this approach in turn.

3   Schedule Phase

The schedule phase might be solved in a number of ways, depending on the exact requirements on the schedule. In the absence of constructive techniques, like those of Russell [16], we need a computational method for finding schedules. Ideally, such a method would be flexible enough to allow alternative evaluation rules, and allow such extensions as fixing particular games. We will concentrate on the issue of fixed games and conclude with some comments on the carry-over model.

Given n teams (labeled $1 \ldots n$), and a series of triplets (i, j, t), find a round-robin schedule for the teams such that i plays j in slot t. For six teams, the input might look like

```
Slot 1 2 3 4 5
A:   B F   C
B:   A   F
C:   D   E A
D:   C E
E:   F D C
F:   E A B
```

One output would be

```
Slot 1 2 3 4 5
A:   B F D C E
B:   A C F E D
C:   D B E A F
D:   C E A F B
E:   F D C B A
F:   E A B D C
```

This problem is an example of a "completion problem" in combinatorial design. Colbourn [6] showed that many of these problems, including the one we address, are NP-complete (de Werra [24] gives some instances related to sports scheduling that can be solved quickly). Our goal is to create a reasonable computational method, rather than solve the problem for extremely large instances. Therefore, we explore two alternative constraint programming approaches for generating solutions to these problems.

We begin with the case that there are no required game placements. In this case, we are only trying to generate tournament schedules, a problem closely related to finding latin squares. Of course, many construction techniques are known for finding a schedule, but these typically find only one or a small number of schedules, so are unsuitable for our purposes.

Our first formulation for this is perhaps the most natural. A variable is set for each (team, time) pair which gives the opposing team in that time slot. Constraints are added to ensure that every team plays every other team and that symmetry holds: if i plays j then j plays i. This formulation can be strengthened by including the constraint that for every time slot, every team plays some opponent. We can also break the symmetry of the formulation by labeling team 1's opponents $2, 3, \ldots, n$ in order. We call this the opponent formulation. The OPL (ILOG [10]) code for this is given in Figure 1.

```
int n=20;
range Teams 1..n;
range Time 0..n-2;
range Opponent 1..n;
var Opponent opponent[Time,Teams];

solve {
   forall (i in Teams) {
      alldifferent(all (t in Time)(opponent[t,i]));
      forall (t in Time) {
       opponent[t,opponent[t,i]] = i;
      opponent[t,i] <> i;
      };
   };
   forall (t in Time) {
      alldifferent(all (i in Teams) (opponent[t,i]));
      opponent[t,1] = t+2;
   };
};
```

Fig. 1. Opponent formulation

An alternative formulation is to give, for each pair of teams, the slot in which the teams play. Each team must play only one game in a slot, and symmetry can be broken as above. This is the slot formulation. The OPL code is in Figure 2.

The formulations act very similarly (and an appropriate search strategy can make them act nearly identically), but some constraints are easier to write in one model than the other. Each model can generate a 20 team schedule in less than 1 s and generate 500 20 team schedules in around a minute. OPL has the capability of using different propagation algorithms for the **alldifferent** constraint: the strongest setting of **onDomain** gave the best results for these tests.

Rosa and Wallis [15] explored the concept of "premature sets": assignments of games to slots in such a way that the schedule could not be completed. They showed that such premature sets could occur and placed bounds on the size of the sets. Perhaps surprisingly, neither formulation for this problem had any difficulty with these premature sets: no backtracking was needed for any of the instances up to size 20. This is not true for larger instances.

The behavior of these formulations is not changed significantly by the addition of game requirements (of course, the problem is difficult, so there must exist difficult instances, but they do not seem to occur in the 10–20 team instances of interest). Empirically, it seems that adding game requirements makes this phase solve more quickly. Other types of constraints may be more difficult to handle, but forced pairings seem easy.

It should be noted that the opponent formulation could be modified to find the balanced-carry-over schedules of Russell [17]. Simply adding the constraint

```
int n=20;
range Teams 1..n;
range Slots 0..n-1;
var Slots slot[Teams,Teams];

solve {
   forall (i in Teams) {
      alldifferent(all (j in Teams)(slot[i][j]));
      alldifferent(all (j in Teams)(slot[j][i]));
      slot[i,i] = 0;
      forall (j in Teams) {
         slot[i,j] = slot[j][i];
   slot[1,j] = j-2;
      };
   };
};
```

Fig. 2. Slot formulation

```
alldifferent(all (t in time: t>0)
   (opponent[t-1,opponent[t,i]]));
```

prevents a team from having more than one carry-over effect on team i. The resulting formulation solves quickly for eight teams, and very slowly (on the order of a day) for 16 teams. Russell (1980) also addresses the question of minimizing the variance of the carry-over effects (equivalently the sum of squared carry-over counts). Again, this is simply the addition of a constraint to either scheduling formulation. For 6 teams, he found a schedule with a value of 60, and we can confirm that is indeed optimal. For 10 teams, his value of 138 is suboptimal: there exists a schedule of value 122. The computational time for a straightforward modification of our formulations is very high (it took one day to find the 10 team solution), so this is an opportunity for further work.

4 Break Phase

The second phase of our approach is to assign the home team (and away team) to each game in a schedule to create a timetable. We wish to assign home/away teams in such a way as to minimize the number of breaks (consecutive home games or consecutive away games). This problem turns out to be much more difficult than the first phase.

Régin [13], [14] has examined this problem in detail using constraint programming. Using a natural formulation (with 0–1 variables) and extensive symmetry breaking and dominance rules, his best results take 5.2 s for a 16 team instance, 80 s for an 18 team instance, and 5603 s for a 20 team instance (Régin's timings are on a 200Mz Pentium computer, so the "normalized" timings for a 266 Mz machine would be 3.9, 60 and 4213 s respectively).

We present an integer programming model for this problem that is at least competitive with Régin's constraint programming model. While the model is not difficult, it has a number of pieces, all of which are needed in order to have the model work effectively.

The variables for this formulation could be chosen in many ways. We begin with three major variables:

`start[i]`: 1 if team i starts at home, and 0 otherwise.
`to_home[i,t]`:1 if team i goes home after slot t, and 0 otherwise
`to_away[i,t]`:1 if team i goes away after slot t, and 0 otherwise

From these variables, it is easy to determine if team i is home in slot t or not:

`at_home[i,t]= start[i]+sum(t1<t)(to_home[i,t1]-to_away[i,t1])`

is 1 if team i is at home in time t, and 0 if team i is away in time t.

We limit the at_home variables to be 0 or 1. The at_home variables are not needed in the formulation (we can replace them with the identity above), but they make the exposition cleaner.

The primary constraint is that for every pair of teams i, j, if they play in slot t then

`at_home[i,t]+at_home[j,t] = 1`

We also need a constraint that for every team i and time t, the team cannot go both home and away:

`to_home[i,t]+to_away[i,t] <= 1`

This formulation is a sufficient integer program. Unfortunately, it works extremely poorly: even for an 8 team example, the program takes 81 s, while a 10 team example would not solve in 1200 s.

To improve this formulation, we begin by breaking symmetry as Régin did: we assume that team 1 begins at home. Since clearly any schedule can be "flipped", reversing home and away, such an assignment does not affect the optimal solution. This has only a marginal effect, decreasing the 8 team time to 75 s without allowing us to solve the 10 team example (in 1200 s).

We then try to strengthen the constraints to avoid unwanted fractional values that slow down the branch and bound search. We can strengthen the use of to_home and to_away by noting that the requirement on at_home does not prohibit setting both `to_home[i, t]` and `to_away[i, t]` to each be 0.5 no matter what `at_home[i, t]` is. We can therefore strengthen the formulation by adding the constraints

`at_home[i,t]+to_home[i,t] <= 1;`
`at_home[i,t]-to_away[i,t] >= 0;`

for all i and t. This formulation is much better, solving the 8 team example in 4 s and the 10 team example in 75 s.

To improve performance, we need to add additional constraints that force breaks to be taken when the teams need them. Consider three teams A, B, and C such that A plays B in slot 1, B plays C in slot 2, and A plays C in slot 3. Then, it is easy to see that either A has a break after slots 1 or 2, or B has a break after slot 1, or C has a break after slot 2. To see this, suppose A is home to B in slot 1. If there are no breaks, then B is home to C in slot 2, so C is home in slot 3. But A is also home in slot 3. Therefore, there must be a break involving one of the three teams. This generalizes to any three teams in any three slots: if we have teams i, j and k who play in slots slot[i, j]<slot[j, k]<slot[i, k], then either

- i has a break between slot[i, j] and slot[i, k], or
- j has a break between slot[i, j] and slot[j, k], or
- k has a break between slot[j, k] and slot[i, k].

This triangle constraint is very powerful because it links teams and forces one or more of them to take a break when the fractional solution might not do so. The resulting OPL code for this formulation is shown in Figure 3.

We can improve on this formulation in a number of different ways. In general, we will try to exploit knowledge of what solutions must look like in order to avoid unnecessary branching. Two improvements are

- Breaks come in pairs (if one team goes AA, then another team must go HH). Therefore, once the gap in the branch and bound tree is less than 2, the search can stop.
- In any slot, there are either no breaks or at least two breaks.

We can branch on auxiliary variables that model this latter requirement. In the tests below, we simply created two subproblems: the first with no breaks in the first period, and the second with at least two. This alone was a great improvement over just branching on the natural variables. The instances we were looking at now become trivial: the 8 team example solves in 0.4 s while the 10 team example takes 3.5 s. Table 1 shows values for larger instances:

Table 1. Sample computation times

Size	Régin	IP
16	4	22
18	60	43
20	4213	1092
22		2293

It should be noted that this is not a serious computational test: it is only on the size 20 instance that we are even solving the same instance as Régin.

```
int n= ...;
range Teams 1..n;
range Time [1..n-1];
range TweenTime [1..n-2];
range Binary [0..1];
range Slot [0..n-1];
Slot slot[Teams,Teams] = ...;//slot[i,j] is the slot when i plays j
var Binary to_away[Teams,TweenTime];
var Binary to_home[Teams,TweenTime];
var Binary start[Teams];

range Obj [0..n*n];
var Obj obj;

maximize
   obj
subject to {
file://Minimizing breaks is the same as maximizing non-breaks
   obj = sum (i in Teams, t in TweenTime)(to_away[i,t]+to_home[i,t]);
   start[1] = 1;
   forall (i in Teams, t in TweenTime ) {
      to_away[i,t]+to_home[i,t] <= 1;
   };
   forall (i in Teams, t in [2..n-1]) {
      0<= start[i]+sum(t1 in [1..t-2]) (to_home[i,t1]-
                               to_away[i,t1])+to_home[i,t-1] <= 1;
      0<= start[i]+sum(t1 in [1..t-2]) (to_home[i,t1]-to_away[i,t1])
                               -to_away[i,t-1] <= 1;
   };
   forall (i in Teams, t in [2..n-1]) {
      0<= start[i]+sum(t1 in [1..t-1]) (to_home[i,t1]-to_away[i,t1])
                               <= 1;
   };
   forall (ordered i,j in Teams) {
      start[i]+start[j] + sum(t in [1..slot[i,j]-1])
                               (to_home[i,t]-to_away[i,t])
      + sum (t in [1..slot[i,j]-1]) (to_home[j,t]-to_away[j,t]) = 1;
   };
   forall (i,j,k in Teams : (i<>k) & (j<>k) & (i<>j) &
                   (slot[i,j]<slot[j,k]) & (slot[j,k]<slot[i,k])) {
      sum(t in [slot[i,j]..slot[i,k]-1]) (to_away[i,t]+to_home[i,t]) +
      sum(t in [slot[i,j]..slot[j,k]-1]) (to_away[j,t]+to_home[j,t]) +
      sum(t in [slot[j,k]..slot[i,k]-1]) (to_away[k,t]+to_home[k,t])
      <= (slot[i,k]-slot[i,j])+
              (slot[j,k]-slot[i,j])+
              (slot[i,k]-slot[j,k]) -1;
   };
};
```

Fig. 3. Break formulation

On the smaller problems, the differences are well within the range of variation across individual instances. It is clear, however, that the IP approach is at least comparable to the constraint programming approach of Régin, and seems to be scaling up perhaps a little better.

5   Extensions and Conclusions

We have presented an alternative approach to creating sports timetables in cases where a fixed game assignment is a critical feature. We presented effective solution techniques for the two subproblems and showed how they could be used for leagues with up to approximately 20 teams.

There are a number of extensions that would make this work more applicable to real sports leagues. Perhaps foremost would be extending this work to double round-robin tournaments. Such a schedule is often played in real leagues, and leads to the additional complication that of the two games between a pair of teams, each team must be at home for one of them. This makes the problem much harder to solve.

Another approach would be to apply a more realistic measure of good HAPs. While breaks are important, some breaks are worse than others. For instance, a sequence like AAA is often much worse than AAHAA, although each have two breaks. And many leagues, including many US college basketball conferences and Major League Baseball actually prefer to go two games at home or two away before breaking. How can sequences like AAHHAAHH be encouraged in this model?

Sports scheduling provides a rich field for creating and testing constraint and integer programming formulations. The break phase solution presented was helped by the constraint programming efforts of Régin. Perhaps the biggest question raised by this paper is to determine how integer programming and constraint programming can work together to solve these problems.

References

1. Armstrong, J., Willis R.J.: Scheduling the Cricket World Cup: a Case Study. J. Oper. Res. **44** (1993) 1067–1072
2. Ball, B.C., Webster, D.B.: Optimal Scheduling for Even-Numbered Team Athletic Conferences. AIIE Trans. **9** (1977) 161–169
3. Bean, J.C., Birge, J.R.: Reducing Traveling Costs and Player Fatigue in the National Basketball Association. Interfaces **10** (1980) 98–102
4. Cain, W.O., Jr: A Computer Assisted Heuristic Approach Used to Schedule the Major League Baseball Clubs. In: Ladany, S.P., Machol, R.E. (eds.): Optimal Strategies in Sports. North-Holland, Amsterdam (1977) 32–41
5. Campbell, R.T., Chen, D.-S.: A Minimum Distance Basketball Scheduling Problem. In: Machol, R.E., Ladany, S.P., Morrison, D.G. (eds.): Management Science in Sports. North-Holland, Amsterdam (1976)15–25
6. Colbourn, C.J.: Embedding Partial Steiner Triple Systems is NP-complete. J. Combinat. Theory, Series A **35** 1983 100–105

7. Ferland, J.A., Fleurent, C.: Computer Aided Scheduling for a Sports League. IN-FOR **29** (1991) 14–24

8. Henz, M.: Constraint-Based Round Robin Tournament Planning. Proc. 1999 Int. Conf. on Logic Programming, Las Cruces, NM. (1999)

9. Henz, M.: Scheduling a Major College Basketball Conference: Revisited. Oper. Res. (2000) to appear

10. ILOG OPL Studio, User's Manual and Program Guide. (2000)

11. Lovasz, L., Plummer, M.D.: Matching Theory. North-Holland, Amsterdam (1986)

12. Nemhauser, G.L., Trick, M.A.: Scheduling a Major College Basketball Conference. Oper. Res. **46** (1998) 1–8

13. Régin, J.-C.: Minimization of the Number of Breaks in Sports Scheduling Problems using Constraint Programming. DIMACS Workshop on Constraint Programming and Large Scale Discrete Optimization (1999)

14. Régin, J.-C.: Modeling with Constraint Programming. Dagstuhl Seminar on Constraint and Integer Programming (2000)

15. Rosa, A., Wallis, W.D.: Premature Sets of 1-factors or How Not to Schedule Round Robin Tournaments. Discrete Appl. Math. **4** (1982) 291–297

16. Russell, K.G.: Balancing Carry-Over Effects in Round Robin Tournaments. Biometrika **67** (1980) 127–131

17. Russell, R.A., Leung, J.M.: Devising a Cost Effective Schedule for a Baseball League. Oper. Res. **42** (1994) 614–625

18. Schaerf, A.: Scheduling Sport Tournaments using Constraint Logic Programming. Constraints **4** (1999) 43–65

19. Schreuder, J.A.M.: Constructing Timetables for Sport Competitions. Math. Program. Study **13** (1980) 58–67

20. Schreuder, J.A.M.: Combinatorial Aspects of Construction of Competition Dutch Professional Football Leagues. Discrete Appl. Math. **35** (1992) 301–312

21. Wallis, W.D.: A Tournament Problem. J. Austral. Math. Soc., Series B **24** (1983) 289–291

22. Walser, J.P.: Integer Optimization by Local Search: A Domain-Independent Approach, Springer Lecture Notes in Artificial Intelligence, Vol. 1637. Springer-Verlag, Berlin Heidelberg New York (1999)

23. de Werra, D.: Geography, Games, and Graphs. Discrete Appl. Math. **2** (1980) 327–337

24. de Werra, D.: Some Models of Graphs for Scheduling Sports Competitions. Discrete Appl. Math. **21** (1988) 47–65

25. de Werra, D., Jacot-Descombes, L., Masson, P.: A Constrained Sports Scheduling Problem. Discrete Appl. Math. **26** (1990) 41–49

26. Willis, R.J., Terrill, B.J.: Scheduling the Australian State Cricket Season Using Simulated Annealing. J. Oper. Res. Soc. **45** (1994) 276–280

27. Wright, M.: Timetabling County Cricket Fixtures Using a Form of Tabu Search. J. Oper. Res. Soc. **45** (1994) 758–770

Three Methods to Automate the Space Allocation Process in UK Universities

E.K. Burke[1], P. Cowling[1], J.D. Landa Silva[1*], and Barry McCollum[2]

[1] Automated Scheduling, Optimisation and Planning Research Group,
School of Computer Science and IT, University of Nottingham,
{ekb,pic,jds}@cs.nott.ac.uk
http://www.asap.cs.nott.ac.uk/ASAP/
[2] School of Computer Science,
The Queen's University of Belfast, UK
B.McCollum@queens-belfast.ac.uk

Abstract. The space allocation problem within UK universities is highly constrained, has multiple objectives, varies greatly among different institutions, requires frequent modifications and has a direct impact on the functionality of the university. As in every optimisation problem, the application of different advanced search methodologies such as local search, metaheuristics and evolutionary algorithms provide a promising way forward. In this paper we discuss three well known methods applied to solve the space allocation problem: hill climbing, simulated annealing and a genetic algorithm. Results and a comprehensive comparison between all three techniques are presented using real test data. Although these algorithms have been extensively studied in different problems, our major objective is to investigate the application of these techniques to the variants of the space allocation problem, comparing advantages and disadvantages to achieve a better understanding of the problem and propose future hybridisation of these and additional methods.

Keywords: space allocation, neighbourhood search, metaheuristics.

1 Introduction

The space allocation problem in academic institutions is described as *the allocation of resources to areas of space such as rooms, satisfying as many requirements and constraints as possible.* Resources are staff, students, meeting rooms, lecture rooms, special rooms, etc. Requirements are certain conditions to be fulfilled such as the amount of space needed for each resource. Constraints (see Section 2.3) are rules that cannot be violated (hard constraints) or ones that can be broken but penalised (soft constraints).

The aims of our research are to carry out a complete and detailed investigation of the space allocation problem, to produce a model of this problem and to propose a set of well studied techniques to find solutions for the different forms of the space allocation problem not only in academic institutions, but also in commercial and industrial areas. Developing hybrid metaheuristic techniques and focusing on initialisation, decomposition and multicriteria decision making, we expect to provide fast and high-

---

[*] JDLS acknowledges support of Universidad Autónoma de Chihuahua and PROMEP, México.

E. Burke and W. Erben (Eds.): PATAT 2000, LNCS 2079, pp. 254–273, 2001.

quality solutions to large space allocation problems in universities and other environments.

In its simplest form, space allocation can be regarded as a bin packing or knapsack problem [2]. These two optimisation problems are frequently used to describe a wide range of industrial and commercial problems. Finding a new set of metaheuristics to solve the space allocation problem may well benefit these related applications. The space allocation problem is also related to scheduling, which is defined by Wren [19] as "the arrangement of objects into a pattern in time or space in such a way that some goals are achieved, or nearly achieved, and that constraints on the way the objects may be arranged are satisfied, or nearly satisfied". Research work within the Automated Scheduling, Optimisation and Planning group has demonstrated that the use of hybrid metaheuristic approaches in real applications of scheduling-related problems offers a significant opportunity of success [4], [5], [6], [9], [10].

Some approaches have been proposed for solving space allocation and space planning problems related to teaching facilities [1], [3], [13], [18]. In [7], the results of a survey on the space allocation problem within UK universities were published. A detailed description of the variety, complexity, characteristics of the problem and available solutions in each institution was obtained. Later, in [8] it was stated that the implementation of metaheuristic methodologies is a promising way to tackle the space allocation problem in universities and that the more highly constrained a real situation is, the less likely it is that we can ensure an acceptable level of space utilisation.

In this paper, we summarise the problem domain and define what a good solution is in terms of our evaluation function. Then we discuss the performance of three well known techniques applied to the space allocation problem (hill climbing, simulated annealing and genetic algorithms) and present a detailed comparison between these three approaches. Finally, some conclusions are established and future research directions are suggested.

2   Problem Description

2.1 Problem Domain

The problem of allocating resources into rooms in UK universities can be summarised as follows: *the process of assigning rooms or areas of space for specific resources, ensuring the efficient utilisation of the space and satisfying as many requirements and constraints as possible.* Types of rooms considered here are non-residential, i.e. focusing on academic-related space. Resources are considered to be staff, students, laboratories, storage areas, common rooms, lecture theatres, etc. Requirements and constraints vary from one university to another, so for each problem instance different requirements and constraints exist. However, most of those requirements and constraints are considered here as a result of our previous work [7]. Solving real instances of the space allocation problem is a multicriteria decision-making process because to determine the quality of an allocation it is necessary to consider different objectives such as: achieve an efficient space utilisation, maximise the satisfaction of constraints, minimise costs and guarantee people's satisfaction.

2.2 Phases and Modes of the Process

The process of allocating rooms in UK universities can be performed in three stages:

1. The centralised office allocates space to faculties and assigns common areas,
2. Faculties assign areas to schools and departments,
3. Departments allocate specific rooms to resources.

During these three phases, the problem can be solved in different ways at each stage:

- Fitting all resources into a limited amount of space,
- Minimising the amount of space required to allocate a set of resources,
- Reorganising because of the addition or removal of space and/or resources,
- Reorganising/optimising the current allocation due to the possible variation of requirements and/or constraints.

2.3 Types of Constraint

Constraints considered so far in the domain of this problem, can be any of the following classes:

- Sharing restrictions: e.g., head of department does not share a room;
- Proximity/adjacency requirements: e.g., secretary must be adjacent to the head of school;
- Grouping requirements: e.g., people in a research group must be in the same room;
- Requirements and limits for wastage and overuse of space: e.g., research students require 6 m^2, but it is acceptable to assign 15% more (6.9 m^2) or less (5.1 m^2) space;
- Requirements for staff sharing between different departments: e.g., a lecturer working for two different departments should share a room;
- Resource specific location: e.g., network technician must be adjacent to the networking laboratory or in a specific room.

These constraints are divided into two groups. The first and basic group consists of space overuse, space wastage, unallocated resources, sharing and grouping restrictions. The second group consists of constraints that are required to be satisfied in each specific case: e.g., technical services coordinator in the School of Computer Science and IT at the University of Nottingham must be in a non-shared room in the 2nd floor, adjacent to other members of the technical services group, and all of them should be close to the networking laboratory. Any constraint can be either hard or soft according to the real problem. For example, in some universities it is strongly required that no member of staff shares an office, while in others this requirement is only desirable. Additional constraints can be added as required.

2.4 Fitness Evaluation of an Allocation

The allocation of all resources may be a hard constraint (a feasible solution must have all resources allocated) or a soft constraint (some resources may be unallocated but a penalty is applied). A feasible solution must satisfy all the hard constraints in the specific space allocation problem. The quality of a feasible allocation is measured using the aggregating function (1). This function is a sum of the penalty due to unallocated resources, the penalty due to inefficient space utilisation and the penalty due to unsatisfied soft constraints. If any of these is a hard constraint or requirement in the problem instance, the corresponding penalty in a feasible solution must be equal to zero. The lower the total penalty value, the higher the quality of the allocation.

$$\text{total penalty} = \sum_{i=1}^{N} UP(r_i) + \sum_{i=1}^{M} [WP(s_i) + OP(s_i)] + \sum_{i=1}^{N} SCP(r_i). \quad (1)$$

UP is the penalty applied to the resource r_i if it has not been allocated, WP is the penalty applied to the room s_i if there is space wastage, OP is the penalty applied to the room s_i if there is space overuse, SCP is the penalty applied if there is a soft constraint violation for the resource r_i, N is the total number of resources to be allocated in the problem and M is the total number of rooms to be used in the allocation process.

We calculate the penalties for violated soft constraints using weights and exponents according to each specific scenario (for our experiments these values are included in the test data sets available). The penalty for each violated soft constraint is equal to $(violation\ level \times weight)^{exponent}$, where the $violation\ level$ is a measure of the soft constraint violation. Suppose we have a space allocation problem in which the allocation of all resources is a soft constraint and a feasible solution has the following constraint violation levels: six resources are not allocated, three rooms have space wasted (4.6, 0.6 and 2.7 m² respectively), one room has space overuse equal to 2.4 m², two sharing restrictions and five adjacency constraints are not satisfied. Assume the following values for weights and exponents:

Constraint	Weight	Exponent
wastage	2	1
overuse	2	2
unallocated	5000	1
sharing	2000	1
adjacent to	500	1

For the example described above the total penalty is calculated using (1) as follows:

$$\text{total penalty} = (6\times5000)^1 + ((7.9\times2)^1 + (2.4\times2)^2) + ((2\times2000)^1 + (5\times500)^1) = 36538.84.$$

The weight is a measure of the impact in the penalty value of the unsatisfied constraints, while the exponent penalises the degree to which the soft constraints are violated.

3   Three Methods to Automate Space Allocation

3.1 Neighbourhood Exploration

The methods we have implemented to approach the space allocation problem are hill climbing, simulated annealing and a genetic algorithm. The three algorithms attempt to find the global optimum in the solution space, but while the first one is well known as a search heuristic that may become stuck in poor local optima, simulated annealing and genetic algorithms attempt to avoid this by performing a wider exploration of the solution space [14], [15], [16], [17].

An allocation is represented using the structure shown in Figure 1. A solution is coded using a string that contains one element for each resource in the problem. Each resource is associated with the room to which the resource has been allocated. If unallocated resources are permitted in a feasible solution, those resources have a *bin* room associated. If the same room is associated to more than one resource then those resources are sharing the specified room.

Lab B	Mr Lee	Store	Director	Catering	Ms Shang	Lab A	Mr Khan
1B01	1B04	1B08	1B17	1B10	1B07	bin	1B04

Fig. 1. The structure used to represent an allocation in the space allocation problem

Three moves are used to modify an allocation and therefore explore the search space: ALLOCATE, RELOCATE and SWAP. The ALLOCATE move selects an un-allocated resource and finds a room to allocate to it. The RELOCATE move changes the assigned room for one allocated resource. Finally, the SWAP move selects two rooms and interchanges the allocated resources between them. The construction of an initial solution is done by means of the ALLOCATE move. During the construction of the initial solution and also during the space exploration, the following parameters are used to modify the searching process: resource search, room search, space deviation and termination criteria. In our experiments (see Section 4) these parameters were investigated to determine the appropriate neighbourhood exploration in each algorithm.

- **Resource search.** The selection of the resource for the ALLOCATE and RELOCATE moves can be: random or the worst offender. In the first case, the resource to be allocated or relocated is randomly selected from the corresponding list (unallocated or allocated resources). Selecting the worst offender means that the move is evaluated for each resource in the corresponding list and the resource that causes the least penalty is chosen. Obviously, the second option takes more time to select the resource because it performs a wider search attempting to make a better resource selection.
- **Room search.** To select the room for the ALLOCATE and RELOCATE moves, or a pair of rooms for the SWAP move, two options are possible: random or the best of *NB* rooms. In random selection we choose at random one resource (ALLOCATE or RELOCATE moves) or two rooms (SWAP move). In the second

case, *NB* random rooms (ALLOCATE or RELOCATE moves) or *NB* random pairs (SWAP move) of rooms are evaluated and the best room or pair of rooms is finally chosen to implement the move. If *NB* equals the total number of rooms *M*, then all rooms are tested and the best is used. Random selection permits faster construction and neighbourhood exploration, but the second strategy performs a more thorough search.

- **Space deviation.** When selecting the room for an ALLOCATE or RELOCATE move or the pair of rooms for the SWAP move, it is possible to perform or skip an evaluation of the percentage of space that can be wasted or overused. If this space deviation is not evaluated, the selected room will be used even if it is too big or too small for the selected resource. If this space deviation is evaluated, then the percentage of space wastage or space overuse in the selected room must be within the problem requirements.
- **Termination criteria.** To investigate the performance of the three algorithms, two termination criteria are available: a fixed number of iterations or no improvement in the allocation after a certain number of iterations.

3.2   Hill Climbing and Simulated Annealing

The standard hill climbing strategy is based on the inspection of the neighbourhood in the solution space, so that by means of moves in the existing solution, progressive improvements can be achieved to reach the local optima. The most important part of this algorithm is the heuristic used to explore the neighbourhood using the three possible moves: ALLOCATE, RELOCATE and SWAP. This strategy is shown below:

> If all N resources are allocated
>> Select a random move between RELOCATE and SWAP
> If not all N resources are allocated
>> If NA ≥ MA
>>> If last move was ALLOCATE
>> Select a random move between RELOCATE and SWAP
>>> If last move was not ALLOCATE
>> Select ALLOCATE move
>>> NA ← 0
>> If NA < MA
>>> If last move was not ALLOCATE
>>>> Select a random move between RELOCATE and SWAP.

where *N* is the total number of resources, *NA* is the number of failed (i.e. non-improving) move attempts and is incremented after one move attempt has failed, *MA* is the maximum number permitted of failed move attempts, and there is an equal probability of choosing either the RELOCATE or the SWAP move.

The strategy shown above to select a move takes into account the current state of the allocation and the viability of accomplishing a certain type of move. In this sense, the type of move that is undertaken in each iteration depends on the number of allocated resources and the number of prior failed attempts to find a feasible move.

When all resources in the current problem are already allocated, the algorithm explores the neighbourhood using the moves RELOCATE and SWAP to improve the solution. In the case that not all resources are allocated, a certain number of attempts (*MA*) normally set to *N/5*, is given to either the ALLOCATE or the RELOCATE and SWAP moves. (Our experiments have shown that it is likely to find a move when one-fifth of the number of resources is evaluated for the required move.) The heuristic tries to ALLOCATE as many resources as required to produce a feasible solution, but also attempts to avoid getting stuck by examining the RELOCATE and SWAP moves. For example, suppose that in the current solution there are still five unallocated resources from a total of 100 in the allocation problem. Then, if after 20 failed attempts none of these resources have been successfully allocated, the algorithm examines the feasibility of modifying the solution using the RELOCATE and SWAP moves up to a maximum of 20 failed attempts. The number of failed modification attempts is set to zero when an improving move has been found.

The simulated annealing algorithm is a well known method in which new solutions are accepted during the process with a probability that varies according to a temperature parameter [16], [17]. Our simulated annealing and hill climbing algorithms use the same heuristic to select the type of move to improve the current solution. The temperature is reduced slowly starting from a random search at high temperature and carrying out pure hill climbing at zero temperature. The goal of the temperature variation process is to combine random selection with the local search heuristic to find global optima. When the current allocation is improved by trying the moves ALLOCATE, RELOCATE or SWAP, a high temperature corresponds to random movements and other solutions are visited even if their fitness is not better than the current solution. Low temperature corresponds to little randomness and worse solutions are not visited. The temperature is set to a high value when the algorithm starts, then it is decreased after a fixed number of iterations. The parameters used in our simulated annealing algorithm are explained in Section 3.4. The acceptance or rejection of the selected move in the current solution is controlled as follows:

> If the selected move improves the current solution
> >   Accept move and new solution
> Else
> >   If current temperature = 0
> >   >   Reject move and new solution
> >   If current temperature > 0
> >   >   Probability of acceptance = $\exp^{(-delta\,/\,current\,temperature)}$
> >   >   If probability of acceptance ≥ random number
> >   >   >   Accept move and new solution
> >   >   Else
> >   >   >   Reject move and new solution.

Delta is the fitness variation due to the proposed move and a value greater than zero means an improvement in the existing solution (decrease in the total penalty value).

3.3 Genetic Algorithm

The genetic algorithm that was implemented for this problem is shown below, where each chromosome is a possible allocation as shown in Figure 1:

Create Initial Population
Calculate Fitness (Initial Population)
Current Population = Initial Population
while Termination Criteria Not Satisfied
        For OffspringNo = 0 to OffspringNo = PopulationSize do
            Parent1 = Roulette_Wheel_Selection (Current Population)
            Parent2 = Roulette_Wheel_Selection (Current Population)
            Heuristic_Crossover (Parent1,Parent2,New Population)
        Mutate Population (New Population)
        Calculate Fitness (New Population)
        Replace_Population (Current Population, New Population)

The construction of each individual in the initial population is carried out using the heuristic explained in Section 3.2 with the moves ALLOCATE, RELOCATE and SWAP. Our genetic algorithm evaluates the fitness of each solution using the penalty function (1) presented in Section 2.4. Using Roulette_Wheel_Selection, two parents are selected from the current population. In the roulette wheel operator, the probability of selecting each individual is proportional to its fitness [12]. Here, the sum of the fitness (F_{sum}) for all chromosomes is obtained, then a random number n between 0 and F_{sum} is generated. The first individual whose fitness added to the fitness of the preceding population members is greater than or equal to n is selected as a parent.

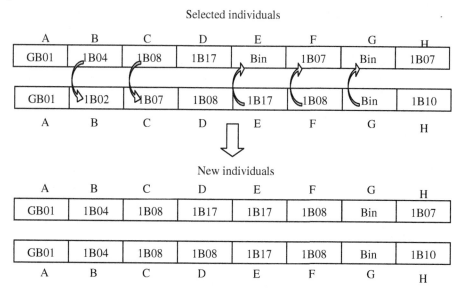

Fig. 2. In the four-point crossover operator, the section in the chromosome with the highest penalty is chosen, so the size of this section varies accordingly

Our crossover operator works as illustrated in Figure 2 using the chromosome string representation shown in Figure 1. This four-point crossover strategy identifies, in each parent, the chromosome section that contains the group of resources (adjacent in the chromosome representation) whose penalty values are the greatest. After detecting these sections, the group of highest penalty in each parent is replaced with the corresponding substring in the other parent.

The mutation operator consists of a random change of the assigned room for a randomly selected resource. In Replace_Population, the new population replaces the current population and elitism is applied to guarantee the selection of the fittest individual so that this solution is preserved between generations. Here, elitism consists of substituting the worst individual in the new population with the best individual in the previous generation.

3.4 Selection of Search Parameters

Making modifications to the searching parameters described in Section 3.1, we obtained the variants of the algorithms shown in Table 1. These 12 variants are different heuristics that were tested to find the set of parameters that produce the best solutions in each type of space allocation problem. Our goal is to investigate the effect of these parameters to design a heuristic for neighbourhood exploration in the space allocation problem.

Table 1. Variants of the three algorithms. Parameters for simulated annealing are: initial temperature 2000, decrement value 100 and decrement interval 300. Parameters for the genetic algorithm are: population size 20, crossover probability 80%, mutation probability 5%. *NB*, the neighbourhood size, is replaced by a number according to each problem instance in the results described in Section 4.

Algorithm variants	Algorithm			Searching options			
	HC	SA	GA	Random room	NB best room	Worst resource offender	Space deviation check
HCRand	√			√			
HCRandChk	√			√			√
HCRandWrst	√			√		√	
HCNBRms	√				√		
HCNBRmsChk	√				√		√
SARand		√		√			
SARandChk		√		√			√
SARandWrst		√		√		√	
SANBRms		√			√		
SANBRmsChk		√			√		√
GANBRms			√		√		
GARandWrst			√	√		√	

After examining the performance of each of the 12 variants of algorithms presented in Table 1, the parameters for our algorithms are those that produce the best results in terms of the penalty function (1) described in Section 2.4. The values shown for these parameters have been set up according to the problem size in our problem instances. For simulated annealing we found that the best decrement interval size is around twice the number of resources (150 resources in average for these test problems). Decrement value is the best when it is set to 1/10 of delta, the fitness variation after implementing a move in the current solution (delta is in the range of 800 to 1500 on these problems). The initial temperature of 2000 is the value that produced the best fitness in these problems. For the genetic algorithm we tested the range of parameters that are proposed by Goldberg [14], and we found that with a population of about 1/10 of the number of resources, 80% crossover probability and 5% mutation probability we obtain the best results.

In the next section we give a description of the five data sets that were used in our experiments. We show how the conditions differ from one problem to another due not only to the fact that each university imposes its own requirements and standards, but also to the different available information to specify the problem and then construct the solution. For example, while some universities provide information about space requirements for resources, adjacency and proximity between rooms and constraints to be satisfied, others simply do not use any standard data and the acceptance/rejection of the solution depends only on space utilisation and some vague sharing restrictions.

4   Results

4.1 The Experiments

All variants of the three algorithms were tested with different real data obtained from three universities in the UK. As we stated in Section 2.2, space allocation can be applied in four ways. We use two of them that represent real situations in academic institutions: optimisation/reorganisation and construction of a complete allocation. Optimisation is when the existing allocation, with all the resources already allocated, must be improved using the same set of rooms and constraints. Reorganisation means that a subset of the resources (some specific rooms, like laboratories, common rooms or strategic offices) has been previously allocated and then all remaining resources are allocated to construct a solution. A complete allocation refers to the situation where all resources are unallocated, and a solution involving all resources must be found. Each algorithm variation was tested 40 times with all data sets, and then we selected the ones that produced the best results for each case. Since in this paper we attempt to determine an efficient strategy for the neighbourhood exploration, the best performance of each algorithm is compared. The computational times required in our tests are shown as a reference to compare them with the time taken to construct a manual solution (weeks or months). A manual solution is constructed by the space officers and varies from university to university [7]. The best solutions shown in Tables 2–6 were selected according to the total penalty value obtained using a PC Pentium 300MHz with 64MB RAM. In all tables, the last column shows the total penalty for the manually constructed allocation as implemented in each case. We present the solutions in tables to facilitate the analysis of the algorithms' performance not only according to total penalty but also for each different evaluation criteria (space utilisation, unallocated resources penalty and soft constraints penalty).

4.2 The University of Nottingham Data

The School of Computer Science and IT recently moved to a new building, so it was necessary to obtain a new allocation. There are 90 rooms of different sizes and 117 resources distributed according to their level, indicating sharing and space requirements: 6 professors, 9 laboratories, 9 meeting rooms, 10 technical staff, 5 storage rooms, 1 teaching assistant, 3 senior lecturers, 7 secretaries, 47 researchers, 19 lecturers and 1 visiting lecturer. For these problems we have the 5 basic constraints, 8 specific groups of people, 30 specific resource locations and 8 particular proximity/adjacency requirements. The three data sets used are as follows.

CSBuildingAllocatedIdeal. This is the real allocation at the Computer Science and IT Building in this university. All 117 allocated resources and 90 rooms are used. The goal is fitness improvement using all constraints specified by the problem, i.e. an optimisation problem. Results obtained for this case are shown in Table 2 compared with the real allocation.

Table 2. Results for optimising the current allocation, University of Nottingham data

Fitness statistics	CSBuldingAllocatedIdeal					Real allocation
	HC30Rms	HCRand	SARandWrst	SA30Rms	GARandWrst	
Resources allocated	117	117	117	117	103	117
Rooms used	90	90	90	90	78	90
Space utilisation	82.45%	81.56%	81.32%	78.27%	63.69%	77.99%
Constraints penalty	714.87	1094.93	4221.15	1264.21	71284.91	1264.21
Space wastage penalty	479	487.4	535.9	624.8	859.6	639.8
Space overuse penalty	403.26	1314.51	2225.76	14666.52	8290.07	17400.27
Total penalty	**1597.13**	2896.84	6982.82	16555.53	80434.48	19304.28
Time taken (h:min:s)	0:29:53	0:19:46	0:23:04	0:14:30	0:45:06	-----
Iterations	20000	50000	5000	5000	15	-----

CSBuildingReorganiseIdeal. This has 21 allocated resources (laboratories, meeting rooms, storage rooms), 96 resources to be allocated (researchers, secretaries, lecturers, senior lecturers, professors, technical staff, teaching assistants, visiting lecturers) and 21 used rooms. The goal is to reorganise the allocation using all requirements for the problem and focusing on staff accommodation. Table 3 shows results for this data set.

CSBuildingNewIdeal. All 117 resources are to be allocated and all rooms are available. The goal is to fit all resources into the limited amount of space, using all constraints for the problem and studying the impact of allocating all resources in one

Table 3. Results for reorganising the current allocation, University of Nottingham data

Fitness statistics	CSBuldingReorganiseIdeal					
	HCRandWrst	HC30Rms	SARandWrst	SA30Rms	GA117Rms	Real Allocation
Resources allocated	117	117	117	117	117	117
Rooms used	90	90	90	90	90	90
Space utilisation	79%	79.99%	81.65%	89.16	75.52%	77.99%
Constraints penalty	8500	13000	13343.146	6500	62083.78	1264.21
Space wastage penalty	477.2	464	594.2	476.6	521.1	639.8
Space overuse penalty	403.26	99.51	2293.9	404.07	12133.97	17400.27
Total penalty	9380.45	13563.51	16231.24	**7380.67**	74738.85	19304.28
Time taken (h:min:s)	0:26:19	0:41:36	0:06:18	0:34:54	3:15:23	-----
Iterations	5000	20000	1000	20000	13	-----

stage. The current allocation in the CSBuilding has been evaluated with the penalty function and compared with the allocations obtained with the algorithms, as shown in Table 4.

Table 4. Results for creating a new allocation, University of Nottingham data

Fitness statistics	CSBuldingNewIdeal					
	HCRandWrst	HC30Rms	SARandWrst	SA30Rms	GA117Rms	Real allocation
Resources allocated	113	115	116	115	117	117
Rooms used	86	89	87	90	90	90
Space utilisation	62.67%	62.98%	70.56	77.65%	60.27%	77.99%
Constraints penalty	21979.086	32309.801	40526.35	29895.5	53530.54	1264.21
Space wastage penalty	1529.4	1354.6	1247.8	1098.4	512	639.8
Space overuse penalty	336.41	4617.88	7514.32	11346.32	8573.22	17400.27
Total penalty	**43962.48**	48309.28	54369.49	52340.22	62615.76	19304.28
Time taken (h:min:s)	0:03:15	0:57:47	0:08:27	0:05:30	2:34:07	-----
Iterations	1000	50000	5000	10000	11	-----

4.3 University of Wolverhampton Data

The Estates department at the University of Wolverhampton provided us with information about the SC Building in the Telford University Campus. In this case there are 115 rooms and 115 resources, which are classified in 13 different levels but not all of them with standard defined space, sharing or special requirements. In this university, the Estates department labels each room with a specific use (for example, staff working room) then, depending on the actual size of the room, its shape and the resource standard space requirement, the capacity is determined for that room. The types of resources are laboratories, staff working rooms, computer rooms, teaching rooms, store rooms, common rooms, toilets, etc. The interest here is to automate allocation of staff working rooms, teaching rooms and some specific laboratories or computer rooms, and to improve the distribution of these resources. We have a set of eight constraints: the five basic ones and three that specify grouping requirements. An important note is that there is no available information about proximity/adjacency between rooms. This condition gives us the opportunity to evaluate the algorithm's performance with missing information.

WolverhamptonReorganiseIdeal. There are 71 allocated resources (special purpose rooms like laboratories, computer rooms, store rooms, common rooms, toilets, etc.), 44 resources to be allocated (staff working rooms, teaching rooms, some specific laboratories or computer rooms) and 71 used rooms. The goal is to fit all resources into the available space, using specified requirements for the problem and focusing on academic related room's accommodation. Table 5 shows results for this case and as with the previous data set, the current allocation in the SC Building at the University of Wolverhampton has been evaluated with the penalty function and compared with the allocations obtained with the three algorithms.

Table 5. Results: reorganising the current allocation, Wolverhampton University data

Fitness statistics	WolverhamptonReorganiseIdeal					
	HC30Rms	HCRand	SARandChk	SA20Rms	GA117Rms	Real allocation
Resources allocated	115	113	114	114	115	115
Rooms used	96	102	103	100	114	115
Space utilisation	65.34%	58.21%	64.52%	61.54%	65.34%	65.33%
Constraints penalty	6171.316	11297.45	51726.85	11384.28	15915.08	16044.82
Space wastage Penalty	1959.994	2369.063	2083.78	2419.06	2218.15	2407.37
Space overuse penalty	0	1815.066	6349.94	212.95	0	0
Total penalty	**8359.67**	25824.67	65541.09	19268.68	18133.23	18452.20
Time taken (h:min:s)	0:14:07	0:01:16	0:01:00	0:02:10	0:56:42	-----
Iterations	10000	20000	50000	10000	12	-----

4.4 Nottingham Trent University Data

This data set is the one with the least information available about requirements for each different resource level, and there is no information available about proximity/adjacency between rooms. Initially, the University did not specify standard sharing, space or grouping requirements. We have 151 resources classified in 7 levels, 74 rooms and the basic constraints. There are 32 administrative assistants, 7 administrators, 9 coordinators, 81 lecturers, 7 managers, 6 professors and 9 technicians. After defining some space requirements and evaluating the actual allocation, the goal here is to improve it using our heuristics.

TrentAllocatedBasic. This is the real allocation at the Chaucer Building in Nottingham Trent University. All 151 allocated resources and 74 rooms are used. The goal is optimisation using only the basic constraints for this problem.

Table 6. Results for optimising the current allocation, Trent University data

Fitness statistics	TrentAllocatedBasic					
	HCRand	HC20Rms	SARand	SA20RmsChk	GARandWrst	Real allocation
Resources allocated	151	151	151	151	95	151
Rooms used	74	74	74	74	64	74
Space utilisation	80.6%	80.6%	80.6%	80.6%	65.65%	75.36%
Constraints penalty	0	0	0	4000	324800	58000
Space wastage penalty	573	573	573	573.4	3467.78	727.88
Space overuse penalty	0	0	0	0.015	68768.30	220738.90
Total penalty	573	573	573	4573.41	397036.08	279466.75
Time taken (h:min:s)	0:01:59	0:00:24	0:05:43	0:12:59	1:47:12	------
Iterations	50000	5000	10000	20000	15	-----

4.5   Selection of the Search Strategy

For Tables 2–6 presented in the last sections, we selected the best options for each algorithm using the information obtained from graphs like the ones in Figures 3–7. In the graphs we indicate the best performance obtained by the selected variants of the hill climbing algorithm included in Tables 2–6. We observe that hill climbing variants produce the best results when applied to optimisation problems (Figures 3 and 7), i.e. when there is an existing allocation and it should be improved. In these cases (CSBuildingAllocatedIdeal and TrentAllocatedBasic problems) all variants obtain substantial improvement over the real allocation. We observe from Figures 3 and 7 that the variants HCRandChk, HCNBRmsChk and HCRand provide poor solutions in the first iterations, but find considerable improvement after 5000 iterations, while the variants HCRandWrst and HCNBRms produce high-quality solutions even with just a few iterations. This means that all our hill climbing heuristics effectively take an existing allocation provided by the user and find good local optima, obtaining a substantial improvement measured with the penalty function (1) described in Section 2.4.

On the other hand, when our hill climbing is applied to reorganisation problems (CSBuildingReorganiseIdeal and WolverhamptonReorganiseIdeal), three variants produce competitive results. This can be noted in Figures 4 and 6, where the variants HCRandChk and HCNBRmsChk offer poor solutions compared with the existing one.

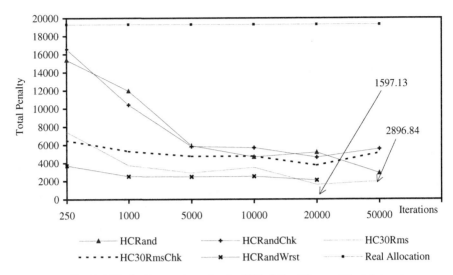

Fig. 3. Hill climbing variants for the CSBuildingAllocatedIdeal data set

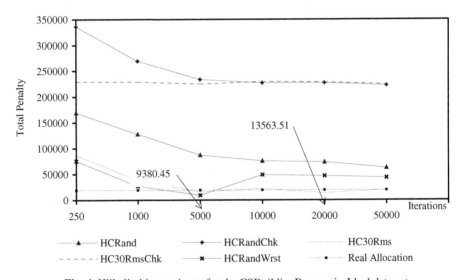

Fig. 4. Hill climbing variants for the CSBuildingReorganiseIdeal data set

HCRand again produces low-quality solutions at the beginning, but after 5000 iterations the allocation obtained is highly competitive. Here, HCNBRms and HCRandWrst are the best options since both are equal to or even improve the current solution in the first iterations.

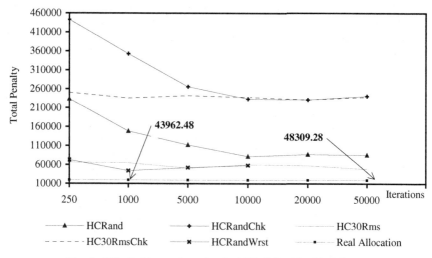

Fig. 5. Hill climbing variants for the CSBuildingNewIdeal data set

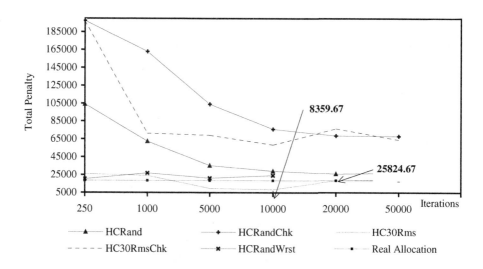

Fig. 6. Hill climbing variants for WolverhamptonReorganiseIdeal data set

If our hill climbing heuristics are applied to construct a completely new solution (CSBuildingNewIdeal) then none of the variants produce a better solution than the solution generated by a human expert measured with the penalty function described in Section 2.4. In Figure 5 we observe that HCRandChk and HCNBRmsChk offer their best performance after 5000 iterations but the solutions produced have low fitness (high total penalty value).

If HCRand after 5000 iterations together with HCRandWrst and HCNBRms are compared with the existing solution, then we can say that these variants produce allocations with a total penalty that is slightly greater. If we also consider that the existing solution was constructed by the experts using all their knowledge and that this allocation is the best one using the traditional and non-automated method, then allocations provided here by hill climbing achieve a reasonable quality. In all test problems, the variants of the hill climbing algorithm that produce the best results are HCNBRms and HCRandWrst. The HCRand variant offers an interesting option while HCRandChk and HCNBRmsChk are the worst of all.

Note that in these graphs, comparison has been made using only the total penalty of an allocation, but there are several aspects to consider before establishing final conclusions about our heuristic's performance. Further analysis with information from the best variants of the three algorithms and additional fitness measures is presented in Section 4.6 using Tables 2–6. Similar analyses were achieved for simulated annealing and genetic algorithm variants, but as we stated before, only the best results are presented here in Section 4. In the next section we analyse the searching strategy and the performance of our implemented algorithms for the space allocation problem.

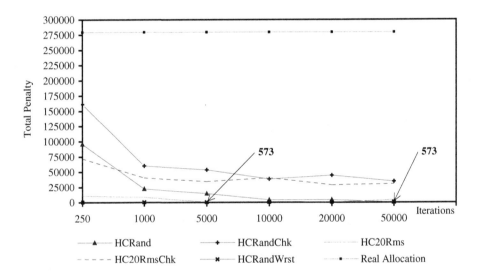

Fig. 7. Hill climbing variants for the TrentAllocatedBasic data set

4.6   Discussion on the Algorithms' Performance

We observe from Tables 2–6 that the best results in terms of the allocation quality measured with the aggregating penalty function (1) are produced by the hill climbing variants. (We are comparing our variants of these algorithms on the space allocation problem but there are other successful implementations of these three algorithms: see [16], [17].) The simulated annealing variants produce good results when the problem is not highly constrained (TrentAllocatedBasic). The genetic algorithm implemented here did not produce improvements over the current allocation. Both hill climbing and simulated annealing strategies, reach the goal of improving an existing allocation.

When reorganising an allocation, hill climbing and simulated annealing variants obtain the best results, as can be noted in Tables 3 and 5. Our genetic algorithm has a good performance in reorganising problems (WolverhamptonReorganiseBasic) if there are only basic constraints. An important observation here is that in reorganising problems (CSBuildingReorganiseIdeal and WolverhamptonReorganiseBasic) where the allocation process is centred on the staff and certain specific and conflicting rooms were allocated previously, both hill climbing and simulated annealing are capable of allocating all resources and improving the manual solution, i.e. these algorithms find a locally optimal solution that is better that the manual solution (here the quality of solutions is measured with the penalty function described in Section 2.4. Our genetic algorithm performs better (allocates all resources) in the situation in which the initial population is originated from a partially constructed allocation.

In Table 4 we observe that if the problem is to construct a completely new allocation for all resources, none of our algorithms produced a better solution than the manual approach. The hill-climbing and simulated annealing strategies constructed allocations that correspond to local optima, which do not match the quality of the manual solution. For example, HCRandWrst produced solutions with four unallocated resources and SA30Rms obtained allocations with two unallocated resources. In the same case, our genetic algorithm produced a set of solutions which have a competitive quality compared with hill climbing and simulated annealing, but neither provided a better solution than the one obtained manually. This genetic algorithm performs well for this type of problem compared with the optimisation and reorganisation cases, because the algorithm constructs all individuals from scratch and is then capable of accomplishing a wide exploration of the solution space.

In all variants of our algorithms, a *completely random* searching strategy can be seen as one that uses only random selection of rooms and resources without any space deviation check. A *steepest descent* searching strategy would be one that always selects the pair resource/room that provides the highest improvement in the current solution. We observe from Tables 2–6 and Figures 3–7 that the three algorithms achieve the best performance when the searching strategy is partially *heuristically directed* (i.e. a random selection of the resource with *NB* rooms evaluated) and space deviation checking is not performed.

5   Conclusions

The problems that space managers face most often are the reorganisation and optimisation of the current allocation. The time required for constructing an allocation varies from weeks to months [7]. Our heuristics offer a promising alternative to automate the space allocation process in a shorter time. From the approaches investigated so far, hill climbing appears to be the best for optimisation problems, using the strategy of selecting the best among *NB* rooms in the neighbourhood exploration heuristic. For reorganising situations, both simulated annealing and hill-climbing strategies produce their best performance using the strategy of selecting the best among *NB* rooms. The reason why these strategies have a good performance in optimising and reorganising problems might be that the most conflicting resources are already allocated and that the improvement of these solutions can then be accomplished using these local search strategies. In constructing a complete allocation, our hill climbing and simulated annealing variants construct good solutions but do not match the quality of the manually constructed allocation. Constructing a completely new allocation is not a frequently needed task, but the experts spend days, even months, on it, while our heuristics produce competitive initial solutions in minutes or hours. The implemented genetic algorithm is capable of producing acceptable results in terms of time when constructing complete allocations. It produces a set of solutions that can be improved using a local search heuristic. We observe that the neighbourhood exploration in these problems produces the best results using our algorithms when a random selection of the resource is performed, NB rooms are evaluated and the best of them is chosen and no space deviation checking is done.

One future research direction is to modify the neighbourhood search heuristic to construct a completely new solution allocating the most conflicting resources at the beginning of the process. It is also important to investigate the hybridisation of genetic algorithms and local search operators in order to produce a robust solution. The effect of the evaluation method to establish the quality of an allocation will also be considered in future research work. This paper shows that the space allocation process in UK universities can effectively be developed in a better way using the algorithms presented. We have studied how some modifications to three well known approaches can be used to tackle the different instances of the space allocation problem within universities, helping us to construct both a comprehensive model for the problem and a well studied set of techniques to solve it.

References

1. Benjamin, C., Ehie, I., Omurtag, Y.: Planning Facilities at the University of Missouri-Rolla. J. Interfaces **22** (1992) 95–105
2. Baase, S., Computer Algorithms Introduction to Design and Analysis. 2nd edn. Addison-Wesley, Reading, MA (1988)
3. Bland, J.A.: Space-Planning By Ant Colony Optimisation. Int. J. Comput. Appl. Technol. **12** (1999) 320–328
4. Burke, E.K., Newall, J.P., Weare, R.F.: A Memetic Algorithm for University Exam Timetabling, Selected Papers from PATAT'95 (Edinburgh). Lecture Notes in Computer Science, Vol. 1153. Springer-Verlag, Berlin Heidelberg New York (1996) 241–250
5. Burke, E.K., Clark, J.A., Smith, A.J.: Four Methods for Maintenance Scheduling. Proc. 3rd Int. Conf. on Artificial Neural Networks and Genetic Algorithms (University of East Anglia, Norwich), Vol. 1. Springer-Verlag, Berlin Heidelberg New York (1997) 264–269
6. Burke, E.K., Smith, A.J.: A Memetic Algorithm for the Maintenance Scheduling Problem. Proc. ICONIP'97 Conf. (Dunedin, New Zealand) Springer-Verlag, Berlin Heidelberg New York (1997) 469–474
7. Burke, E.K., Varley, D.B.: Space Allocation: An Analysis of Higher Education Requirements. Selected Papers from PATAT'97 (Toronto, Canada). Lecture Notes in Computer Science, Vol. 1408. Springer-Verlag, Berlin Heidelberg New York (1998) 20–33
8. Burke, E.K., Varley, D.B.: Automating Space Allocation in Higher Education. Proc. 2nd Asia Pacific Conf. on Simulated Evolution and Learning (Canberra, Australia) (1998) 66–73
9. Burke, E.K., Smith, A.J.: A Memetic Algorithm to Schedule Planned Grid Maintenance. Proc. Int. Conf. on Computational Intelligence for Modelling Control and Automation (Vienna, Austria) (1999) 122–127

10. Burke, E.K., Newall, J.P., A Multi-stage Evolutionary Algorithm for the Timetable Problem. IEEE Trans. Evolut. Comput. **3** (1999) 1085–1092
11. Burke, E.K., Landa Silva, J.D.: The Space Allocation System – Test Data. Available at www.asap.cs.nott.ac.uk/ASAP/space/spacedata.html [March 1st 2001]
12. Coley, D.A.: An Introduction to Genetic Algorithms for Scientists and Engineers. World Scientific, Singapore (1999) 23–34
13. Giannikos, J., El-Darzi, E., Lees, P.: An Integer Goal Programming Model to Allocate Offices to Staff in an Academic Institution. J. Oper. Res. Soc. **46** (1995) 713–720
14. Goldberg, D.E.: Genetic Algorithms in Search, Optimisation and Machine Learning. Addison-Wesley, Reading, MA (1989)
15. Man, K.F., Tang, K.S., Kwong, S.: Genetic Algorithms. Springer-Verlag, Berlin Heidelberg New York (1999)
16. Osman, I.H., Kelly J.P.: Meta-heuristics: Theory and Applications. Kluwer, Dordrecht (1996)
17. Rayward-Smith, V.J., Osman, I.H., Reeves, C.R., Smith, G.D.: Modern Heuristic Search Methods. Wiley, New York (1999)
18. Ritzman, L., Bradford, J., Jacobs, R.: A Multiple Objective Approach to Space Planning for Academic Facilities, J. Manage. Sci. **25** (1980) 895–906
19. Wren, A.: Scheduling, Timetabling and Rostering, a Special Relationship? Selected Papers from PATAT'95 (Edinburgh). Lecture Notes in Computer Science, Vol. 1153. Springer-Verlag, Berlin Heidelberg New York (1996) 46–75

Practical Considerations
and General Issues

Resource-Constrained Project Scheduling and Timetabling

Peter Brucker and Sigrid Knust

Universität Osnabrück, FB Mathematik/Informatik,
Albrechtstr. 28, D-49069 Osnabrück, Germany
{peter, sigrid}@mathematik.uni-osnabrueck.de

1 Introduction

Crew or staff scheduling is a complex combinatorial problem. It consists of assigning tasks to persons which have to perform these task and to schedule the performance of the tasks. Railway and airline crew scheduling problems, audit scheduling problems, school and university course scheduling and staff scheduling in hospitals are typical examples of such combinatorial problems.

Usually, the solution process for these problems is decomposed into two stages: in one stage the assignment is done while in the other stage scheduling decisions are taken. For railway and airline crew scheduling the scheduling problem is solved first by fixing the timetable and the routing of trains or airplanes. In a second stage tasks are derived (pairing) and the tasks are assigned to crew members (rostering). For school and university course scheduling the decomposition is done in the reverse order. First it is negotiated which teacher takes which courses. The second stage in which a schedule for the courses (and teachers) is produced is called timetabling. Timetabling is a difficult process due to the fact that many constraints have to be satisfied.

Timetabling problems are usually modelled by zero-one linear programs (see Schaerf [16] for a survey on timetabling). More recently, constrained logic programming languages have been used to model and to solve timetabling problems (Goltz and Matzke [12]). In this paper we model basic timetabling problems and combined assignment and basic timetabling problems as variants of the resource-constrained project scheduling problem (RCPSP) with time-dependent resource profiles or the multi-mode RCPSP.

We describe concepts which are useful to develop local search heuristics and enumerative methods for solving these problems. One of these concepts is to represent a dominant set of schedules by sequences of the activities (jobs, tasks, courses) which are to be scheduled and a method to calculate schedules associated with these sequences. Thus, local search methods can be performed by navigating through an appropriate set of sequences. Such representations are also useful in combination with enumerative methods. Another important concept is constraint propagation which allows one to reduce the search space. This concept is mainly used in connection with enumerative methods based on branchings.

The paper is organized as follows. In Section 2.1 the RCPSP with time-dependent resource profiles and some of its generalizations which are useful to

E. Burke and W. Erben (Eds.): PATAT 2000, LNCS 2079, pp. 277–293, 2001.

model timetabling problems are introduced. In Section 2.2 timetabling examples
are discussed. Section 3 deals with representations of schedules by sequences and
with earliest start schedules which can be constructed using these sequences.
Section 4 introduces the main concepts of constraint propagation. In Section 5
we discuss how local search heuristics can be implemented. In Section 6 two
branching concepts for enumeration methods are presented. The last section
contains some concluding remarks.

2   The Resource-Constrained Project Scheduling Problem

In this section we introduce the resource-constrained project scheduling problem
in a general way and illustrate with some examples how timetabling problems
can be modelled in this context. The models described in these examples are
simplifications of the real world. Depending on the actual situation usually other
constraints have to be added.

2.1   Problem Formulation

The *resource-constrained project scheduling problem* (RCPSP) with time-dependent resource profiles is a general scheduling problem which may be formulated
as follows.

We are given a time horizon T, n *activities* (tasks, jobs, operations, courses)
$j = 1, \ldots, n$ and r *resources* $k = 1, \ldots, r$, where the resource availabilities depend
on the time. The *availability* of resource k is given by a function $R_k(t)$, where
t enumerates the time periods $[t, t + 1[$ for $t = 0, \ldots, T - 1$. This includes the
nonavailability of resource k if we set $R_k(t) := 0$ for the relevant t-values. Activity
j must be processed for p_j time units without interruption. Thus, if we denote by
S_j the starting time of activity j, it completes at time $C_j = S_j + p_j$. During this
time period $[S_j, C_j[$ a constant amount of r_{jk} units of resource k $(k = 1, \ldots, r)$ is
occupied. Furthermore, *precedence constraints* C are defined between activities.
They are given by relations $i \rightarrow j \in C$, where $i \rightarrow j$ means that activity j cannot
start before activity i is completed, i.e. $S_i + p_i \leq S_j$ must hold.

A schedule $S = (S_j)_{j=1}^{n}$ is called *feasible* if

- in each time period t the total resource demand is less than or equal to the
  availability $R_k(t)$ of each resource type k, and
- the given precedence constraints are fulfilled.

We call a problem of finding a feasible schedule with completion times C_j such
that $C_j \leq T$ for $j = 1, \ldots, n$ a search problem or *feasibility problem*. A search
problem with threshold value T has a solution if and only if a schedule S exists
such that the *makespan*

$$C_{\max} := \max_{j=1}^{n} \{C_j\} = \max_{j=1}^{n} \{S_j + p_j\} \tag{1}$$

is not greater than T.

The RCPSP is usually formulated as the problem of finding a feasible schedule which minimizes the makespan. Other important objective functions besides (1) are based on cost functions $f_j(t)$ for the activities. One has to find a feasible schedule which minimizes the total costs $\sum_{j=1}^{n} f_j(C_j)$.

Associated with each activity j may also be a set $\mathcal{T}_j$ of τ_j disjoint time windows $[s_j^\nu, e_j^\nu[$ with $0 \leq e_j^\nu \leq s_j^{\nu+1} \leq T$ for $\nu = 1, \ldots, \tau_j - 1$ in which j can be processed. This can be modelled by an additional resource j' with $R_{j'}(t) = 1$ if t belongs to $\mathcal{T}_j$ and 0 otherwise. The single unit of this resource is only needed by activity j and ensures that j is processed in $\mathcal{T}_j$ only.

Since it may be very hard to find schedules respecting all constraints, we augment the time horizon T by overflow periods $T, T + 1, \ldots, T + p_{\max}$, where $p_{\max}$ denotes the largest processing time of all activities. Furthermore, we assume that in $[T, T + p_{\max}[$ sufficient resources are available, i.e. a schedule with starting times $S_j = T$ for all activities j is feasible. In this case a schedule is feasible for the original problem if and only if its makespan is not greater than T. If we consider a sum objective function and an activity is scheduled in $[T, T + p_{\max}[$, we penalize this infeasibility by adding a job-dependent positive penalty term (which is usually very large) to the objective function. Then a schedule is feasible if and only if the sum of all penalties is zero.

A resource k is called *disjunctive* if $R_k(t) \in \{0, 1\}$ for all $t = 0, \ldots, T - 1$. Otherwise it is called *cumulative*. If k is disjunctive, two activities i, j with $r_{ik} = r_{jk} = 1$ cannot be processed at the same time.

The assumption that each activity j must be processed within an interval $[S_j, S_j + p_j[$ may be relaxed by allowing preemptions. *Preemption* of an activity means that processing may be interrupted and resumed at a later time. In this case we allow preemptions and continuations of activities at integer times only.

A generalization of the RCPSP is the *multi-mode RCPSP* (MRCPSP). In the multi-mode case a set M_j of modes (i.e. processing alternatives) is associated with each activity j. In mode m the processing time of activity j is p_{jm}, it needs r_{jkm} units of resource k, and the costs are given by c_{jm}. One has to assign a mode to each activity and to schedule the activities in the assigned modes.

The described constraints may be generalized or further constraints may be added to the (multi-mode) RCPSP by introducing the following concepts.

Parallelity constraints. Two activities may be forced to be processed in parallel for at least one time unit.

Minimal and maximal time-lags. The precedence relations $S_i + p_i \leq S_j$ may be replaced by $S_i + d_{ij} \leq S_j$ where d_{ij} is an arbitrary integer. If $d_{ij} \geq 0$ holds, activity j cannot be started earlier than d_{ij} time units after the start of activity i (minimal start–start time-lag). On the other hand, if $d_{ij} < 0$ holds, activity j cannot start earlier than $|d_{ij}|$ time units before the starting time of activity i (maximal time-lag). Activities i and j may be processed in parallel if $d_{ij} < p_i$ and $d_{ji} < p_j$.

Separating times between activities. Given two activities i and j, the condition $S_i + d_{ij} \leq S_j$ may be replaced by the weaker condition that $S_i + d_{ij} \leq S_j$ or

$S_j + d_{ji} \leq S_i$ has to be satisfied. If $d_{ij} = p_i + l$ and $d_{ji} = p_j + l$ with $l \geq 0$, then we have the condition that between the completion time of one of the two activities i, j and the starting time of the other activity we must have a minimal separating time of l time units.

Bounding restrictions. Let B be a set of activities and U be a set of time periods. Then we may require that the activities in B are processed for at least l and for at most u time units within the time periods defined by U, where l and u are integers with $l \leq u$. We also may set $l = 0$ or $u = \infty$ which imposes no restriction.

Throughout this paper we assume that all data defining an instance of the RCPSP are integers.

2.2   Timetabling Examples

In the following we will illustrate how some problems related to timetabling can be modelled as resource-constrained project scheduling problems.

Example 1 : A high-school timetabling problem (de Werra [19], Schaerf [17]). There are m classes $i = 1, \ldots, m$, n teachers $j = 1, \ldots, n$, and T time periods $t = 0, \ldots, T - 1$. Each teacher and each class is available only in prespecified time periods. Furthermore, for each class i and each teacher j the number r_{ij} of lectures which teacher j must give to class i is known in advance. Each lecture takes one time unit.

One has to schedule the lectures in such a way that

- the availability constraints are satisfied,
- no teacher is involved in more than one lecture at a time, and
- no class is involved in more than one lecture at a time.

This problem is equivalent to an open-shop problem (see, for example, Brucker [2]), where the teachers correspond to the machines and the classes to the jobs. If all teachers and classes are available over the whole time horizon, this problem can be solved in polynomial time using bipartite matchings (Even et al. [11]). As soon as teachers or classes are unavailable in certain time periods, the problem becomes NP-hard [11].

In a corresponding RCPSP model, lectures correspond to activities and teachers and classes correspond to disjunctive resources. These resources are not available at certain time periods. A lecture can be given in period t only if the corresponding class and teacher are available in t. All activities have equal processing lengths (corresponding to the length of the time periods) and have to be scheduled without preemption.

This is only a basic high-school timetabling problem. In practice, many additional constraints have to be satisfied. Next we discuss some of these additional constraints.

(a) For a given set U_i of periods each class i must be necessarily involved in one lecture per period $t \in U_i$. Thus we have the bounding restriction that exactly $|U_i|$ lectures of class i must be scheduled in the time periods of U_i.

(b) For each teacher (or class) the number of teaching periods per day is bounded from above and below by given numbers. These restrictions can be modelled by bounding restrictions as well.

(c) Certain lectures must be taught in special rooms. To model this, one may introduce additional resources (one for each special room type). If additionally some lecture can be taught in different room types, this can be modelled by introducing multiple modes for the corresponding lecture. In this situation each mode corresponds to a different room type.

(d) Pairs of lectures require to be scheduled simultaneously. These requirements take care of various features that are usually present in many schools like laboratory assistants, shared gymnastic rooms, bilingual classes and others. This can be modelled by introducing corresponding parallelity constraints. □

Example 2: A university course timetabling problem (de Werra [19], Schaerf [16]).
There are q courses $i = 1, \ldots, q$. Course i consists of n_i lectures $L_{i1}, \ldots, L_{in_i}$. A curriculum S_k $(k = 1, \ldots, r)$ is a subset of courses which have common students. This means that the lectures of courses in S_k must be scheduled at different time slots. We assume that $\bigcup_{k=1}^{r} S_k = \{1, \ldots, q\}$, i.e. all courses are covered by the curricula. The total number of time slots is T and u_t is the maximal number of lectures which can be scheduled simultaneously in period t (i.e. the number of rooms available at period t).

In a corresponding RCPSP model the activities are the lectures L_{ij}. We have one cumulative resource "classroom" with a time-dependent resource profile u_t $(t = 0, \ldots, T-1)$. Furthermore, for each curriculum we introduce a disjunctive resource which avoids that different lectures of the same curriculum are scheduled in the same time period. The objective is to find a feasible schedule within the time horizon T. As in the high-school timetabling problem many additional constraints concerning the rooms or other features may be given.

Examination scheduling may be seen as a special case of university course timetabling. For each course exactly one examination must be scheduled. Thus, we have essentially a problem with $n_i = 1$ for $i = 1, \ldots, q$. However, the additional constraints of both problems are usually different. To avoid a clustering of examinations to be taken by the same students, one may introduce separating times between the examinations. □

Example 3: An audit-scheduling problem (Brucker and Schumacher [6]).
A set of jobs $J_1, \ldots, J_g$ must be processed by auditors $A_1, \ldots, A_m$. Job J_l consists of n_l tasks $(l = 1, \ldots, g)$. The tasks of all jobs are enumerated in the form

$$1, \ldots, n_1, n_1 + 1, \ldots, n_1 + n_2, \ldots, n_1 + n_2 + \cdots + n_g =: n.$$

There may be precedence constraints $i_1 \rightarrow i_2$ between tasks i_1, i_2 of the same job. Associated with each job J_l is a release time r_l, a due date d_l and a weight

w_l. Each task $i = 1, \ldots, n$ must be processed by exactly one auditor. If task i is processed by auditor A_j, then its processing time is p_{ij} and the corresponding mismatching costs are c_{ij}. If a task i cannot be processed by an auditor A_j, then we set $c_{ij} = \infty$. Auditor A_j is only available during disjoint time intervals $[s_j^\nu, e_j^\nu[$ $(\nu = 1, \ldots, m_j)$ with $e_j^\nu \leq s_j^{\nu+1}$ for $\nu = 1, \ldots, m_j - 1$. Furthermore, the total working time of A_j is bounded from below by a value H_j^- and from above by H_j^+ with $H_j^- \leq H_j^+$ $(j = 1, \ldots, m)$.

A task i assigned to auditor A_j can be preempted only in the following sense: if i is not finished at the end of an interval $[s_j^\nu, e_j^\nu[$, it must be continued at the beginning $s_j^{\nu+1}$ of the next interval of A_j.

We have to assign all tasks to the auditors and for each auditor to schedule the assigned tasks. A schedule for all auditors A_j is feasible if

- the tasks assigned to A_j are processed in the given time windows where in each time period at most one task is processed,
- the total workload of A_j is bounded by H_j^- and H_j^+,
- tasks are preempted in the restricted sense only,
- the precedence constraints are not violated, and
- tasks of J_l do not start before r_l.

Given a schedule (for all auditors), the finishing time C_l of J_l is the latest finishing time of all tasks of J_l. The objective is to assign each task i to an auditor $A_{\alpha(i)}$ and to construct a feasible schedule such that the sum of mismatching costs $c_{i\alpha(i)}$ and weighted sum of tardiness $T_l = \max\{0, C_l - d_l\}$, i.e.

$$\sum_{i=1}^n c_{i\alpha(i)} + \sum_{l=1}^g w_l T_l$$

is minimized.

The problem may be formulated as a MRCPSP with bounding restrictions and restricted preemptions.   □

3   Representation of Solutions

In order to apply an optimization algorithm to a problem at first a suitable problem representation has to be chosen. In this section we will describe list scheduling techniques known for project scheduling problems which may be used to generate non-preemptive or preemptive schedules for the RCPSP with time-dependent resource profiles.

We assume that the time-dependent resource profile of resource k is represented by pairs (t_k^μ, R_k^μ) for $\mu = 1, \ldots, m_k$, where $0 = t_k^1 < t_k^2 < \cdots < t_k^{m_k} = T$ and R_k^μ denotes the resource capacity in the time interval $[t_k^\mu, t_k^{\mu+1}[$. For each activity j let $\mathcal{K}(j)$ be the index set of all resources needed by j (including the disjunctive resource j' which takes into account the time window constraints).

We represent solutions by sequences (lists) of all activities which are compatible with the precedences, i.e. in the list i is placed before j if $i \to j \in C$

holds. With each sequence we associate an "earliest start schedule" by planning the activities in the order induced by the list. For each activity j we calculate the earliest time t where all predecessors are finished and sufficient resources are available. Then we schedule j in the interval $[t, t + p_j[$ and update the resource profiles. This process is summarized in the following procedure which calculates the corresponding non-preemptive earliest start schedule for a given sequence (say $(1, \ldots, n)$) of all activities.

PROCEDURE Earliest Start Schedule $(1, \ldots, n)$
1. FOR $j := 1$ TO n DO BEGIN
2. $t := \max_{i \to j \in C} \{S_i + p_i\}$;
3.   WHILE a resource $k \in \mathcal{K}(j)$ with $r_{jk} > R_k(\tau)$ for a $\tau \in [t, t + p_j[$ exists DO
4.   Calculate the smallest time $t_k^\mu > t$ such that j can be scheduled in
                     the interval $[t_k^\mu, t_k^\mu + p_j[$ if only resource k is considered and set $t := t_k^\mu$;
5.   Schedule j in the interval $[S_j, C_j[:= [t, t + p_j[$ and update the current
            resource profiles for all $k \in \mathcal{K}(j)$ in $[t, t + p_j[$;
     END

In each iteration we have to check the resource profiles at most once. Scheduling activity j creates at most two new intervals where the resource profile changes.

Thus, procedure Earliest Start Schedule can be implemented in such a way that it needs at most $O(\sum_{j=1}^{n} \sum_{k \in \mathcal{K}(j)} (m_k + 2(j - 1)) + |C|) = O(n^2 r + n \sum_{k=1}^{r} m_k)$ time, i.e. it runs in polynomial time. It remains to show that with this procedure a dominant set of schedules is achieved. We have to show that if a feasible solution exists, such a solution always is contained in the set of schedules which is obtained by applying the procedure to all possible sequences.

Theorem 1. *If a feasible solution for the considered problem exists, a sequence of all activities exists such that the procedure* Earliest Start Schedule *provides a feasible schedule.*

In order to prove the theorem we first observe that the described procedure, like the so-called "serial generation scheme" for the RCPSP (cf. Kolisch [13]), calculates only active schedules, i.e. schedules in which no activity can be started earlier without delaying some other activity. Furthermore, if a feasible schedule exists, we may take the sequence in which the activities are ordered according to non-decreasing starting times in this feasible schedule. Applying procedure Earliest Start Schedule to this sequence obviously provides a feasible schedule too.

The procedure can easily be extended to some generalizations of the described model such that Theorem 1 remains valid.

If minimal time-lags $d_{ij} \geq 0$ are given, we only consider sequences which are compatible with the induced precedences. In the scheduling procedure we only have to replace the values $S_i + p_i$ in Step 2 by the values $S_i + d_{ij}$. Separating times can be taken into account in a similar way. In the case of maximal time-lags it is an open question whether a set of dominant list schedules can be calculated in polynomial time.

Parallelity constraints often only occur in the case of equal processing times. In this case the corresponding activities can be combined to one activity with adapted resource requirements (either by adding the corresponding resource requirements or by taking the maximum of the requirements).

Bounding restrictions can only be incorporated to a certain extent. Assume that some sets of time intervals with upper bounds for the total processing time of activities scheduled in these intervals are given. A schedule satisfying these constraints can be obtained by updating the bounds for all intervals in each step of the scheduling procedure. If we have bounding restrictions with lower bound values greater than zero, obviously not necessarily a feasible active schedule exists since it may be advantageous to shift activities to the right. Thus, in this case Theorem 1 does not hold.

Another list scheduling procedure which is often used for scheduling problems also produces a dominant set of schedules for the described situation. Sprecher and Drexl [18] used such a procedure in a branch and bound algorithm (generating a so-called "precedence tree") for the RCPSP; Carlier and Neron [7] used it for multi-processor flow-shop problems. This procedure generates schedules where the starting times are ordered according to the given list, i.e. $S_1 \leq S_2 \leq \cdots \leq S_n$ holds. The resulting schedules are in general not active, but again it can be shown that if a feasible solution exists, it can be provided by the procedure. To implement this procedure, in Step 2 of the procedure `Earliest Start Schedule` we only have to set t equal to the maximum of $\max_{i \to j \in C} \{S_i + p_i\}$ and the starting time S_{j-1} of the previous activity. The complexity of the procedure reduces to $O(\sum_{k=1}^{r} (m_k + n) + |C|) = O(nr + n^2 + \sum_{k=1}^{r} m_k)$ since we never go back to previous time periods when we check the resource profiles.

Procedure `Earliest Start Schedule` can also be adapted to the case with restricted preemptions (like in the audit-scheduling problem from Example 3). In this situation an activity may be interrupted at the end of a time window $[s_j^\nu, e_j^\nu[$ and continued at the beginning $s_j^{\nu+1}$ of the next interval. To calculate a schedule with restricted preemption we modify Steps 4 and 5 of procedure `Earliest Start Schedule` in such a way that if it is not possible to plan activity j completely in the interval $[t, t + p_j[$, then we schedule j in a longest interval of the form $[t, t'[$ and continue starting at the smallest $s_j^\nu > t'$. This process is repeated until all p_j time units of j are processed.

4   Constraint Propagation

Constraint propagation techniques are used to reduce the search space in connection with local search procedures or branch-and-bound methods. They are also used to calculate lower bounds. In this section we apply such techniques to the RCPSP with time-dependent resource profiles.

4.1   Basic Concepts and Notations

We consider a resource-constrained project scheduling problem with r resources and corresponding resource profiles $R_k(t)$ for $k = 1, \ldots, r$. The objective is to find a feasible non-preemptive schedule for n activities $j = 1, \ldots, n$ defined by their processing times p_j and resource requirements r_{jk} such that $C_{\max} \leq T$, where T is a given threshold value (or to conclude that such a schedule does not exist). Each resource is either disjunctive or cumulative. It is convenient to introduce a *dummy start-activity* 0 and a *dummy end-activity* $n + 1$ requiring no resources with processing times $p_0 = p_{n+1} = 0$. Furthermore we have $0 \to j \to n + 1$ for all activities $j = 1, \ldots, n$. We represent a relevant schedule by the starting times $S = (S_j)_{j=0}^{n+1}$ of all activities where $S_0 = 0$ and $S_{n+1} \leq T$. In this case activity j is processed in the time interval $[S_j, S_j + p_j[$ which covers the time periods $[t, t+1[$ for $t = S_j, S_j + 1, \ldots, S_j + p_j - 1$. A *domain* Δ_j of activity j is a set of potential starting times of activity j. We may assume that

$$\Delta_j \subseteq D_j = \{t \mid r_{jk} \leq R_k(t) \text{ for all } k = 1, \ldots, r\}.$$

Δ_j can be represented by a set of disjoint intervals: a *first* and a *last interval* (which may be identical) and some *inner intervals*.

The purpose of constraint propagation is to eliminate time periods $[t, t+1[$ from the domains Δ_j. This is achieved by proving that no feasible schedule $S = (S_j)_{j=0}^{n+1}$ exists with $S_j = t$.

One method to reduce the domains Δ_j is to calculate an *earliest starting time* est_j and a *latest completion time* lct_j for activity j, i.e. a lower bound est_j for the starting time S_j and an upper bound lct_j for the completion time $S_j + p_j$ of activity j. In this case est_j is also called *head* and $T - lct_j$ is called *tail* of activity j. All periods $[t, t+1[$ with $t < est_j$ and $t > lct_j - p_j$ can be eliminated from Δ_j.

In Section 4.2 methods for calculating heads and tails are presented.

Another useful concept for constraint propagation are relations $i \to j$, $i \parallel j$, or $i - j$ between activities i, j. We write $i \to j \in C$ if in any feasible schedule with $C_{\max} \leq T$ activity j does not start before i is finished, i.e.

$$S_i + p_i \leq S_j \tag{2}$$

holds. $i \to j$ is called a *conjunction*.

The meaning of $i \parallel j$ is that activities i and j are processed in parallel for at least one time unit. This is the case if and only if neither $i \to j$ nor $j \to i$ holds, i.e. if both inequalities

$$S_i + p_i > S_j \text{ and } S_j + p_j > S_i \tag{3}$$

hold. $i \parallel j$ is called a *parallelity relation*.

The negation of $i \parallel j$ is a relation called *disjunction* and is denoted by $i - j$. $i - j$ means that either $i \to j$ or $j \to i$ holds.

An initial set of conjunctions is given by the set of precedence relations $i \to j \in C$. An initial set of disjunctions may be induced by resource constraints for pairs of activities. We set $i - j$ if for any $S_i \in \Delta_i, S_j \in \Delta_j$ with $[S_i, S_i + p_i[\cap[S_j, S_j + p_j[\neq \emptyset$ at $t \in [S_i, S_i + p_i[\cap[S_j, S_j + p_j[$ and a k exists with $r_{ik} + r_{jk} > R_k(t)$.

In Section 4.3 we discuss how additional relations can be derived from given ones. Finally, in Sections 4.4–4.6 we will show how interval inconsistency tests can be used to reduce domains.

4.2   Calculating Heads and Tails

In this section we describe how heads and tails can be calculated for the activities. A head r_j is a lower bound for the starting time S_j of activity j and a tail q_j defines a lower bound for the time period between the completion time C_j of j and the optimal makespan. Obviously, heads and tails can be adjusted in such a way that $r_j \geq r_i + p_i$ and $q_i \geq p_j + q_j$ hold for all conjunctions $i \rightarrow j \in C$. These constraints can be strengthened by regarding the resource constraints. An activity cannot start before all predecessors are completed and sufficient resources are available for processing. This observation can be used in a procedure similar to the procedure **Earliest Start Schedule** described in Section 3. We consider a sequence which is compatible with all precedence constraints and calculate heads for all activities in the order induced by the list. In each iteration the head r_j of the considered activity j is given by the smallest time $t \geq \max\limits_{i \rightarrow j \in C} \{r_i + p_i\}$ where all resources $k \in \mathcal{K}(j)$ are available to process j in the interval $[t, t + p_j[$. Contrary to the procedure **Earliest Start Schedule** we may not update the resource profiles since the order of all activities which are not linked by a conjunction is not fixed (i.e. they may be scheduled simultaneously for the lower-bound calculations). This procedure needs at most $O(\sum\limits_{j=1}^{n} \sum\limits_{k \in \mathcal{K}(j)} m_k + |C|) = O(n^2 + n \sum\limits_{k=1}^{r} m_k)$ time. Symmetrically, tails can be calculated by a backward recursion.

4.3   Start–Start Distance Matrix

Let $S = (S_j)_{j=0}^{n+1}$ be a feasible schedule with $S_0 = 0$ and $S_{n+1} \leq T$. Then we have $i \rightarrow j$ if and only if (2), i.e. $S_j - S_i \geq p_i$ holds. Furthermore, we have $i \parallel j$ if and only if (3) holds, which is equivalent to

$$S_j - S_i \geq -(p_j - 1) \text{ and } S_i - S_j \geq -(p_i - 1) \tag{4}$$

because all data are integers. Additionally, for arbitrary activities i, j we have

$$S_i + p_i \leq S_{n+1} \leq T \leq T + S_j \text{ or } S_j - S_i \geq p_i - T.$$

If we define

$$d_{ij} = \begin{cases} 0 & \text{if } i = j \\ p_i & \text{if } i \rightarrow j \\ -(p_j - 1) & \text{if } i \parallel j \\ p_i - T & \text{otherwise,} \end{cases} \tag{5}$$

then the entries of d_{ij} $(i, j = 0, \ldots, n+1)$ of the $(n+2) \times (n+2)$-matrix $D = (d_{ij})$ are lower bounds of the differences $S_j - S_i$, i.e.

$$S_j - S_i \geq d_{ij} \text{ for all } i, j = 0, \ldots, n+1. \tag{6}$$

We call D a *start–start time-lag matrix* (SSTL matrix).

If additionally time-lags d'_{ij} are given, i.e. $S_i + d'_{ij} \leq S_j$ or equivalently $S_j - S_i \geq d'_{ij}$ has to be satisfied, then we may incorporate these constraints by replacing d_{ij} by

$$d_{ij} := \max\{d_{ij}, d'_{ij}\}. \tag{7}$$

In this case all derived results are also valid for the RCPSP with time-lags.

The relation $S_j - S_i \geq d_{ij}$ has the following transitivity property:

$$S_j - S_i \geq d_{ij} \text{ and } S_k - S_j \geq d_{jk} \text{ imply } S_k - S_i \geq d_{ij} + d_{jk} \tag{8}$$

(we obtain the third relation by adding the first and second).

Due to (8) the matrix $D = (d_{ij})$ may be replaced by its transitive closure, which can be derived from D in time $O(n^3)$ by applying the Floyd–Warshall algorithm (see Ahuja et al. [1]).

The first row of an SSTL matrix represents heads because $S_i = S_i - S_0 \geq d_{0i}$ for each activity i. Similarly, the last column of an SSD matrix contains information on the tails of the activities. More precisely, $d_{i,n+1} - p_i$ is a tail of activity i because we have

$$S_{n+1} - (S_i + p_i) = (S_{n+1} - S_i) - p_i \geq d_{i,n+1} - p_i.$$

From (2) and (4) we derive

$$i \rightarrow j \text{ if and only if } d_{ij} \geq p_i \text{ holds}, \tag{9}$$

and

$$i \parallel j \text{ if and only if both } d_{ij} \geq -(p_j - 1) \text{ and } d_{ji} \geq -(p_i - 1) \text{ hold}. \tag{10}$$

If $d_{ii} > 0$ for some activity i, then no feasible schedule with $C_{\max} \leq T$ exists because $d_{ii} > 0$ implies the contradiction

$$0 = S_i - S_i \geq d_{ii} > 0.$$

There is another way to detect infeasibility. Assume that for some entry of the SSTL matrix we have $d_{ij} > 0$. Furthermore, let $d_{ji} \leq -d_{ij}$ (otherwise by transitivity we have $d_{ii} > 0$). In this case activity j must be started at least d_{ij} time units after S_i and cannot be started more than $-d_{ji}$ time units after S_i. Thus, we can eliminate from Δ_i all periods $[t, t + 1[$ with the property that due to the resource requirements activity j cannot be started in the interval $[t + d_{ij}, t - d_{ji}[$ if i is started at time t. If after such an elimination process Δ_i gets empty, then no feasible schedule exists.

Calculating the transitive closure of an SSTL matrix is the simplest form of constraint propagation. By increasing the d_{ij} values we may increase heads and tails, we may induce new relations by (9) or (10) and we may decrease domains. Other constraint propagation methods are discussed in the following sections.

4.4 Symmetric Triples

In this section we show how relations can be derived from given ones and some additional conditions. First we have to introduce some terminology.

An activity k is called *inconsistent* with activities i and j if for each $t \in [S_k, S_k + p_k[$ with $S_k \in \Delta_k$ a resource l exists such that condition $r_{il} + r_{jl} + r_{kl} > R_l(t)$ holds. i, j, k are inconsistent if one of these activities is inconsistent with the other two activities.

A triple (i, j, k) of activities is called *symmetric* if

- $k \parallel i$ and $k \parallel j$, and
- i, j, k are inconsistent.

For a symmetric triple (i, j, k) we must have $i - j$. This result can be derived easily.

Additional relations can be deduced in connection with symmetric triples (i, j, k):

1. If $l \parallel i$ and j, k, l are inconsistent, then $l - j$. If additionally $i \to j$ ($j \to i$) then $l \to j$ ($j \to l$).
2. Assume that $p_k - 1 \le p_i$, $p_k - 1 \le p_j$, $p_k - 1 \le p_l$. If i, k, l are inconsistent and j, k, l are inconsistent, then $k - l$ is a disjunction.
3. Suppose that the conditions $p_k - 1 \le p_i$ and $p_k - 1 \le p_l$ hold. Furthermore, assume that the precedence relations $l \to j$ and $i \to j$ are given and i, k and l are inconsistent. Then the additional conjunction $l \to k$ can be fixed.
4. Suppose that the conditions $p_k - 1 \le p_j$ and $p_k - 1 \le p_l$ hold. Furthermore, assume that the precedence relations $i \to j$ and $i \to l$ are given and j, k and l are incosistent. Then the additional conjunction $k \to l$ can be fixed.
5. Suppose that the conditions $p_k - 1 \le p_i$ and $p_k - 1 \le p_j$ hold. Furthermore, assume that the precedence relations $l \to i$ and $l \to j$ ($i \to l$ and $j \to l$) are given. Then we may fix the conjunction $l \to k$ ($k \to l$).
6. Let the parallelity relations $l \parallel i$ and $l \parallel j$ be given. Then the additional parallelity relation $l \parallel k$ can be fixed.

The proofs of these results are identical with the proofs in Brucker [3].

4.5 Disjunctive Resources

Let k be a disjunctive resource, i.e. $R_k(t) \in \{0, 1\}$ for all t. Then we denote by $A_k := \{i \mid r_{ik} > 0\}$ the set of all activities requiring resource k. Furthermore, we set $P(A) := \sum_{j \in A} p_j$ for $A \subseteq A_k$ and define for $u, v \in A_k$

$$d(u, v) := lct_v - est_u - |\{t \mid [t, t+1[\subseteq [est_u, lct_v[; R_k(t) = 0\}|$$

which is an upper bound on the number of time periods available for scheduling a set $A \subseteq A_k$ of activities if $u \in A$ is scheduled first and $v \in A$ is scheduled last. The following theorem generalizes a result in Dorndorf et al. [9].

Theorem 2. *(Sequence consistency): Let $A', A'' \subseteq A \subseteq A_k$. If*

$$\max_{\substack{u \in A\setminus A' \\ v \in A\setminus A'' \\ u \neq v}} d(u,v) < P(A), \tag{11}$$

then an activity in A' must start first or an activity in A'' must end last in A in any feasible schedule.

Proof. If no activity in A' starts first and no activity in A'' ends last, then the left-hand-side of (11) is an upper bound on the number of periods available for processing all activities in A. Thus, if (11) holds, no feasible schedule exists. □

By choosing different subsets A' and A'' we get different tests.

Input test. If condition (11) holds for $A' = \{i\} \subseteq A$ and $A'' = \emptyset$, then activity i must start first in A, i.e. we must have

$$i \rightarrow j \text{ for all } j \in A\setminus\{i\}.$$

In this case i is called *input* of A.

Output test. If condition (11) holds for $A' = \emptyset$ and $A'' = \{i\} \subseteq A$, then activity i must start last in A, i.e. we must have

$$j \rightarrow i \text{ for all } j \in A\setminus\{i\}.$$

In this case i is called *output* of A.

Input-or-output test. If condition (11) holds for $A' = \{i\}$ and $A'' = \{j\}$ with $i, j \in A$, then activity i must start first in A or activity j must start last in A, i.e. i is input for A or j is output for A.

Input/output negation test. If condition (11) holds for $A' = \{i\}$ and $A'' = A\setminus\{i\}$ with $i \in A$, then i must not start last in $A\setminus\{i\}$ (output negation).

The input-or-output test may be combined with the input negation test or with the output negation test leading to the conclusion $i \rightarrow j$ for all $i \in A\setminus\{j\}$ or $i \rightarrow j$ for all $j \in A\setminus\{i\}$.

Additional conjunctions $i \rightarrow j$ may be derived by these tests, which generally leads to smaller domains.

4.6 Cumulative Resources

In this section we generalize the concepts derived in the last section for disjunctive resources to the general case which includes cumulative resources as well. Let k be some resource. As before we define $A_k = \{i \mid r_{ik} > 0\}$. Then $w_i = r_{ik} p_i$ is the *work* needed to process activity $i \in A_k$. For $A \subseteq A_k$ we define $W(A) := \sum_{i \in A} w_i$. Using this terminology we can generalize Theorem 2.

Theorem 3. *Let $A', A'' \subseteq A \subset A_k$. If*

$$\sum_{\substack{\min_{u \in A\setminus A'} est_u \leq t < \max_{v \in A\setminus A''} lct_v}} R_k(t) < W(A), \tag{12}$$

then an activity in A' must start first or an activity in A'' must end last in A in any feasible schedule.

The proof of this theorem is similar to the proof of the previous theorem, because the left-hand side of (12) is an upper bound on the total work available for the activities in A if no activity in A' starts first and no activity in A'' ends last in A. Note that we dropped the restriction $u \neq v$ because, due to parallelity, an activity may start first and end last in A.

Again, input tests, output tests, input-or-output tests, and input/output negation tests can be derived.

4.7 Shaving

Let Δ_i be the (current) domain of activity i and let $[t_1, t_2[\subseteq \Delta_i$ be one of the intervals of Δ_i. If for some integers t'_1, t'_2 with $t_1 \leq t'_1 < t'_2 \leq t'_2$ we can show that no feasible schedule with $S_i \in [t'_1, t'_2[$ exists, then we may replace Δ_i by $\Delta_i \backslash [t'_1, t'_2[$. This process, which may be applied iteratively, is called *shaving*. Shaving has been introduced in connection with the job-shop scheduling problem by Martin and Shmoys [15].

5   Local Search Methods

In this section we describe how local search methods can be used to calculate heuristic solutions for the RCPSP with time-dependent resource profiles. As explained in Section 3 solutions can be represented by sequences which are compatible with all conjunctions (the given precedences and the relations derived by constraint propagation). By applying the procedure Earliest Start Schedule to a sequence a corresponding schedule is calculated. To cope with the hard problem of finding feasible solutions in local search procedures it should be allowed to search through infeasible regions. Infeasibilities are penalized by an objective function and the goal is to find a solution where all penalties are zero.

In order to search systematically through the search space all neighbourhoods can be used in which sequences are changed. The simplest one is the adjacent pairwise interchange (API) neighbourhood where adjacent activities in the list are interchanged (if they are not linked by a conjunction). A larger neighbourhood contains shifts of activities to other positions in the sequence (when the shift is compatible with the conjunctions).

Based on the representation by sequences several local search algorithms like simulated annealing, tabu search or genetic algorithms can be applied. For an overview on heuristic methods for the RCPSP and a computational comparison of different algorithms see Kolisch and Hartmann [14]. A disadvantage of the representation by sequences is the fact that many sequences represent the same schedule, i.e. the same schedule can be generated several times.

For the more general situation in which different modes are given for the activities (MRCPSP), as well as the scheduling problem an assignment problem has to be solved. This problem can be solved by a two-stage local search procedure where on the first level the assignment is changed and on the second level the resulting scheduling problem is treated. Such an approach has been applied to the audit-scheduling problem from Example 3 (Brucker and Schumacher [6]).

6   Enumeration Procedures

In this section we review enumeration procedures for constructing a feasible schedule or to prove that no such schedule exists. We restrict our attention to the basic RCPSP model with time-dependent resource profiles. Additional restrictions can be incorporated without difficulties. Thus, a schedule is feasible if and only if the following conditions are satisfied:

- the precedence constraints are fulfilled,
- its makespan is not greater than T, and
- for each period $t = 0, \ldots, T - 1$ the resources needed in period t are not greater than $R_k(t)$ for $k = 1, \ldots, r$.

We will discuss in more detail two procedures based on two different branching schemes:

- disjunctive branching, and
- time-oriented branching.

The disjunctive branching scheme can only be applied if all resources are disjunctive. In both schemes the nodes of the enumeration tree correspond to sets of schedules which are split into disjoint subsets by the branching process.

In each node of the enumeration tree, constraint propagation techniques are applied to reduce the search space or to prove that no corresponding feasible solution exists. In the latter case no further investigations for the node are necessary.

6.1   Disjunctive Branching

The disjunctive branching scheme has been successfully applied to job-shop and open-shop problems (Carlier and Pinson [8], Brucker et al. [5], Dorndorf et al. [10]).

A node of the enumeration tree corresponds to a set C of conjunctions $i \rightarrow j$ which contains the conjunctions given by the precedence constraints and additional conjunctions fixed during the branching process. Furthermore, additional conjunctions which are derived by constraint propagation procedures are added to C if no contradiction is derived. If a contradiction is derived, the node is abandoned. Otherwise a disjunction $i - j$ (i.e. $i \neq j$ with $r_{ik} = r_{jk} = 1$ for some resource k) with $i \rightarrow j \notin C$ and $j \rightarrow i \notin C$ is chosen and either $i \rightarrow j$ or $j \rightarrow i$ is added to C. The child nodes $C \cup \{i \rightarrow j\}$ and $C \cup \{j \rightarrow i\}$ are treated recursively. If no disjunction $i - j$ with $i \rightarrow j \notin C$ and $j \rightarrow i \notin C$ exists and the directed graph induced by C has no cycles, a corresponding schedule is constructed by applying the Procedure `Earliest Start Schedule` to a sequence of all activities which is compatible with C. If this schedule is feasible, we are finished. Otherwise we have to continue the search at some other branch of the enumeration tree. If all branches lead to infeasibility, then no feasible solution exists.

6.2   Time-Oriented Branching

The time-oriented branching applies to general RCPSPs with time-dependent resource profiles. A node of the enumeration tree is defined by the domains Δ_j of all activities $j = 1, \ldots, n$. Branching is done by splitting the domain of one activity into disjoint sets. At the beginning one should split the domains into their intervals. Thereafter, when all domains consist of intervals, these intervals should be further split. The leaves of the enumeration tree correspond to one-element domains Δ_j for all $j = 1, \ldots, n$. They define a feasible or infeasible schedule. The procedure stops if a feasible schedule is found or all schedules represented by the nodes are infeasible.

6.3   Further Methods

Both disjunctive branching and time-oriented branching can be mixed. Other search methods developed for the RCPSP try to build a (feasible) schedule from left to right by scheduling step-by-step one activity (see Brucker et al. [4]).

7   Concluding Remarks

We described a general model of a resource-constrained project scheduling problem with time-dependent resource profiles and showed how basic timetabling problems can be formulated within this framework. However, the presented timetabling examples are simplifications of the real world. Usually, additional constraints have to be considered depending on the actual situation. Some of these constraints can be covered by considering the more general multi-mode RCPSP. Others should be handled as soft constraints which can be incorporated into the objective function in many cases.

We also presented concepts to derive exact and heuristic methods for these types of problems and discussed how constraint propagation, a very promising tool, can be applied in connection with these methods.

References

1. Ahuja, R., Magnanti, T., Orlin, L.: Network Flows: Theory, Algorithms and Application. Prentice-Hall, Englewood Cliffs, NJ (1993)
2. Brucker, P.: Scheduling Algorithms. Springer-Verlag, Berlin Heidelberg New York (1998)

3. Brucker, P.: Complex Scheduling Problems Osnabr"ucker Schriften zur Mathematik, Reihe P, Heft 214 (1999)
4. Brucker, P., Drexl, A., M"ohring, R., Neumann, K., Pesch, E.: Resource-Constrained Project Scheduling: Notation, Classification, Models and Methods. Eur. J. Oper. Res. **112** (1999) 3–14
5. Brucker, P., Jurisch, B., Sievers, B.: A Branch & Bound Algorithm for the Job-Shop Problem. Discrete Appl. Math. **49** (1994) 107–127
6. Brucker, P., Schumacher, D.: A New Tabu Search Procedure for an Audit-Scheduling Problem. J. Scheduling **2** (1999) 157–173
7. Carlier, J., Neron, E.: An Exact Method for Solving the Multi-Processor Flow-Shop. RAIRO Oper. Res. **34** (2000) 1–25
8. Carlier, J., Pinson, E.: An Algorithm for Solving the Job-Shop Problem. Manage. Sci. **35** (1989) 164–176
9. Dorndorf, U., Phan Huy, T., Pesch, E.: A Survey of Interval Capacity Consistency Tests for Time- and Resource-Constrained Scheduling. In: Węglarz J. (ed.): Handbook on Recent Advances in Project Scheduling. Kluwer, Dordrecht (1999) 213–238
10. Dorndorf, U., Pesch, E., Phan Huy, T.: Solving the Open Shop Scheduling Problem. J. Scheduling (1999) (to appear)
11. Even, S., Itai, A., Shamir, A.: On the Complexity of Timetabling and Multicommodity Flow Problems. SIAM J. Comput. **5** (1976) 691–703
12. Goltz, H.-J., Matzke, D.: University Timetabling Using Constraint Logic Programming. In: Gupta G. (ed.): Practical Aspects of Declarative Languages. Lecture Notes in Computer Science, Vol. 1551. Springer-Verlag, Berlin Heidelberg New York (1999) 320-334
13. Kolisch, R.: Serial and Parallel Resource-Constrained Project Scheduling Methods Revisited: Theory and Computation. Eur. J. Oper. Res. **90** (1996) 320–333
14. Kolisch, R., Hartmann, S.: Heuristic algorithms for the Resource-Constrained Project Scheduling Problem: Classification and Computational Analysis. In: Węglarz J. (ed.): Handbook on Recent Advances in Project Scheduling. Kluwer, Dordrecht (1999) 147–178
15. Martin, P.B., Shmoys, D.B.: A New Approach to Computing Optimal Schedules for the Job Shop Scheduling Problem. Proc. 5th Int. IPCO Conference (1996)
16. Schaerf, A.: A Survey of Automated Timetabling. CWI-Report CS-R 9567. Amsterdam (1995)
17. Schaerf, A.: Tabu Search Techniques for Large High-School Time-Tabling Problems. CWI-Report CS-R 9611. Amsterdam (1996)
18. Sprecher, A., Drexl, A.: Solving Multi-mode Resource-Constrained Project Scheduling Problems by a Simple, General and Powerful Sequencing Algorithm. Eur. J. Oper. Res. **107** (1998) 431–450
19. de Werra, D.: An Introduction to Timetabling. Eur. J. Oper. Res. **19** (1985) 151–162

Graph Colouring by Maximal Evidence Edge Adding*

Barry Rising[1], John Shawe-Taylor[1], and Janez Žerovnik[2]

[1] Department of Computer Science, Royal Holloway, University of London, Egham, Surrey TW20 0EX, UK
{barry,jst}@dcs.rhbnc.ac.uk
[2] Faculty of Mechanical Engineering, University of Maribor, Smetanova 17, 2000 Maribor, Slovenia and IMFM, Department of Theoretical Computer Science, Jadranska 19, Ljubljana, Slovenia
janez.zerovnik@uni-lj.si

Abstract. Graph colouring is one of the most studied NP-hard problems. Many problems of practical interest can be modelled as colouring problems. The basic colouring problem is to group items in as few groups as possible, subject to the constraint that no incompatible items end up in the same group. Classical examples of applications include timetabling and scheduling [25]. We describe an iterative heuristic algorithm for adding new edges to a graph in order to make the search for a colouring easier. The heuristic is used to decide which edges should be added by sampling a number of approximate colourings and adding edges which have fewest conflicts with the generated colourings. We perform some analysis of the number of approximate colourings that might be needed to give good bounds on the probability of including an edge which increases the chromatic number of the graph. Experimental results on a set of "difficult graphs" arising from scheduling problems are given.

Keywords: graph colouring, maximal evidence, edge adding, scheduling problems

1 Introduction

The problem of creating an educational timetable for examinations or lectures and classrooms has been studied over a number of years with the aim of devising efficient practical algorithms that work effectively for the sizes of instances arising in schools or universities. Since it is known that even in its simplest form the timetabling problem is NP-complete, this search has focused on heuristic approaches frequently imported from related problem areas (see, for example, Schaerf [31] for a survey). Timetabling is also related to the more general problem of scheduling.

---

* Work supported by a British Council Exchange Grant with Slovenia, ALIS No. 27

E. Burke and W. Erben (Eds.): PATAT 2000, LNCS 2079, pp. 294–308, 2001.

Some of the earliest methods drew on techniques from graph colouring, see for example Welsh and Powell [36], Wood [37] or Burke et al. [4], who show an application to university timetabling. Meta-heuristics were also used to give further improvements in timetabling automation [6]. Burke et al. [5] subsequently hybridized meta-heuristics with graph colouring approaches.

The current paper introduces the meta-heuristic of adding additional constraints based on probabilistic inference from partial solutions. The effectiveness of this approach is tested on difficult graph colouring instances arising from large scheduling problems that are therefore closely related to timetabling problems.

A *colouring* of a graph $G = (V(G), E(G))$ is any mapping $c : V(G) \rightarrow C$, where C is the set of "colours", for convenience usually a subset of the natural numbers. Unless otherwise stated we will assume that $V(G) = \{1, \ldots, n\}$. We say that a colouring c is *proper*, if there is no pair $(u, v) \in E(G)$ of adjacent vertices of the same colour $(c(u) = c(v))$. A graph G is *k-colourable* if there is a proper colouring with $card(C) \leq k$ colours. The minimal k for which there is a proper k-colouring is the *chromatic number* of G, denoted $\chi(G)$. The decision problem of k-colouring is:

INPUT: graph G, integer k

QUESTION: is G k-colourable?

The problem is NP-complete for general graphs [18], but also for many special classes of graphs including planar graphs [17], regular graphs [40], etc. It is also known that there is no polynomial approximation algorithm with a guaranteed performance bound unless P = NP [26].

Graph colouring is one of the most studied problems of combinatorial optimization. Many problems of practical interest can be modeled as colouring problems. The general form of these applications involves forming a graph with vertices representing items of interest. The basic colouring problem is to group items in as few groups as possible, subject to the constraint that no incompatible items end up in the same group. In special cases, there may be some additional constraints or costs. Classical examples of applications include timetabling and scheduling: see, for example, [10], [25], [27], [28], [32].

Experimental studies include [7], [22], [9], [30], [21]. The techniques reported in [30] have been extended in a series of papers [38], [39], [34], [2], [3], which have both analysed and improved the heuristics. In [2] the energy landscape for mean field annealing is altered by adapting weights on the graph edges. The current paper investigates an extension of this idea by introducing additional edges into the graph. This may seem a counter-intuitive strategy as it might appear to make the graph colouring problem more difficult. We will see in the next section, however, that this is not necessarily the case.

One way of viewing the strategy is in terms of the way that humans often attempt to solve problems by further constraining them based on probabilistic reasoning from known constraints, hence reducing the size of the search space, while hopefully not excluding all feasible solutions.

2   Background to the Algorithm

Edwards [11] showed that when restricted to the class of graphs with lowest vertex degree greater than αn for arbitrary $\alpha > 0$, the decision problem of 3-colouring is polynomial.

Further experimental support for the assertion that dense graphs are easier to colour comes from [30], [9]. A random k-colourable graph $G_{n,p,k}$ graph is an n vertex graph, whose vertices are divided into k independent sets of size n/k. Edges are included between pairs of vertices in different sets independently with probability p. When applying a graph colouring algorithm to the random graphs $G_{n,p,k}$, a so-called critical region of density was detected. Experimental results show that graphs with

$$\frac{2np}{k} \sim 4.8 \tag{1}$$

are difficult to colour [30], [39], [34]. The algorithm used in locating this difficult region was due to Petford and Welsh [30]. The algorithm works by starting from a random colouring and repeatedly updating vertices selected at random until a proper colouring has been found. A selected vertex is updated by counting for each colour i the number of neighbours $S(i)$ coloured with that colour. The new colour for the vertex is then selected to be colour i with probability proportional to

$$4^{-S(i)} = \exp(-S(i)/0.72).$$

Note that the value $T = 0.72$ is referred to as the standard temperature. In the experimental section this value will be treated as a parameter to the algorithm. More recently, further independent experimental confirmation of this result has been obtained using very different adaptive evolutionary algorithmic techniques [12]. The $G_{n,p,k}$ graphs are, however, readily colourable with an adaptive temperature version of the Petford–Welsh algorithm [35], which we now explain briefly. Initially, a distribution of temperatures is specified from which temperatures are drawn at random. For each chosen temperature the Petford–Welsh algorithm is run for a specified number of iterations. The number of miscoloured edges is used to give an exponentially decaying weighting to the chosen temperature. As the number of restarts increases, the distribution used to generate temperatures becomes centred on the weighted average of the temperatures used so far, while the deviation is gradually reduced. At the same time, the length of each run is increased by 1% for each newly chosen temperature. This small adaption to the Petford–Welsh algorithm ensures that all the ·randomly generated graphs $G_{n,p,k}$ with $n = 450$ are coloured in relatively few restarts and hence proved uninteresting for testing the algorithm proposed in this paper.

Although the variable temperature version appears more effective, the temperature learned will differ for different graphs. If we were to take the union of two difficult graphs for which efficient colouring requires different temperatures, the learning algorithm may fail since it cannot run one temperature for one part of the graph and a different temperature for the second part. Indeed, the results in Figure 2 show that the effectiveness of the standard algorithm can be sensitive to temperature, so that choosing a compromise temperature for the two parts may not be a feasible option.

In this paper we investigate an alternative strategy, which appears to overcome the difficulty of choosing the correct temperature. We use the constant temperature annealing algorithm as a subroutine to a main program that attempts to "boost" the performance by combining the information gained in separate runs. We give experimental evidence to show that by using this technique the effect of the temperature parameter becomes far less severe, and that over a wide temperature range the performance matches that of the best a posteriori chosen temperature.

It seems clear that random graphs of very low density are easy to colour, since they are with high probability not connected, so every component can be coloured separately. Edwards's result suggests why dense random graphs can usually be coloured quickly.

There is some theoretical evidence that repeated trials of various random heuristics (including simulated annealing [24], mean field annealing [29] and the Petford–Welsh algorithm [30]) on average perform better than one long run [14].

However, at each restart we lose all information obtained by previous unsuccessful trials, which may have obtained reasonably good near-optimal colourings.

Therefore, a possible way of accelerating the repeated trials algorithm is to somehow accumulate the information obtained by the near optimal colourings obtained so far. The idea of using the information of previous local searches is not new. In a different way it is explored by heuristics such as tabu search [19], [20], reactive tabu search [1], and genetic algorithms [15].

The two observations above give an intuitive explanation for the approach we use. First, a number of near optimal colourings are generated by a randomized algorithm (in our case, this is the Petford–Welsh generalized algorithm). After some time, when the information obtained so far is sufficient according to some measure, we decide to add edge(s) to the graph which join pairs of vertices that are least likely to be coloured the same and try to colour the new graph. There are two things we wish to point out: since edges are added the graph gets denser and hence in view of Edwards's result potentially "easier" to colour. However, it is possible that by unlucky edge adding the chromatic number of the new graph is larger and in this case the procedure fails.

3   The Algorithm

In the algorithm described below we make use of the following quantities which are defined for the various colourings generated. Each colouring c is assigned a weight $w_G(c)$ given by

$$w_G(c) = \exp(-b(c)/T_w),$$

where $b(c)$ is the number of edges in G that join vertices coloured with the same colour by c, and T_w is a parameter which defines the particular weighting function. Note that $w(c) \leq 1$ for all c, while $b(c) = 0$ and $w(c) = 1$ for proper colourings. Colourings with $b(c) > 0$ are called approximate colourings, where we assume that the number of available colours has been fixed beforehand.

The complement of a graph $G = (V, E)$ is the graph

$$\bar{G} = (V, (V \times V) \setminus (E \cup \{(i, i) | i \in V\})).$$

For a set of colourings C, we define the evidence for a non-edge $(x, y) \in \bar{G}$ to be

$$E_{G,C}((x, y)) = \sum_{c \in C, c(x) \neq c(y)} w_G(c).$$

We also define the evidence of a set of colourings as

$$E_G(C) = \sum_{c \in C} w_G(c).$$

We use a generalized annealing algorithm as a subroutine which returns a colouring for our algorithm. We denote a call to this procedure by

$$c := PW(G, T, M),$$

where T is the (fixed) temperature used in the updates [34] and M is a specified number of iterations to be performed. In some experiments this routine will be the Petford–Welsh algorithm, while in others it will be a hardware approximation to it [3]. When a number of calls have been executed, the evidence in favour of the non-edges and the non-edge with maximal evidence is inserted into the graph – see Figure 1. In the first version, we take the first edge among edges with the same maximal evidence (the first according to the "lexicographic order" in some arbitrary numbering of the vertices). This was often found to improve the performance of the algorithm when compared with choosing an edge at random. There appeared to be some benefit in concentrating on one part of the graph. The iteration condition in these experiments was $E_G(C) < E_0$.

In the second version, all edges with maximal evidence are added and the iteration condition was a fixed number of 25 iterations.

4   Estimating the Probabilities for Edge Adding

In this section we perform some analysis of the number of approximate colourings that might be needed to give good bounds on the probability of including an edge which increases the chromatic number of the graph.

Throughout this section we make the natural assumption that approximate colourings are drawn independently according to some fixed distribution. A colouring c will be called incompatible with adding the edge (u, v), if $c(u) = c(v)$. For two vertices u, v which are not adjacent in the graph G, let a_{uv} denote the probability that an approximate colouring is incompatible with the edge (u, v).

By $G \cup \{(u, v)\}$ we denote the graph obtained from G by adding the edge (u, v). If $\chi(G \cup \{u, v\}) > \chi(G)$ we say that (u, v) is a bad (non) edge, otherwise (u, v) is a good edge. We denote the set of bad edges by $B(G)$. Note that the increase in chromatic number can be at most one.

```
C := ∅
while G ≠ Kₙ do begin
    while iteration condition do begin
        c = PW(G, T, 200 * n);
        if wG(c) = 1 then success
        else C := C ∪ {c};
    end;
    e := argmax e∈Ḡ{EG,C(e)};
    G := (V, E ∪ {e});
end;
failure
```

Fig. 1. Basic edge adding algorithm

Lemma 1. *Fix $\alpha > 0$. The probability over a sample of m random colourings that the fraction of incompatible colourings is less than $a_{uv} - \alpha$ for any pair of vertices u, v not adjacent in G is bounded by*

$$n(n-1)\exp(-\alpha^2 m/(2\alpha_{uv}))/2 \leq n(n-1)\exp(-\alpha^2 m/2)/2,$$

where $n = |V(G)|$ is the number of vertices in the graph G.

Proof. The probability of a rate β or better being observed for a particular pair u, v (i.e. the number of colourings for which u and v are assigned the same colour is less than or equal to βm) is given by

$$\sum_{i=0}^{\beta m} \binom{m}{i} (a_{uv})^i (1 - a_{uv})^{m-i} \leq \exp(-(a_{uv} - \beta)^2 m/2a_{uv})$$

using Chernoff bounds [8] to obtain the inequality. Hence, if $\beta = a_{uv} - \alpha$ we obtain the bound
$$\exp(-\alpha^2 m/2a_{uv}) \leq \exp(-\alpha^2 m/2).$$
The union bound of probabilities over all possible non-edges gives the result. □

Corollary 1. *Suppose that for a graph G with $\chi(G) = k$ the probability that an approximate colouring is incompatible with a bad edge is A times the probability for a random colouring, then with probability greater than $1 - \delta$ adding any edge for which the observed relative frequency is less than $1/(2k)$ will not increase the chromatic number, provided*

$$m \geq \frac{8Ak}{(2A-1)^2} \left\{ \ln \binom{n}{2} + \ln \frac{1}{2\delta} \right\}.$$

If the observed relative frequency is 0 *then with probability greater than* $1 - \delta$
adding any edge will not increase the chromatic number provided

$$m \geq \left\{ \ln \binom{n}{2} + \ln \frac{1}{\delta} \right\} / \ln \left(\frac{k}{k - A} \right).$$

Proof. The frequency of incompatible random colourings is $1/k$, since this is the fraction of random assignments that give the same colour to both ends of an edge. By assumption we have $a_{uv} \geq A/k$ for all bad edges (u, v). Therefore, we can bound the probability that a bad edge has an observed relative frequency of incompatibility less than $1/(2k)$ using Lemma 1 with α satisfying $\alpha_{uv} - \alpha \leq 1/(2k)$. Hence,

$$\frac{\alpha_{uv} - \alpha}{\alpha_{uv}} \leq \frac{1}{2A} \quad \Rightarrow \quad \frac{\alpha}{\alpha_{uv}} \geq 1 - \frac{1}{2A}$$

$$\Rightarrow \quad \frac{\alpha^2}{\alpha_{uv}^2} \geq \left(1 - \frac{1}{2A} \right)^2 \quad \Rightarrow \quad \frac{\alpha^2}{\alpha_{uv}} \geq \frac{(2A - 1)^2}{4Ak}.$$

We require m such that $\binom{n}{2} \exp(-m(2A - 1)^2/(8Ak)) \leq \delta$. Solving for m gives the first result. If $\alpha_{uv} \geq A/k$, then the probability that the observed relative frequency is 0 for a bad edge can be bounded by

$$\binom{n}{2} \left(1 - \frac{A}{k} \right)^m.$$

Setting the right-hand side equal to δ and solving gives the second result.   □

In practice we observe that many edges are compatible with all approximate colourings. Corollary 1 indicates that, provided the approximate colourings do increase the incompatibility of bad edges by twice that of random colourings, then with probability at least $1 - \delta$ such edges can safely be added provided

$$m \geq \left\{ \ln \binom{n}{2} + \ln 1/\delta \right\} / \ln \left(\frac{k}{k - 2} \right).$$

Table 1 gives some relevant values of these numbers. Note how few extra colourings are needed to give very significant increases in confidence or to apply to very much larger graphs.

We have so far considered the case where there is assumed to be a bound on the probabilities a_{uv} for all bad edges (u, v) and the observed relative frequency of incompatibility on the added edges are very low. This will be the case if there are at least some good edges for which a_{uv} is also low. These restrictions may not hold in all cases, and the strategy of only including the maximal evidence edges is intended to extend the applicability of the technique. The next corollary shows that using the maximal evidence criterion does indeed make it possible to weaken the conditions on the values of a_{uv}.

Table 1. Numbers of colourings required to give confidence in edge adding

n	k	δ	$m(n,k,\delta)$
450	5	0.1	28
450	5	0.01	32
450	5	0.001	37
450	15	0.1	97
450	15	0.01	113
450	15	0.001	129
100 000	5	0.001	58

Corollary 2. *If there is at least one good edge (u,v) for which*

$$a_{uv} < \min_{(u',v')\in B(G)} \{a_{u'v'}\} - \alpha,$$

for $\alpha > 0$, then with probability $1 - \delta$ the maximal evidence edge will be a good edge provided the number m of approximate colourings satisfies

$$m \geq \frac{8}{\alpha^2}\left\{2\ln n + \ln\frac{1}{2\delta}\right\}.$$

Proof. If the difference between $a_{u'v'}$ and the corresponding observed relative frequency is less than $\alpha/2$ for all edges (u',v'), then the maximal evidence edge will be a good edge, because no bad edge will have observed relative frequency better than (u,v) as $a_{uv} + \alpha/2 < a_{u'v'} - \alpha/2$, for all bad edges (u',v'). By Lemma 1 the probability that the difference between observed relative frequency and $a_{u'v'}$ is less than $\alpha/2$ for all non-edges is bounded by

$$n(n-1)\exp(-\alpha^2 m/8)/2.$$

By choosing m as in the corollary statement this value is made less than δ as required. □

Hence, the heuristic will work provided that there is at least one edge which can be separated from the bad edges by a polynomially small gap. This leads to the following further corollary.

Corollary 3. *Assume that $RP \neq NP$. For any polynomial time-randomized approximate colouring algorithm there exists a sequence of graphs G_n, with $|V(G_n)| = n$ for which the difference*

$$\min_{(u,v)\in B(G)} \{a_{uv}\} - \min_{(u,v)\in E(G)} \{a_{uv}\}$$

decays faster than $1/P(n)$ for any polynomial $P(n)$.

Proof. If no such sequence existed for some randomised colouring algorithm $\mathcal{A}$ then we could produce a randomised polynomial algorithm for the NP-complete problem of graph colouring as follows. Let $P(n)$ be a polynomial which satisfies the condition that

$$\min_{(u,v)\in B(G)} \{a_{uv}\} - \min_{(u,v)\in E(G)} \{a_{uv}\} > 1/P(n)$$

for all graphs G with $|V(G)| = n$. Note that for a graph with no good edges we assume that the colouring algorithm always delivers a correct colouring. This is a trivial addition to the algorithm if it does not already implement it, since there is no choice of different colours up to permutation of their values. Set $\alpha = 1/(2P(n))$ and set m according to the estimate of Corollary 2 with $\delta = 1/n^2$:

$$m = 8P(n)^2 \{4\ln n - \ln 2\}.$$

Perform approximate colourings before adding the single maximal evidence edge. Continue until either a colouring is found or too many edges have been added to the graph (e.g. the graph is complete). If the graph is coloured pronounce the original graph k-colourable, otherwise answer not k-colourable. Since the separation between bad and good edges is at least $1/P(n)$, with probability $1 - \delta$ the edges added were consistent with a k-colouring provided the original graph was k-colourable. Hence, with probability at least $1/2$ the final graph remains k-colourable if the original graph was k-colourable. On the other hand, if the original graph was not k-colourable the final graph will also not be. The running time of the algorithm is polynomial since a polynomial number of edges is added and at each stage a polynomial number of approximate colourings is generated. □

This final result suggests that an interesting avenue of research might be to identify classes of graphs for which a polynomial separation does exist for some well studied simple graph colouring algorithms.

The results described in this section show that under apparently mild assumptions about the quality of the algorithm generating the sample colourings, a very modest number of colourings is sufficient to ensure that with high confidence the edges added by the edge adding algorithm will not in fact increase the chromatic number. The next section attempts to test how effective the approach is in practical colouring problems.

5   Experimental Results

As mentioned earlier, the random graphs with known chromatic number are very efficiently coloured already by the Petford–Welsh algorithm and its variants [30], [39], [38], [34], [35].

We have also investigated the possibility of using the graph tre-s-92 described in [13]. For the unweighted version the standard Petford–Welsh algorithm without edge adding proved successful for a range of temperatures $T = 0.3$ to 0.46, slightly lower than the standard $T = 0.72$. Hence, for this graph the edge adding technique was not required. We further adapted the Petford–Welsh algorithm to respect the additional constraint of 655 on the size of the independent set. Again, the algorithm succeeded in finding an optimal 25-colouring over a wide range of temperatures, including the standard $T = 0.72$.

In view of the success of the standard Petford–Welsh algorithm for these problems, we have only reported experiments for a set of four five-colourable Leighton graphs. We should emphasize that the aim of the paper is not to contend that the presented algorithm is necessarily the best over a range of graphs, but rather to demonstrate that the edge adding technique can significantly improve the performance and robustness of an already very competitive algorithm.

5.1   Five-Colourable Leighton Graphs

The algorithms were run on a series of four "Leighton" graphs [25] which have chromatic number equal to 5. Colouring these graphs corresponds to solving a large scheduling problem. They were proposed as part of the DIMACS Challenge in graph colouring [23] as examples of graphs that are hard to colour.

In the first set of experiments the parameters T, T_w and E_0 were set to 1.0, 3.0 and 1.0 respectively and the Petford–Welsh algorithm was implemented. The update temperature of 1.0 had been chosen as giving very good performance on these graphs and hence represents an a posteriori choice. As a comparison we also ran the Petford–Welsh algorithm in two regimes. In the first instance we performed repeated short runs of the algorithm with the same number of iterations as used in the edge adding algorithm. In the second regime one long run was performed. Note that in both cases the same fixed temperature was used. The numbers in Table 2 indicate the total number of vertex updates performed whether or not the colour of the vertex was changed. For each application of the algorithms to each of the four graphs 10 experiments were performed starting with different random seeds. At the bottom of each section a total number of iterations is given as well as a final total at the bottom of the table.

The average degrees of the four graphs are also given in the first column of the table. It can be seen that they are all above the average degree of 19.2 predicted by equation (1) to correspond to hard graphs.

We have also included the standard deviation of the number of iterations over the 10 runs as this gives some indication of the reliability with which a particular algorithm will deliver a result without using many more iterations than the average. The edge adding algorithm is also competitive when measured by this criterion. For the parallel version, a larger total number of iterations was needed. But it is interesting to note that in all cases, though a large number of edges was added, an optimal colouring was found. This experimentally confirms that the edges with maximal evidence are not in conflict with optimal colouring.

The results for the Petford–Welsh algorithm in Table 2 represent a benchmark against which the multiple edge adding algorithm will be compared. Note, however, that even for the single edge adding there are improvements over standard restarts, particularly for graph 5a and 5b, with the overall average number of iterations being fewer by 78%. In the experiments presented here, long runs on average did better than restarts, but worse than the single edge adding regime.

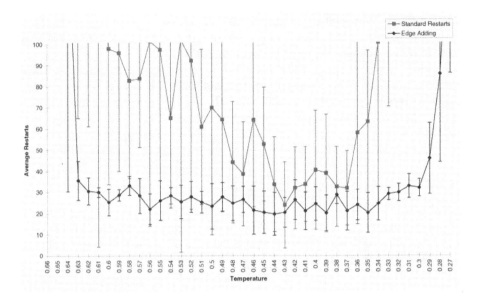

Fig. 2. Average numbers of restarts for standard MSBSN algorithm and edge adding at different temperatures

The main thrust of this paper is, however, to show that edge adding can make the underlying algorithm less sensitive to the choice of temperature. In order to investigate this question the multi-state bit-stream network was used as the base colouring algorithm. This hardware device behaves similarly to the Petford–Welsh algorithm by performing a hard summation of the conflicting neighbours but probabilistically excluding a subset of these vertices [3]. As with the probabilistic update of the Petford–Welsh algorithm, this has the effect of introducing some randomness into the choice of colour. The temperature of the machine determines the probability that a neighbour is ignored, so that in this case the lower the temperature the more randomness is introduced. A restart consists of 10 000 parallel iterations (80 000 iterations in total on 8 processors – note that in Table 2 iterations refers to the total number and not the number of parallel iterations). The network was checked at every 1000 parallel iterations.

The algorithm was run for a wide range of temperatures in two regimes. Firstly, no edge adding was performed, but in the second regime after 25 standard restarts all the maximal evidence edges were added at once. After this, standard restarts were continued with the extended graph but no more edges were added. Figure 2 plots the average number of iterations taken to find a solution in the two regimes against the temperature for the graph 5d. The error bars correspond to plus/minus one standard deviation. Figure 2 shows how, on average, the standard restart algorithm performs well between the temperatures 0.44 and 0.37. In contrast, the edge adding adaptation performs comparably to the best temperature (0.43) restart runs over the much wider range between 0.63 and 0.3, and furthermore the standard deviation is in all cases much smaller than that obtained for standard restarts.

It is interesting to note that despite using only 25 initial colourings, a value that would appear to be at the limit of reliability according to Table 1, there was not one occasion in all of the experiments when a bad edge was added and the graph failed to be coloured with 5 colours. This suggests that our estimate of bad edges having a probability of being incompatible with a colouring of at least twice the probability for a random colouring was conservative. The factor A may in fact be much larger, though the analysis is complicated by the fact that more than one edge is being added at once.

It is clear that for the graph 5d the edge adding technique has succeeded in reducing the sensitivity of the underlying colouring algorithm to the choice of temperature.

In Table 2, the total number of iterations and the number of edges added are given for one non-optimal setting of the temperature parameter. This version is referred to as the parallel edge adding algorithm with MSBSNs. The results are given for all four 5-colourable Leighton graphs and they show that the MSBSN is competitive with the Petford–Welsh algorithm run with a temperature that has been chosen a posteriori. The comparison is particularly favourable if we discount the initial 25 iterations used to accumulate evidence about which edges to add – these values are given in the column headed "iterations after 25".

6 Conclusions

The paper has described an iterative heuristic algorithm for adding new edges to a graph in order to make the search for a colouring easier. The heuristic is used to decide which edges should be added by sampling a number of approximate colourings and adding edges which have fewest conflicts with the generated colourings. Hence, the algorithm can be viewed as a bootstrapping method for improving the quality of an approximation algorithm.

Experimental results demonstrate that the method is effective on a set of difficult-to-colour graphs arising from scheduling problems. The average degree of the graphs is above that predicted by equation (1) to correspond to the hardest graphs. The important property of the technique highlighted in this paper is that multiple edge adding appears to reduce the sensitivity of the underlying algorithm to the choice of temperature.

The results presented here have shown how the meta-heuristic of adding constraints suggested by a probabilistic analysis of approximate solutions proved effective in the case of graph colouring. We believe that the approach holds promise in other constraint satisfaction problems and in particular for timetabling.

Some adaptations may be needed to take advantage of the type of constraints that can arise in these problems, so that some tuning of the approach will undoubtedly be required. That said, the proposed approach has a clear intuitive appeal as adding constraints that are likely to be required can significantly reduce the search space around those solutions that warrant investigation.

Table 2. Results of the algorithms on the 5-colourable Leighton graphs (PW: Petford–Welsh)

	Try	Edge adding (PW) Iterations (×10³)	No. of edges	Parallel adding (MSBSN) Iterations (×10³)	No. of edges	Iterations after 25 (×10³)	Standard restarts (PW) Iterations (×10³)	No. of restarts	One run (PW) Iterations (×10³)
le450_5a:	1 –	73	0	2328	36849	328	73	0	73
Average	2 –	698	3	2168	36222	168	862	9	3730
degree: 25.40	3 –	304	0	2032	38413	32	304	3	352
	4 –	73	0	2024	37468	24	73	0	73
	5 –	104	0	5528	35640	3528	104	1	265
	6 –	407	0	2232	35656	232	407	4	773
	7 –	177	0	2016	38448	16	177	1	139
	8 –	2063	12	2784	37498	784	1555	17	831
	9 –	168	0	2384	36662	384	168	1	1202
	10 –	1735	12	2080	37969	80	3546	39	129
	Total –	**5802**		**25576**		**5576**	**7268**		**7568**
	S.D. –	687		10697			1039		1056
le450_5b:	1 –	1783	16	3064	39200	1064	5640	62	1256
Average	2 –	5217	49	3304	37822	1304	8628	95	3841
degree: 25.48	3 –	1323	0	2008	36958	8	1323	14	2292
	4 –	2720	15	4192	34571	2192	3775	41	415
	5 –	2061	15	2808	37456	808	4947	54	1478
	6 –	4072	37	4264	36838	2264	2571	28	298
	7 –	873	0	3872	38086	1872	873	9	372
	8 –	3127	25	3736	36503	1736	2737	31	841
	9 –	1219	0	2552	34915	552	1219	13	206
	10 –	1287	4	3056	35633	1056	2495	27	1069
	Total –	**23682**		**32856**		**12856**	**34209**		**12068**
	S.D. –	1341		7323			2286		1072
le450_5c:	1 –	25	0	768	0	—	25	0	25
Average	2 –	387	0	1384	0	—	387	4	172
degree: 43.57	3 –	148	0	1240	0	—	148	1	2130
	4 –	1284	5	2968	41954	968	1269	14	6780
	5 –	4209	28	376	0	—	2095	23	3596
	6 –	667	0	2168	43130	168	667	7	2131
	7 –	819	1	1936	0	—	819	9	1101
	8 –	345	0	2728	42642	728	345	3	153
	9 –	666	0	1792	0	—	666	7	779
	10 –	1022	4	1504	0	—	1022	11	1079
	Total –	**9572**		**20248**		—	**7444**		**17946**
	S.D. –	1144		9040			579		1970
le450_5d:	1 –	4083	41	2008	42345	8	1676	18	4668
Average	2 –	220	0	1224	0	—	220	2	2451
degree: 43.36	3 –	2357	6	392	0	—	7612	84	453
	4 –	390	0	352	0	—	390	4	2653
	5 –	348	1	672	0	—	1383	15	215
	6 –	575	1	784	0	—	575	6	990
	7 –	64	0	2488	41480	488	64	0	64
	8 –	237	0	2032	41953	32	237	2	1303
	9 –	2515	0	2416	42862	416	2515	27	1002
	10 –	250	0	2008	41851	8	250	2	308
	Total –	**11038**		**14376**		—	**14921**		**14108**
	S.D. –	1309		8430			2177		1377
Grand total:		**47579**		**93056**		—	**60663**		**51690**

References

1. Battiti, R., Tecchiolli, G.: The Reactive Tabu Search. ORSA J. Comput. **6** (1994) 126–140

2. Burge, P., Shawe-Taylor, J.: Adapting the Energy Landscape for MFA. J. Artif. Neural Networks **2** (1995) 449–454

3. Burge, P., Shawe-Taylor, J.: Bitstream Neurons for Graph Colouring. J. Artif. Neural Networks **2** (1995) 443–448

4. Burke, D.K., Elliman, D.G., Weare, R.F.: A University Timetabling System Based on Graph Colouring and Constraint Manipulation. J. Res. Comput. Ed. **26** (1993)

5. Burke, E., Newall, J.: A Multi-stage Evolutionary Algorithm for the Timetabling Problem. IEEE Trans. Evol. Comput. **3** (1999) 63–74

6. Carter, M.W., G. Laporte, G.: Recent Developments in Practical Course Timetabling. In: Burke, E., Ross, P. (eds): Lecture Notes in Computer Science, Vol. 1153. Springer-Verlag, Berlin Heidelberg New York (1996) 3–21

7. Chams, M., Hertz, A., de Werra, D.: Some Experiments with Simulated Annealing for Colouring Graphs. Eur. J. Oper. Res. **32** 260–266

8. Chernoff, H.: A Measure of Asymptotic Efficiency of Tests of a Hypothesis Based on the Sum of Observations. Ann. Math. Stat. **23** (1952) 493–507

9. Culberson, J.C.: Iterated Greedy Graph Coloring and the Difficulty Landscape. Technical Report TR 92-07, University of Alberta, Canada (1992) (http://web.cs.ualberta.ca/ joe/index.html/)

10. de Werra, D.: An Introduction to Timetabling. Eur. J. Oper. Res. **19** (1985) 151–162

11. Edwards, K.: The Complexity of Colouring Problems on Dense Graphs. Theor. Comput. Sci. **43** (1986) 337–343

12. Eiben, A.E., van der Havl, J.K., van Hemert, J.I.: Graph Coloring with Adaptive Evolutionary Algorithms. J. Heuristics **4** (1998) 25–46

13. Erben, W.: A Grouping Genetic Algorithm for Graph Colouring and Exam Timetabling. Proc. 3rd Int. Conf. on the Practice and Theory of Automated Timetabling (2000) 397–421

14. Ferreira, A.G., J. Žerovnik, J.: Bounding the Probability of Success of Stochastic Methods for Global Optimization. J. Comput. Math. Appl. **25** (1993) 1–8

15. Fleurent, C., Ferland, J.A.: Genetic and Hybrid Algorithms for Graph Coloring. Ann. Oper. Res. **63** (1996) 437–461

16. Garey, M.R., Johnson, D.S.: The Complexity of Near-Optimal Graph Colouring. J. Assoc. Comput. Machinery **23** (1976) 43–49

17. Garey, M.R., Johnson, D.S., Stockmeyer, L.: Some Simplified NP-complete Graph Problems. Theor. Comput. Sci. **1** (1976) 237–267

18. Garey, M.R., Johnson, D.S.: Computers and Intractability. Freeman, San Francisco, CA (1979)

19. Glover, F.: Tabu Search – Part I. ORSA J. Comput. **1** (1989) 190–206

20. Glover, F.: Tabu Search – Part II. ORSA J. Comput. **2** (1990) 4–32

21. Hertz, A., de Werra, D.: Using Tabu Search Techniques for Graph Coloring. Computing **39** (1991) 378–406

22. Johnson, D.S., Aragon, C.R., McGeoch, L.A., Schevon, C.: Optimization by Simulated Annealing: an Experimental Evaluation. II: Graph Coloring and Number Partitioning. Oper. Res. **39** (1991) 378–406

23. Johnson, D.S., Trick, M. (eds.): Cliques, Coloring and Satisfiability: 2nd DIMACS Implementation Challenge. American Mathematical Society (1996). DIMACS colouring benchmarks available via ftp at rutgers.dimacs.edu

24. Kirckpatrick, S., Gelatt, C.D., Vecchi, M.P.: Optimization by Simulation Annealing. Science **220** (1983) 671–680

25. Leighton, F.T.: A Graph Coloring Algorithm for Large Scheduling Problems. J. Res. Natl Bur. Standards **84** (1979) 489–506

26. Lund, C., Yanakakis, M.: On the Hardness of Approximating Minimization Problems. J. Assoc. Comput. Machinery **41** (1994) 960–981
27. Neufeld, G.A., Tartar, J.: Graph Coloring Conditions for the Existence of Solutions to the Timetable Problem. Commun. ACM **17** (1974) 450–453
28. Opsut, R.J., Roberts, F.S.: On the Fleet Maintenance, Mobile Radio Frequency, Task Assignment and Traffic Phasing Problems. In: Chatrand, G., Alavi, Y., Goldsmith, D.L., Lesniak-Foster, L., Lick, D.R. (eds.): The Theory and Applications of Graphs. Wiley, New York (1981) 479–492
29. Peterson, C., Anderson, J.R.: A Mean Field Annealing Theory Learning Algorithm for Neural Networks. Int. J. Neural Systems **1** (1987) 995–1019
30. Petford, A., Welsh, D.: A Randomised 3-colouring Algorithm. Discrete Math. **74** (1989) 253–261
31. Schaerf, A.: A Survey of Automated Timetabling. Artif. Intell. Rev. **13** (1999) 87–127
32. Schmidt, G., Ströhlein, T.: Timetable Construction – an Annotated Bibliography. Comput. J. **23** (1979) 307–316
33. Shawe-Taylor, J., Žerovnik, J.: Boltzmann Machine with Finite Alphabet. In: Artificial Neural Networks 2, Vol. 1. Int. Conf. on Artif. Neural Networks (Brighton) (1992) 391–394.
34. Shawe-Taylor, J., Žerovnik, J.: Analysis of the Mean Field Annealing Algorithm for Graph Colouring. J. Artif. Neural Networks **2** (1995) 329–340
35. Shawe-Taylor, J., Žerovnik, J.: Adapting Temperature for some Randomized Local Search Algorithms. Preprint 614. Dept Math., University of Ljubljana (1998)
36. Welsh, D.J.A., Powell, M.B.: An Upper Bound for the Chromatic Number of a Graph and its Applications to Timetabling Problems. Comput. J. **10** (1967) 85–86
37. Wood, D.C.: A System for Computing University Examination Timetables. Comput. J. **11** (1968) 41–47
38. Žerovnik, J.: A Randomised Heuristical Algorithm for Estimating the Chromatic Number of a Graph. Inform. Process. Lett. **33** (1989) 213–219
39. Žerovnik, J.: A Randomized Algorithm for k-colorability. Discrete Math. **131** (1994) 379–393
40. Žerovnik, J.: Regular Graphs are "Difficult" for Colouring. Informatica **17** (1993) 59–63

Modelling Timetabling Problems with STTL

Jeffrey H. Kingston

Basser Department of Computer Science,
The University of Sydney, 2006,
Australia
jeff@cs.usyd.edu.au

Abstract. STTL is a language for specifying and evaluating timetabling problems, instances and solutions. An interpreter for STTL is freely available on the Internet. After a brief overview of the language, this paper explores the issues involved in the application of STTL to real-world problems, in particular the modelling of timetabling resources, time and meetings, and the evaluation of solutions. High school timetabling is used as a case study.

1 Introduction

The timetabling research community needs a standard format for specifying and evaluating timetabling problems, instances and solutions. As discussed at length in a previous paper [3], such a format must be *general* (able to express all kinds of timetabling problems), *complete* (able to express them in full detail, including the criteria by which solutions are to be evaluated), and *accessible* (easy to translate to and from).

In Section 2 we briefly describe such a format, or language, which we call STTL. It could be used for modelling any complex assignment-type problem, not just timetabling. An interpreter for STTL has been written by the author [4]. It is freely available on the Internet, in both Eiffel and C versions. STTL is very close to a previous proposal [3], with some cosmetic changes that reflect the author's preferences, and some refinements suggested by practical experience.

The author has used STTL to specify the high school timetabling problem following [2], and to specify one sample instance and solution. This case study is the basis of Sections 3–6, which explore the use of STTL in modelling timetabling entities, and in evaluating solutions.

2 The STTL Language

This section is a brief overview of STTL. Space precludes a definitive account here; this may be found in a separate User's Guide [5].

STTL is an object-oriented functional language. Objects are the natural means of defining timetabling entities such as times, resources and meetings, while functions allow arbitrary requirements to be expressed clearly. STTL also offers a few specialized features for defining assignment-type problems.

STTL has the usual *BOOLEAN, INTEGER, REAL,* and *STRING* built-in types, with the usual operations and the usual syntax. There are also two built-in types, *SET[T]* and *SEQ[T]*, denoting sets and sequences of any type *T*. There are set comprehensions in the

E. Burke and W. Erben (Eds.): PATAT 2000, LNCS 2079, pp. 309–321, 2001.

usual notation, such as $\{1, 3, 5, 7\}$, and sequence comprehensions also, using brackets [] instead of braces.

Functions in STTL have a familiar form:

contains(*s*: *SET*[*INTEGER*], *x*: *INTEGER*): *BOOLEAN* =
(
   if *s* = [] **then**
     false
   elseif *x* = **head** *s* **then**
     true
   else
     contains(**tail** *s*, *x*)
   end
)

This function determines whether or not *s* contains *x*. In practice one would use the built-in **member** operator to do this.

There are *scan expressions* which avoid recursion in simple cases. For example,

concatenation *x* **in** [1, 2, 3, 4, 5] **suchthat** *x* >= 3 **of** [*x* * *x*]

has value [9, 16, 25]. For each *x* in the set or sequence after **in** such that the condition after **suchthat** evaluates to **true**, the expression after **of** is evaluated and the values accumulated according to the initial keyword (in this case, concatenated). Concatenation, union, sum, product and a few other operators may be used to accumulate the values.

STTL offers the classes of object-oriented programming, and they are extremely useful for defining timetabling entities. In fact, our first example is *ENTITY*:

class *ENTITY*

   name: *STRING*

functions

   defects: *SEQ*[*DEFECT*] = []

end

Each entity has a name, which is a string, and can calculate a sequence of timetabling defects concerning itself (Section 6).

Class attributes can be of three kinds: *creation variables*, *solution variables* and *functions*. Creation variables appear first in a class declaration; so in *ENTITY*, *name* is a creation variable, and *defects*, appearing as it does after the **functions** keyword, is a function. Solution variables appear after any creation variables and before any functions, preceded by the **solution** keyword. They are like creation variables, except that they are initialized differently: at creation time they receive the special value ?, meaning 'undefined,' and (as we will see below) they are assigned values by solutions.

Inheritance is used extensively in STTL. For example, a resource is a kind of entity, and a teacher is a kind of resource:

class *TEACHER* **inherit** *RESOURCE*

max_class_load: INTEGER
unavailable_times: SET[TIME]
cats: SET[TEACHER_CATEGORY]

functions

defects: SEQ[DEFECT] =
(
    defect_check(TeacherClash, clashes) +
    defect_check(TeacherOverload, overload) +
    defect_check(TeacherDayOverload, day_overload)
)

end

This is only part of the *TEACHER* class from the high school case study, but it shows *TEACHER* inheriting *RESOURCE*, adding three creation variables, and redefining the *defects* function to be the concatenation of three sequences of defects.

To create an object of some class, we write the class name followed by a sequence of expressions in parentheses, whose values are assigned to the creation variables in order of declaration. For example,

TEACHER("Knott", 22, { }, {English})

is an expression whose result is a new *TEACHER* object, with *name* (inherited from *EN-TITY*) set to *"Knott"*, *max_class_load* set to 22, and so on. STTL is a purely functional language, so there is no way to change the values of these variables later.

Although the *TEACHER* class is a part of the high school timetabling problem, particular teachers, such as Knott, are part of some instance of the problem. We would expect to see something like this in an instance:

Knott = TEACHER("Knott", 22, { }, {English})

This defines a global variable, *Knott*, and assigns it the value after the equals sign. Again, this value cannot be changed subsequently. An instance of a timetabling problem is just a sequence of such global declarations, of times, teachers, rooms, meetings, and so on.

The usual dot notation is used: *Knott.max_class_load* returns 22, *Knott.defects* returns a sequence of Knott's defects. The expression

all *TEACHER*

returns the set of all *TEACHER* objects which appear in global declarations, so

sum *t* **in all** *TEACHER* **of** *t.max_class_load*

for example returns the total class load of all teachers. These **all** expressions are the only way that information from instances can be accessed within problems.

We have referred to problems, instances, and solutions; we now make these terms precise, and introduce the features of STTL that support them. A *problem* is a general question to be answered, such as the high school timetabling problem or the examination timetabling problem. An *instance* of a problem is a particular case of that problem, such as the case of Bankstown Girls' High School in 1998, or the University of Sydney examinations in November 1997. A *solution* of an instance is a set of decisions that solves that instance (possibly not very well).

An STTL program begins with a problem, arranged as follows:

STTL problem *HighSchool_CK*

  declarations

evaluation

  expression

end problem

The declarations are an arbitrary mixture of the class declarations and function and variable declarations seen earlier. The **evaluation** clause will be explained shortly.

After the problem, and typically kept in a separate file, comes an instance:

STTL instance *Bankstown*1998 **of problem** *HighSchool_CK*

  HistoryStaffMeeting = STAFF_MEETING(2, {Smith, Jones, Robinson})
  ...

end instance

Instance declarations are limited to global variables, and the expressions are limited to 'immediate' (similar to literal) ones, making instances easier to read than problems. Instances declare objects representing all their times, resources, and meetings.

The specific question posed by the problem and instance together may be expressed as follows: find an assignment of values to the solution variables of the global objects of the instance, which produces a good value of the expression following the **evaluation** keyword in the problem. A proposed solution may appear after the instance, like this:

STTL solution *S*1 **of instance** *Bankstown*1998 **of problem** *HighSchool_CK*

  HistoryStaffMeeting.times := {Mon02, Mon03}
  ...

end solution

The meaning is the obvious one: assign the values on the right-hand sides of the assignment symbols := to the corresponding solution variables named on the left-hand sides. Solutions, like instances, may use only immediate values.

After all these assignments have been made, STTL evaluates the expression following the **evaluation** keyword in the problem, and its value becomes the official assessment of the solution. The author's implementation is able to enter an interactive mode after this, in which the user types expressions which STTL evaluates immediately.

3 Modelling Resources

The participants in a timetable are conventionally divided into *times* and *resources*, where resources are teachers, rooms, students (or groups of students), and so on. It is an ancient observation [1] that the various kinds of resources have similar roles: all attend meetings and must avoid clashes. We prefer not to call time a resource, because the fundamental "no clashes" requirement distinguishes between times and resources.

In the high school model that we use in this paper as a case study, these observations are realized by an inheritance hierarchy. The root class is *ENTITY* from Section 2; it has two child classes: *TIME* which will be discussed in Section 4, and *RESOURCE*. The *RESOURCE* class contributes no variables, but it is a convenient place to put functions for calculating the set of all meetings attended by a resource, the times the resource attends those meetings, and the number of clashes among these times, since these functions are the same for all resources.

The children of *RESOURCE* are *TEACHER*, *ROOM*, and *STUDENT_GROUP* (our high school model timetables groups of students, not individual students). In addition to their name, inherited from *ENTITY*, *TEACHER* objects have three other creation variables: a maximum load, a set of unavailable times and a set of *categories*, which are kinds of classes they are qualified to teach. So instances will declare teachers like this:

Knott = TEACHER("Knott", 22, { }, {English})
Jackson = TEACHER("Jackson", 30, { }, {English, History})

ROOM objects have categories too (Science labs, etc.), but are always available:

r01 = ROOM("r01", {DramaRoom})
r02 = ROOM("r02", {ScienceLab})

Student groups need only their names:

Yr7C = STUDENT_GROUP("Yr7C")
Yr7K = STUDENT_GROUP("Yr7K")

Teacher and room categories are defined by trivial classes:

English = TEACHER_CATEGORY("English")
History = TEACHER_CATEGORY("History")

These are essentially enumerated types, and contain only their own name.

4 Modelling Time

Time seems to be more difficult to model than resources are. Time flows continuously, whereas resources are discrete entities; some timetables are one-off, others repeat periodically, every week or two weeks perhaps. In this section we sketch a fully general model of time, and then present the much more limited one that has been implemented.

The most general model of time possible is one based on intervals of real time, such as 'From 9.30am on 17 November 1999 to 12.30pm on 18 November 1999.' This model applies in the obvious way to one-off timetables, but it also applies to periodic timetables if we consider a weekly meeting to occupy its time interval(s) in all the actual weeks of real time that the timetable is in use. Because this model is the real story, it allows for curiosities such as classes that run every second week in a timetable that has mostly weekly periodicity, or timetables for two semesters where some classes run in just one semester and others run in both.

It would be quite easy to implement this model literally in STTL, using minutes elapsed since some reference moment to represent moments in real time, pairs of moments to represent intervals, and sets of intervals to represent the times occupied by meetings. A function could calculate the amount of overlap between two sets of intervals, for use in clash detection.

If a timetable is one-off or strictly periodic (that is, containing no meetings that do not repeat exactly in every period), we only need to represent one period and the implementation just sketched would be quite efficient. For the periodicity curiosities, though, a literal implementation would require meetings to store all their recurrences, resulting in a highly redundant representation which would be very slow to calculate with, and perhaps not amenable to convenient input and output. In those cases it might be better to use an inheritance hierarchy of time sets; one child class could represent one interval, another could represent a periodic repetition of a set of intervals, and so on. Another idea is to divide time hierarchically, say into semesters, weeks, days, and time intervals within days. A time value might look like

$$TIME(\{2\}, INTERVAL(1, 14), \{Mon\}, \{INTERVAL(t(9, 30), t(12, 30))\})$$

meaning Semester 2 only, weeks 1 to 14, Mondays 9.30 to 12.30.

In view of the complexity in practice of this general model, and the author's lack of experience of complex time models, a much simpler and more limited model has been used in the high school case study. There are *DAY* objects containing a *name* variable inherited from *ENTITY*, and an *ord* field for sequencing the days. A typical instance would declare global variables

$Monday = DAY("Monday", 1)$
$Tuesday = DAY("Tuesday", 2)$

and so on. Times are declared similarly; they contain an ordinal number, their day of the week, and whether or not they are followed by a break:

$Mon1 = TIME(Mon1, 1, Monday, \textbf{false})$
$Mon2 = TIME(Mon2, 2, Monday, \textbf{false})$
$Mon3 = TIME(Mon3, 3, Monday, \textbf{false})$
$Mon4 = TIME(Mon4, 4, Monday, \textbf{true})$

and so on. Each *TIME* represents a particular time interval on some day; the intervals are disjoint, have equal length, and follow each other immediately in the order of the *ord* variables, except where a break is flagged. There are functions in the *TIME* class which determine whether this time is the first or last in the day, whether it forms a double with a given other time:

> *double_with(other: TIME): BOOLEAN = (other.ord = ord +* 1 **and not**
> *break_follows*)

and so on.

The models of time presented in this section sample a spectrum which ranges from comprehensive and complex at one end to limited and simple at the other. Whether this spectrum should be implemented by a single model of time, or whether it will be necessary to offer several, is a question for further experience to answer.

5   Modelling Meetings

A *meeting* is a set of time intervals and a set of resources, whose meaning is that the resources are all occupied (unavailable for other meetings) during the given time intervals.

Alternative definitions are often given which restrict a meeting to just one time interval or just one student group, teacher, and room. However, there are compelling reasons for the more general definition: it makes it easy to require sets of times to be spread through the week in a certain way, to require the same teacher to be assigned at all of the times, and so on.

All meetings require times (at least, those that do not are of no interest in timetabling), but different kinds of meetings (staff meetings, ordinary classes, etc.) require different kinds of resources. Accordingly, in the high school case study there is an inheritance hierarchy whose root class, *MEETING*, is concerned only with naming the meeting and with its times:

> **class** *MEETING* **inherit** *ENTITY*
>
> > *times_required: INTEGER*
> > *times_condition: TIMES_CONDITION*
>
> **solution**
>
> > *times: SEQ[TIME]*
>
> **end**

The classes presented in this section have additional functions devoted to evaluation, which will be discussed later (Section 6) and are omitted here.

This *MEETING* class contains our first solution variable: *times*, a sequence of times (following the simple model of time from Section 4), representing the times when this meeting occurs. The use of *SEQ* rather than *SET* will be justified below. Declaring *times* to be a solution variable means that instances cannot set this variable in the meetings they declare; only solutions can do so, by means of assignments such as

*Year*11*English.times* := [*Mon*01, *Tue*04, *Thu*03, *Fri*06]

However, instances constrain the times chosen by solutions by setting the creation variables, *times_required* and *times_condition*: a timetable will be considered defective if the times chosen do not conform to the requirements expressed by these variables.

In our experience, no computer program is ever asked to choose the duration of a meeting, so we provide the *times_required* variable to allow meetings to ask for the specific number of times they require. In addition, *times_condition* expresses some more complex condition that the meeting's times should satisfy. In the high school case study the particular conditions available are:

FreeChoice	Times may be chosen freely
Usual	No double times, all times on different days, and *last*
Double	One double time, other times on different days, and *last*
TwoDouble	Two double times, other times on different days, and *last*
ThreeDouble	Three double times, other times on different days, and *last*
OneTriple	One triple time, other times on different days, and *last*
SubsetOf(*m*)	Times to be a subset of the times of meeting *m*
SameAs(*m*)	Times to be the same as the times of meeting *m*
FixedTimes(*s*)	Times must equal the fixed sequence *s*; no choice allowed

In this table, *last* means "at most one of the times may be last in its day". It is considered unfair to give a class two last times, since the students are tired and restless at the end of the day. It is very easy to extend this list, since each element is just a child class of *TIMES_CONDITION* containing only its creation variables (if any) plus one function, for evaluating a given sequence of times against the condition. That is fortunate, because more conditions will certainly need to be added: even spread between morning and afternoon, some times fixed but others obeying some condition, and so on. Although we want to avoid the full generality of allowing logic expressions in instances, it may well be that a meeting should hold a *set* of time conditions, not just one, meaning that every condition in the set should be satisfied.

In the high school case study, *MEETING* has two child classes: *STAFF_MEETING*, which models staff meetings by adding a creation variable holding the set of teachers who are to attend, and *CLASS* for ordinary classes:

class *CLASS* **inherit** *MEETING*

> *students*: *SET*[*STUDENT_GROUP*]
> *teachers_requirements*: *SEQ*[*TEACHER_REQUIREMENT*]
> *rooms_requirements*: *SEQ*[*ROOM_REQUIREMENT*]

solution

> *teachers*: *SEQ*[*TEACHER_ASSIGNMENT*]
> *rooms*: *SEQ*[*ROOM_ASSIGNMENT*]

end

The student groups are fixed at creation. A solution must supply (in addition to the inherited *times*) a sequence of teacher assignments conforming to the teacher requirements, and we will now discuss these. Rooms are analogous so will not be mentioned further.

Classes may be large composite affairs, containing many sub-classes constrained to run simultaneously. An example common in Australian high schools is the *elective*, where a heterogeneous mixture of classes is run simultaneously and each student chooses one. Accordingly, the *teachers_requirements* creation variable holds one 'teacher requirement' for each of the teachers required to attend the meeting. For example, a class requiring three teachers might have the following value of *teachers_requirements*:

[*ChooseTeacher*(*History*), *AnyTeacher*, *FixedTeacher*(*Smith*)]

This requires the first teacher to be any History teacher (*History* is a teacher category, as in Section 3), the second to be any teacher, and the third to be Smith. *ChooseTeacher*, *AnyTeacher*, and *FixedTeacher* are child classes of *TEACHER_REQUIREMENT*.

It might be expected that *CLASS* would contain a solution variable holding a sequence of teachers, but in fact a solution can be more complex than that. The solution variable used by *CLASS* to represent the assigned teachers is

teachers: *SEQ*[*TEACHER_ASSIGNMENT*]

Each teacher assignment is supposed to satisfy the corresponding requirement in *teachers_requirements*. For example, one value of *teachers* might be

[*NT*, *T*(*Jones*), *MT*([*Smith, Smith, Robinson, Smith*])]

meaning that no teacher (*NT*) has been allocated to fulfil the first teacher requirement, Jones has been allocated to fulfil the second requirement, and that for the third requirement it was decided to allocate Robinson for the third of the four times of the meeting, and Smith for the other three times (this is why we use sequences of times, rather than sets). *NT*, *MT* and *T* are child classes of *TEACHER_ASSIGNMENT*, representing the three kinds of teacher assignment in this model.

6   Evaluating Solutions

Although STTL is useful as a modelling tool alone, its main value arises from its ability to evaluate solutions to timetabling problem instances against their formal specifications.

Since STTL has equivalent power to a Turing machine, its evaluation of a solution could be any computable function. In this section we present arguments to show that the usual (much more restricted) kind of evaluation is a less arbitrary choice than is often supposed, and we outline its implementation in the high school case study. A similar approach (indeed, identical STTL code) could be used in other problems.

An evaluation function is not likely to be some unstructured, unanalysable function, because timetables affect the entities they are composed of: teachers, students, rooms, meetings, and so on. It is a good timetable if and only if its entities are satisfied.

We do acknowledge that some requirements are not closely associated with one particular entity: a requirement that the average load should be no more than so many hours per week, for example. These are rare and can be handled by associating them with a fictitious entity, such as "the timetable". We associate requirements with entities because it is a useful organizing principle in most cases, but it is not strictly required.

If an entity reports a problem, it must describe that problem in some concrete way: "This teacher is overloaded", or "This meeting needs more times than have been assigned to it" and so on. There can only be finitely many such descriptions. It is true that when a proposed timetable is presented to entities, they sometimes complain about things that had never before been heard of; and also, when a new institution's problem is analysed, a few requirements show up that no previous institution has used. This proves that bringing the finite list to completion requires careful management, but does not invalidate the basic point about its finiteness. We will show below that our evaluation model is easy to extend with new requirements, a key point since, in a spirit of inclusiveness, we would not wish to outlaw any requirement used in practice because of implementation difficulties.

We are thus led to the concept of a *defect* – that is, a concrete point of imperfection in a solution, reported by some entity:

class *DEFECT*

> *entity*: *ENTITY*
> *type*: *DEFECT_TYPE*
> *occurrences*: *INTEGER*

end

A defect records the entity that reported it, the type of defect and the number of occurrences (how overloaded the teacher is, how many times are missing, or whatever).

We have argued that there can be only finitely many defect types. Our *DE-FECT_TYPE* class contains only one variable, a string description of the defect:

> *TeacherClash*: *DEFECT_TYPE* = *DEFECT_TYPE*("teacher clash")
> *TeacherOverload*: *DEFECT_TYPE* = *DEFECT_TYPE*("teacher overloaded")

The high school case study has 24 defect types, covering resource clashes and overloads, defective assignments, and poor distribution of meeting times through the week.

Evaluation consists of finding all the defects, which we do by asking each entity for the defects that affect it. In the high school case study, every entity has a function

> *defects*: *SEQ[DEFECT]*

that returns its defects. An entity can work these out quite routinely using logic expressions. For example, a teacher can calculate the cardinality of the union of the sets of times of the meetings he or she participates in, and compare this to the stated maximum load, to see whether a teacher overload defect exists, like this:

> *load* = **count union** *m* **in** *all_meetings* **of** *m.times*
> *overload* = *max*(0, *load* − *max_load*)

The implementation of *defects* in *TEACHER* entities is

defects: SEQ[DEFECT] = (
 defect_check(TeacherClash, clashes) +
 defect_check(TeacherOverload, overload) +
 defect_check(TeacherDayOverload, day_overload)
)

where + is the concatenation operator. Function *defect_check* checks whether its second parameter (an integer) is greater than zero or not; if it is, it returns a sequence containing one defect of the given type with itself as entity and that many occurrences; if it is not, it returns the empty sequence. This makes defect handling trivial once one has a function (such as *clashes* and *overload* in the example) which calculates how many occurrences of the defect there are. To add a new defect type, one needs only to define its *DEFECT_TYPE* object, write a function to calculate its number of occurrences in affected entities, and add one *defect_check* call to the body of *defects*.

 A scan expression can collect all the defects from all the entities:

concatenation *e* **in all** *ENTITY* **of** *e.defects*

The **evaluation** expression of the high school case study is essentially just this, except that it has extra code to print a neatly formatted table.

 We have argued that defining interpretation to mean finding all defects involves no loss of generality in practice. We are on shakier ground when we come to aggregate the defects to produce an overall score. Aggregation seems to be the only practical way of comparing solutions, but each instance will have its own view of the relative significance of the defects, and how they should be aggregated.

 We now explain how the high school case study aggregates defects into a total score. There is undoubtedly some arbitrariness here, but since all the defects are available from the expression above, other methods could easily be implemented.

 The *DEFECT_TYPE* class contains an *infeasible* function, which returns a Boolean value which is true if the presence of a defect of this type is considered to render a solution infeasible, and a *badness* function, which returns an integer measure of how bad one occurrence of this defect is considered to be. Aggregation is carried out by taking the disjunction of all the *infeasible* values, and the sum of the badness values multiplied by the number of occurrences. Here is an abbreviated version of an actual report:

```
Teachers Description Feasible Cost Count Total
-------------------------------------------------------------------------
Knott teacher overloaded 20 5 100
Miniutti teacher overloaded 20 5 100
Jackson teacher overloaded 20 1 20
...
Classes Description Feasible Cost Count Total
-------------------------------------------------------------------------
Yr7K_English times cover too few days 1 1 1
Yr7A_English times cover too few days 1 1 1
Yr7C_Maths too few double times 2 1 2
...
Yr11_5_12_5_OAS_A times cover too few days 1 1 1
Yr11_5_12_5_OAS_B times cover too few days 1 1 1
Yr12_6 asst split between teachers 10 2 20
-------------------------------------------------------------------------
 Feasible overall: true Badness total: 1688
```

Each line shows one defect: its entity, description, infeasibility (blank if feasible), badness, number of occurrences, and the product of the last two. Aggregate values appear at the end.

Although it would be possible to supply constant values for *infeasible* and *badness* within the definition of *DEFECT_TYPE*, it is better to allow instances to choose these values, since, as we said before, opinions differ over the importance of the defect types. We have implemented this in the high school problem with definitions that allow an instance to contain, for example,

```
Weights = DEFECT_WEIGHTS({
    WEIGHT(TeacherClash, true, 100),
    WEIGHT(TeacherOverload, false, 20),
    WEIGHT(TeacherDayOverload, false, 5),
    WEIGHT(RoomClash, true, 100)
})
```

Each line defines, for one defect type, the values of its *infeasible* and *badness* functions. Any defect types not listed default to **false** for *infeasible* and 0 for *badness*, ensuring that solutions are not penalized for defects that do not appear in the weights list of their instance.

7 Conclusion

The author intends to construct a web site offering a database of problems, instances and solutions, using the STTL format and arranged in the obvious three-level tree structure. Such a site will both encourage the use of STTL and realize its goal of facilitating the storage and communication of timetabling problems.

Although the STTL interpreter will be freely available for download, it should also be possible to upload STTL files to the web site and have them checked there. Instances and solutions that can be read and evaluated by the interpreter without error could then be installed into the web site, at the client's request.

One potential problem is the process of acquiring and standardizing problems. It will be necessary to restrict the installation of problem files, to prevent fragmentation of the space of timetabling problems. On the other hand, the author is opposed to any policy which excludes any researcher's work. Rather, he envisages a process of negotiation in which existing problem files are extended to accommodate specific real-world instances expressed in STTL and offered by researchers. The author volunteers to manage this process.

Another potential problem concerns the speed of the current implementation, which is an interpreter, written in Eiffel and implemented by recursively evaluating expression trees. When compiled by the GNU SmallEiffel compiler with all optimizations on and run on a Sun Sparc Ultra-2 computer using the SunOS operating system, the program takes about 45 seconds to produce the report given earlier. If further optimization is required, as might be the case if, for example, a significantly more complex time model was installed, or university instances containing thousands of students became available, it would be possible to alter the interpreter so that it converts each problem into (say) an Eiffel program, which could then be compiled into a binary that could read and evaluate instances and solutions for that problem only.

The author intends to write a very simple parser for STTL instance files, which researchers can include as a module in their C programs. Given a file name, the module would read that file and return a tree structure with strings such as *"SET"* and *"TEACHER"* as node tags. It would then be very easy to write code which traverses this tree and copies the information in it into the researcher's own format.

We believe that STTL offers a good prospect of making timetabling research much more systematic than it is today. But to be successful it must attract widespread support by meeting its users' needs. The author is accordingly anxious to hear about and address the concerns of timetabling researchers.

Acknowledgement

The author thanks Ben Lin, who implemented the first STTL interpreter, for several ideas carried over from that implementation.

References

1. Appleby, J. S., Blake, D. V., Newman, E. A.: Techniques for Producing School Timetables on a Computer and Their Application to Other Scheduling Problems. Comp. J. **3** (1960) 237–245
2. Cooper, T. B., Kingston, J. H.: The Solution of Real Instances of the Timetabling Problem. Comp. J. **36** (1993) 645–653
3. Burke, E. K., Kingston, J. H., Pepper, P. W.: A Standard Data Format for Timetabling Instances. In: Practice and Theory of Automated Timetabling II. Lecture Notes in Computer Science, Vol. 1408. Springer-Verlag, Berlin Heidelberg New York (1997)
4. Kingston, J. H.: The STTL Interpreter, Version 1.0 (1999). URL

A Language for Specifying Complete Timetabling Problems

Luís Paulo Reis[1,2] and Eugénio Oliveira[1,3]

[1]LIACC – Artificial Intelligence and Computer Science Lab.,
University of Porto, Portugal
lpreis@ufp.pt, eco@fe.up.pt
http://www.ncc.up.pt/liacc
[2]UFP – CEREM, Multimedia Resource Center,
Praça 9 de Abril, 349, 4200 Porto, Portugal
[3]FEUP – DEEC, Rua Dr. Roberto Frias, s/n,
4200-465 Porto, Portugal

Abstract. The timetabling problem consists in fixing a sequence of meetings between teachers and students in a given period of time, satisfying a set of different constraints. There are a number of different versions of the timetabling problem. These include school timetabling (where students are grouped in classes with similar degree plans), university timetabling (where students are considered individually) and examination timetabling (i.e. scheduling of university exams, avoiding student double booking). Several other problems are also associated with the more general timetabling problem, including room allocation, meeting scheduling, staff allocation and invigilator assignment. Many data formats have been developed for representing different timetabling problems. The variety of data formats currently in use, and the diversity of existing timetabling problems, makes the comparison of research results and exchange of data concerning real problems extremely difficult.

In this paper we identify eight timetabling sub-problems and, based on that identification, we present a new language (UniLang) for representing timetabling problems. UniLang intends to be a standard suitable as input language for any timetabling system. It enables a clear and natural representation of data, constraints, quality measures and solutions for different timetabling (as well as related) problems, such as school timetabling, university timetabling and examination scheduling.

1 Introduction

Wren [44] defines timetabling as a special case of scheduling: "Timetabling is the allocation, subject to constraints, of given resources to objects being placed in space-time, in such a way as to satisfy as nearly as possible a set of desirable objectives". In a more common definition, a timetabling problem consists in fixing in time and space, a sequence of meetings between teachers and students, in a prefixed period of time, satisfying a set of constraints of several different kinds. These constraints may include both hard constraints that must be respected and soft constraints used to evaluate the solution quality.

The scientific community has given a considerable amount of attention to automated timetabling during the last four decades. Starting with the works of Appleby

E. Burke and W. Erben (Eds.): PATAT 2000, LNCS 2079, pp. 322–341, 2001.
© Springer-Verlag Berlin Heidelberg 2001

[2] and Gotlieb [24], many papers have been published in conferences and journals, including several surveys [3], [7], [9], [11], [12], [19] ,[29] and annotated bibliographies on the same subject [33], [41]. In addition, several practical timetabling systems have been developed and applied with partial success. The first automated timetabling approaches were based on operational research methodologies such as network flow techniques [20], reduction to graph colouring [43], integer programming [33], direct heuristics [4], simulated annealing [1], tabu search [28] and neural networks [32]. Although timetabling automation may be desirable, there are also a number of clear advantages of manual timetabling over automated timetabling. This has led several authors [42] to advocate the use of interactive methods for timetabling based on decision support systems and human–computer interface techniques. In recent years, research into automated timetabling has included techniques such as logic programming [23], [30], expert systems [25], constraint logic programming [26], [35], [37], [40] and genetic/evolutionary algorithms [6], [17].

A large number of variants of the timetabling problem have been proposed in the literature. Schaerf [38] classifies the timetabling problem using three main classes based on the type of institution involved and the type of constraints:

- **School Timetabling** – Weekly scheduling for all the classes of a high school avoiding teachers and groups of student double booking;
- **University Timetabling** – Scheduling of all the lectures of a set of university degree modules, minimizing the overlaps of lectures having common students and avoiding teachers double booking;
- **Examination Timetabling** – Scheduling the exams of a set of university courses avoiding overlapping exams for the same students and spreading the exams for the students as much as possible.

As was noted by Schaerf [38], this classification is not strict in the sense that some specific problems may fall between two classes and cannot be easily placed in this classification. Carter, in his study about recent developments in practical course timetabling [11] identified five different sub-problems for the course scheduling problem: course timetabling, class teacher timetabling, student scheduling, teacher assignment and classroom assignment.

The recent emergence of the PATAT (Practice and Theory of Automated Time-tabling) series of international conferences [5], [8] and the establishment of the EURO (Association of European Operational Research Societies) Working Group on Automated Timetabling (WATT), indicates that the research interest in this area is increasing dramatically. However, the variety of data formats currently in use and the diversity of existing timetabling problems makes the comparison of research results and exchange of research ideas and data concerning real problems extremely difficult. Several attempts to find a standard to represent timetabling problems have been made in the past but at the present there is no universally accepted language for describing timetabling problems.

This paper introduces UniLang, a new language for representing timetabling problems that can be easily read by computer specialists and school administrators. In Section 2, some timetabling data standards proposed in the past years are briefly reviewed. Eight sub-problems identified in the complete timetabling problem are introduced in Section 3 along with the architecture for a generic timetabling system.

Section 4 presents the requirements for a language to represent timetabling problems. Based on these requirements as well as on the timetabling related sub-problems, we present in Section 5 our proposal of a language for representing timetabling problems. Section 6 briefly shows how a problem represented through this language can be translated to a constraint logic program and solved using a constraint logic programming language. Finally, we give some conclusions together with an outlook to future research.

2   Timetabling Data Standards

At present there is no universally accepted language for describing timetabling problems. Although some data formats have been developed for representing different timetabling problems, they are usually incomplete in some aspect.

Cooper and Kingston [14], [15], [16] propose a formal specification of the problem based on TTL, a timetabling specification language. A TTL instance consists of a time group, a set of resource groups, and a set of meetings. A time group defines the identification of the time slots available for meetings, followed by a specification of the way in which time slots are distributed over the days of the week. For specifying this distribution, a format is proposed, based on the utilization of brackets (to enclose days), colons (meaning breaks), dots (for preferable time slots) and commas (for undesirable days). Resource groups can contain subgroups, which are subsets of the resource set defining functions that they may perform. A resource may be in any number of subgroups. In this specification language, meetings are collections of slots that are to be assigned elements of the various resource groups, under certain constraints. Only a basic set of constraints is defined in the language specification.

Cumming and Paechter propose a standard data format in a discussion paper presented at PATAT'95 conference [18], but not submitted formally to the conference or printed in the proceedings. Their proposal was highly criticized at the conference for lack of generality but generated a big discussion about the subject in which the difficulty of creating such a standard became evident. In their discussion paper [18], Cumming and Paechter propose principles and requirements to guide the creation of the standard. Their standard claims to represent complete and incomplete timetables and preferences. The components used are time slots (using a day.hh:mm representation format), events, staff and students, and rooms. They make no distinction between staff and students, arguing that in some cases students can lecture classes. A list of keywords with different parameters is proposed as the standard. For example, offer.room with parameters event and room represents that a given event must be assigned a room and that the specified room is an option. They also propose a cross-convention (but not as part of the standard) that may be attached to any keyword and enables a Cartesian product between the arguments of that keyword (lists in this case). They also attempt to represent the soft constraints using cost functions. Moreover, they conclude that timetable evaluation is likely to be the most difficult item to standardize. Some important omissions of this work are concerned with the availability of resources, split events, groups of resources, weeks and other type of periods, room types, definition of what is the problem to solve and how to represent the solution.

An interesting work related to standard data format for timetabling problems is included in the GATT (Genetic Algorithm Time Tabler) timetabling system [13] by

Collingwood (now Hart) et al. GATT uses a file format for describing timetabling problems. The format claims to be able to describe any GELTP (general exam/lecture timetabling problem) and also non-educational timetabling problems. The file format is verbose and uses as main components: events, time slots, rooms, students and teachers. The format was essentially devised for exam timetabling problems. Some important omissions include weeks and other periods (useful for staff allocation and university course timetabling problems), room types, event sections (and section duration), continuity and load constraints (useful to achieve good quality schedules for teachers and students), student groups (essential in school timetabling) and teacher groups.

A more recent paper by Burke et al. [10], proposes a different kind of standard for timetabling instances. They include a simple but incomplete description of the data types, keywords and syntax of the language and outline some further facilities to develop. Some concepts and constructs they use are similar to those found in the Z specification language [34]. The format includes as data types: classes, functions, sets, sequences, integers, floats, booleans, chars and strings. Classes include some attributes and functions and an inheritance mechanism is provided. A useful data type of this work are sets since many of the components of timetabling instances involve groups of resources (groups of students, groups of classes, etc.). In addition to the common set operators (member, union, intersection, subset, etc.), they also include operators like forall, exists, sum and prod. All these data types make the specification language close to a kind of programming language and enable the definition of constraints in logic programming language style.

A good extension of Burke et al.'s work could be the use of a constraint logic programming language as the specification language. Logic associated with constraints performs very well in describing and solving timetabling problems. However, logic is not appropriate as an interchange format between the computer specialist and school administrators. Therefore, we here propose a simpler and verbose language as our language for describing timetabling problems.

3   Sub-problems of the Timetabling Problem

Timetabling can be viewed as a multi-dimensional assignment problem [11] in which students and teachers (or invigilators) are assigned to courses, exams, course sections or classes, and events (individual meeting between teachers and students) are assigned to rooms and time slots. This multi-dimensionality indicates that we do not have a single problem designated by timetabling. We can have student assignment, teacher (or invigilator) assignment, room allocation and time allocation problems, all included in a given global timetabling problem. Figure 1 shows a generic timetabling system. The description of a given problem must be pre-processed in order to perform validity checks and decompose the original problem into its associated sub-problems. The sub-problems, after being solved using appropriate algorithms, enable the construction of the timetabling problem final solution.

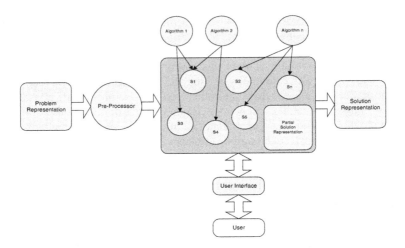

Fig. 1. A generic timetabling system

Extending Carter's definition of the timetabling sub-problems [11], the problem analysis led us to the identification and definition of the following eight classes of inter-related timetabling sub-problems:

- **CTT – Class–Teacher Timetabling:** This is a common problem in most high schools and consists in allocating a time (or set of times) to each lesson of each module in the school (or university). The scheduling unit is a group of students (or class) that has a common programme. It is considered that the assignment of teacher and classes to the events has already been made. Usually, either rooms are used as constraints in this problem or room allocation is performed simultaneously with class–teacher timetabling.
- **CT – Course Timetabling:** Scheduling of all the lectures of a set of university degree modules, minimizing the overlaps of lectures having common students and avoiding teachers double booking. In this case, students are considered individually and are not assumed to belong to some group as in school timetabling. Teachers are usually already assigned (although some flexibility may be included in this assignment). Sometimes, students are not assigned to the event sections before the course timetabling scheduling. Although they are usually already assigned to the events, in some universities this can also be untrue.
- **ET – Examination Timetabling:** Scheduling (in time) of the exams of a set of university courses avoiding overlapping exams having common students and spreading the exams for the students as much as possible. Room assignment and invigilator assignment can be done prior to or after the exam timetabling phase.
- **SD – Section Definition:** This problem occurs within every timetabling problem. It consists in, for each event, defining the number of sections offered. Each section is simply a different occurrence of the same event (for example, the same lecture given to a different group of students). A similar problem occurs in examination timetabling if we consider split examinations (in time).
- **SS – Student Scheduling:** This problem occurs when modules are taught in multiple sections. Once students have selected their modules, they must be

assigned to module sections, trying to provide schedules (for students) without conflicts and balancing section sizes.

- **SA – Staff Allocation (Teacher Assignment):** Assigning teachers to different modules, trying to respect their preferences, including preferred subjects, balance between lecture hours in different periods of the year (semesters or trimesters) and other side constraints. Sometimes this is achieved in two or three consecutive separated steps. The first step consists in selecting teachers responsible for each one of the modules (although in traditional public universities this problem is solved using only small variations from year to year). In the second step, the teachers that will lecture each one of the modules are selected (along with their own workloads). A third step consists in assigning individual sections to teachers. Sometimes this last phase remains open (or at least very flexible) until the timetabling generation phase.

- **IA – Invigilator Assignment:** This is a common problem which is associated to examination scheduling. Each exam needs one or more invigilators. The number of invigilators is related to the number of students having the examination and to the number of rooms needed.

- **RA – Room Assignment:** Usually all real timetabling problems have a room assignment phase. Events must be assigned to specific rooms (or sets of rooms) satisfying the size and type required for the event. Problems with split events, shared rooms and distances between rooms for events may arise.

Other sub-problems could be included in this classification. However, we consider those other problems as unusual or unimportant compared with the described ones. For example, prior to examination timetabling, some institutions may aggregate small exams into sets that will be scheduled at the same time using the same room and supervisors. Other institutions use modules with different workloads throughout the year. Then, the timetables are different every week and the problem becomes a kind of mix between timetabling and job-shop scheduling in which module workloads are defined for each week and its lessons are scheduled in each week. However, we believe that the sub-problems we have taken into consideration include the timetabling problems faced by most schools and universities around the world.

To describe a specific timetabling problem, it is necessary to know in detail which of the sub-problems will be solved and in which order they will be solved. Some flexibility is allowed in this ordering by some universities or schools, while others obey strict and rigid rules.

4   Requirements for the Proposed Language

Many formalisms for timetabling application descriptions have been designed to address the concern of minimizing data storage space or to facilitate fast data processing. Our proposed formalism tries to compromise between generality and simplicity. It is sufficiently general to allow representation of the most common variations of timetabling problems, including school, university and examination-timetabling, while also preserving a very simple syntax. After a series of interviews with timetabling experts, we arrive at the following main requirements associated with the needed language:

- It should be independent of implementation details and timetabling strategies;
- It should be easy to extend, including new concepts and constraints;
- Existing benchmark problems should be easily translated to the new formalism;
- The new formalism should be compatible with most timetabling systems and easily readable by the human user;
- Information not directly concerned with the scheduling problem should not be included in the formalism. Therefore, common information regarding school administration, like student names and addresses, is omitted;
- The formalism should be general enough to enable the representation of school timetabling, university timetabling and exam timetabling problems;
- It should be suitable for representing complete problems and simple related sub-problems (like staff or invigilator assignment, section definition and room assignment);
- It should include a clear definition of all the constraints (both hard and soft) associated with the problems;
- It should be possible to represent both incomplete and complete solutions;
- A timetabling quality evaluation function should be easy to include, enabling one to evaluate directly any proposed solution;
- It should be as concise as possible;
- It should be robust, enabling simple data validation and override of common errors.

5   Language for Representing Timetabling Problems

Based on the identification of the eight timetabling sub-problems given in Section 3 and on the requirements presented in the previous section, we propose a new language called UniLang for representing timetabling problems. UniLang is intended to be a standard suitable as specification language for any timetabling system. It enables a clear and natural representation of data, constraints, quality measures and solutions for different timetabling (as well as related) problems, such as school timetabling, university timetabling and examination scheduling.

5.1 Components

The model behind our proposal includes the definition of different *time periods* composed of a set of *weeks* (or other time period) with equal (or similar) schedules. Each week is composed of a set of *time slots* located at a given *time* of a given *day*.

We prefer to separate resources definition into three main classes: *students*, *teachers* and *rooms*. Although some similarities exist between the three (for example, all the resources have availability constraints), the differences are evident in any timetabling problem. For example, *rooms* can hold (in some timetabling problems) several *events* at the same *time*, have capacity constraints, and variation of types and distances between them. *Students* and *teachers* can be put together in groups (although with different meanings) and have workload constraints. *Teachers* can have maximum and minimum *event* assignment constraints and ability constraints (stating whether they are capable or willing to lecture/supervise a given subject/exam).

Students (and *student groups*) have also associated spreading and continuity constraints. As a consequence, we have considered the following components: *periods, time slots, resources (rooms, teachers* and s*tudents*) and e*vents*.

An *event* is a meeting between a *teacher* (or a set of *teachers*) and a set of *students* (or *student groups*) that takes place in a *room* (or in a set of *rooms*) in a given *time slot* (or set of *time slots*). We do not consider (explicitly) other resources like equipment, because these situations are very rare in timetabling problems.

Our model also includes the specification of both data and constraints in a common format. Since the distinction between data and constraints is not always completely clear, this seems an adequate approach. Some of the types of constraints included have two different versions: a strong version in which the constraint must be respected (hard constraint) and a weak version that should be respected only if possible (soft constraint). The considered set of soft constraints enables the definition of a timetabling quality function. Each one of the soft constraints included has associated a numerical *preference* that represents the cost of not satisfying that constraint.

A solution (partial or global) to a timetabling problem consists in, for each one of the *events* considered, a set of *slots*, a set of *rooms*, a set of *teachers* and, finally, a set of *students* (or *student groups*). Although in most of the timetabling problems found in the literature *students* and *teachers* are already assigned to the *events* (or *event sections*), our model intends to be general enough to also solve those assignment subproblems, (i.e. *teacher assignment, invigilator assignment* and *student scheduling*).

To make the formalism still more robust, and following Hart's idea in the GATT system [13], a list of synonyms (Table 1) may be included for each one of the concepts (keywords) used.

Table 1. Example of a brief list of synonyms for the keywords used in the language

```
[Default|All|Every|Any], [Year|Problem|File], [Schedule|Timetable],
[Event|Module|Lecture|Lesson|Exam|Examination|Tutorial],
[Teacher|Teachers|Invigilator|Supervisor|Lecturer]`, [Student|Students],
[Day|Days], [Time|Times|Hour|Hours|Minute|Minutes], [Slot|Slots|Time
Slot],
[Place|Places|Room|Rooms], [Preference|Weight|Priority|Penalty],
[Consecutive| Continuous], [Simultaneously|Concurrently|Together|At the
same time],
[Teaches|Invigilates|Supervises], [Hold|Holds|Have capacity|Has
capacity|Capacity],
[Specify|Specifies|Require|Requires|Need|Needs], [Last| asts],
[Duration|Length],
[Contains|Comprises|Has|Have], [Is|Are],
[At least|No less than|No fewer than], [At most|No more than],
[Exactly|Precisely],
[Group_teachers|Area|Department], [Group_students|Class|Student_type],
[Room_type|Room_Group|Buildings], [Double_bookings|Clashes|Conflicts]
```

The use of these synonyms makes the description of input data files more easily readable by human users. Also, users can build their own list of synonyms in a separate file and, with the help of a simple pre-processor, that file can be directly converted into our proposed specification language.

5.2 Time Representation

Each file contains a problem description concerning a specific *period of time*: A school *year*. Therefore, the file needs the identification of that specific *year*:

```
this is year <NAME>
```

Each *year* can be divided into smaller *time periods* (semesters, trimesters, exam periods, etc.). UniLang enables defining the number of *periods* used (for the problem) and the names of those *periods*. Each *period* may have a *starting date, starting and ending weekdays* and it is composed of a set of *weeks*. The concept of *week* is not that of a typical *week*. Here we are interested in a concept of *week* as a *period of time* to which the same schedule has been assigned. So, for example, in an examination timetabling problem, a "*week*" may last 15 or 20 *days*, while in a typical university or school timetabling problem, a *week* lasts 5 or 6 *days*:

```
periods are {<PE>}
period {<PE>} contains weeks {<N>}
period <PE> begins on date <DATE>
period <PE> begins|ends on day <D>
```

We then assume that each *week* is composed of a set of consecutive *days* and that each *day* is composed of a set of disjoint and consecutive *time slots*. Each *slot* is located in a given *day* and begins at a given *time*:

```
slots are {<S>}
days are {<D>}
times are {<T>}
slot <S> is on day <D> at time <T>
```

If the *slots*, *days* and *times* are explicitly defined, then the previous sentence can be simplified. For example, if *<S>* is a *slot*, *<D>* is a *day* and *<T>* is a *time* then "*slot <S> is on day <D> at time <T>*" can be described in one of the following ways:

```
<S> is on <D> at <T>
<S> <D> <T>
```

This simplification applies to all keywords that come before any other component in a sentence (like *teacher, student, room, teacher_group, event*, etc.).

Some *slots* cannot be used for timetabling *events*. Examples are *slots* on Sundays or Saturdays and *slots* at night. UniLang allows two ways of stating this impossibility. The first involves simply not defining the impossible *slots*. The second is based on defining the *slots* and stating explicitly the impossibility of the allocation. This enables a coherent time difference measurement between *slots* that may be needed for some timetabling applications:

```
slots {<S>} are unusable
```

We can define *time periods* (like mornings, afternoons, lunch times, etc.) using the concept of *time_period*. Each *time period* is composed of a set of *slots* in a *week*. UniLang includes the keyword '*default*' that enables assignment of values to all elements of a class:

```
time_periods are {<TP>}
time_period {<TP>}|default contains slots {<S>}
```

Using the keyword *default*, the constraints are applied to all the elements belonging to the given class. This can also be applied to any constraint that deals with *time slots*, *rooms*, *teachers*, *students* and *groups* (of *teachers* or *students*).

To each *slot*, a capacity in terms of both *events* and *seats* can be assigned. These capacities are important to enable the definition of timetabling problems that do not have associated *room allocation* problems. Otherwise, if *room allocation* is a part of the complete problem to solve, *slot capacities* are not needed to be stated explicitly:

```
slots {<S>}|default have capacity <N> seats|events
```

To state that the problem of time allocation for the described *events* has to be solved, the following syntax is provided:

```
solve class-teacher timetabling
solve course timetabling
solve examination timetabling
```

These three lines express how to ask for solutions of the three broad classes of timetabling problems. Knowing the timetabling problems to solve, a given time-tabling system may then select the appropriate constraints and quality measures for those problems.

5.3 Space Representation

Our space representation is based on the idea that each *event* usually needs one *room* but may, in some cases, be hosted by more than one *room* or may not even need a *room*. Our formalism enables the definition of the *rooms* available in the following way:

```
rooms are {<R>}
```

Each *room* can hold a given number of *students* at a *time*. In a similar way, each *room* can hold only a given maximum of *events* (usually one) at a *time*. If *preference* is omitted, this becomes a hard constraint. If *preference* is included then the constraint can be violated with a cost *P*:

```
room {<R>}|default holds <N> students|events
[preference <P>]
```

The number of *students* and *events* that a *room* can host simultaneously can be different throughout the *week slots*:

```
room {<R>}|default holds <N> students|events in
slots{<S>} [preference<P>]
```

Usually, *rooms* are not allowed to have *double booking* and may be *unavailable* or preferably usable during some *time slots*:

```
room {<R>}|default cannot have double_bookings
[preference <P>]
room <R>|default specify|excludes in slots {<S>}
[preference <P>]
```

Our formalism also provides *room types*. Each *room* has a given *room_type* or set of *room_types*:

```
Room_types are {<RT>}
room_type {<RT>}|default has rooms {<R>}|default
[preference <P>]
```

We include also the concept of distance between *rooms*. A *room* can be connected (if the preference is omitted) to another *room* or can be close to another *room*, with a given proximity measure:

```
room {<R>} close to room {<R>} [preference <P>]
```

To state that the problem includes the assignment of *rooms* to *events*, the language proposes the following syntax:

```
solve room assignment
```

5.4 Event Description

The concept of *event* is crucial in our specification. An *event* can be a lecture, exam, tutorial, lunch, etc. It has a given duration (in terms of *time slots*), may or may not need space (*rooms*) and resources like *teachers* (or *invigilators*) and *students* (or *student groups*).

The first step is to define the *events* (*event* identifiers) and, eventually, some groups of *events* (like degrees, degree years, etc.):

```
events are {<E>}[in period <PE>]
groups_events are {<GE>}
```

Events may have default duration but may also be given a different individual duration. The *students* that may/will attend the *event* have also to be defined:

```
event {<E>}|default lasts <N> slots
```

```
event {<E>} has students {<ST>} [preference <P>]
```

Besides the number of *students* expected, an *event* can have some (anonymous) external extra *students*. The total number of *students* of the *event* can also be defined directly. This can be useful in cases where *students* are not going to be considered individually. Some *events* may also require a given number of (anonymous) *teachers* or *rooms*. This can be useful if we are not concerned with the *teacher* or *room allocation* problems:

```
event {<E>} has <N> students|extra_students
events {<E>}|default requires <N> teachers [preference
<P>]
events {<E>}|default requires <N> rooms [preference
<P>]
```

An *event* usually needs a given type of *room* and *events* may belong to groups:

```
events {<E>}|default requires room_type {<RT>}
[preference <P>]
group_events {<GE>} has events {<E>} [preference <P>]
```

Events can be split into a number of different *parts* of different duration that will occur in different *days* of the *week*. A typical constraint concerning *event parts* is that they should not be assigned on the same *day*. *Events* can also be divided into multiple *sections*, meaning that different *teachers* can repeat them during the *week* to different *students*. For example, an exam can have two *parts* (a theoretical and a practical part) and three *sections* (one for class A, one for class B and another for class C). A module is usually divided into lectures (*event parts*) that are taught in different days of the week and may have multiple *sections* (taught to different *students*):

```
event {<E>}|default has <N> event_parts of duration
{<N>}
event {<E>}|default has <N> event_sections
event {<E>}|default minimum/maximum <N> students
```

If *event sections* are not defined and the *section definition* problem must be solved, then we have

```
solve section definition
```

5.5 Classes and Students

The number and names (identifiers) of *students* and *student groups* can be declared just in the same way as the names of *slots*, *rooms* and *events*:

```
students are {<ST>}
group_students are {<GST>}
```

The *students* can be grouped into *student groups* (with common, or similar, schedule preferences or degree plans):

```
group_students {<GST>} have students {<ST>}|default
[preference <P>]
group_students {<GST>} has <N> students
```

A *student* can also be a *teacher*. This is the case of postgraduate *students* that also are lecturers to their undergraduate colleagues:

```
student <ST> is teacher <TE>
```

Students or *student groups* can be unavailable in some *time slots* throughout the *week*. This can be a hard or a soft constraint:

```
student|group_students {<ST>|<GST>}|default
specify|excludes slots {<S>}|default [preference<P>]
```

Students and *student groups* may attend *events*:

```
event {<E>}|default has students|group_students
{<ST>|<GST>}|default [preference <P>]
```

A *student* or *student group*, besides being registered in an *event* can also be registered in one of its *sections*. *Preferences* can also be used for allocating *students* and *student groups* to *events*:

```
event_section {<ES>} has students {<ST>} [preference
<P>]
event_section {<ES>} has group_students {<GST>}
[preference <P>]
```

To prevent that *students* (or *student groups*) have *double booking*, we can declare

```
student|group_student {<ST>|<GST>}|default cannot have
double_bookings [preference <P>]
```

Those declarations can be seen either as soft constraints (with a given violating cost *<P>*) or hard constraints (if the keyword *preference* is not used). If the problem of allocating students to *event sections* is considered, then this must be stated using the following:

```
solve student scheduling
```

5.6 Teachers and Invigilators

In the proposed model, *teachers* can be seen both as lecturers and invigilators (and the keyword *teaches*, can also signify *invigilates*). *Teachers* can be grouped in areas or *groups of teachers*. This leads to the definition of departments and scientific areas:

```
teachers are {<TE>}
group_teachers are {<GTE>}
group_teachers {<GTE>}|default has teachers
{<TE>}|default
```

Each *teacher* can be previously assigned to teach a given number of *events* (pre-allocations). The information of each *teacher* is also concerned with his capability of teaching a given *event*:

```
teacher {<TE>} teaches|cannot_teach events {<E>}
[preference <P>]
```

In order to be able to describe *staff allocation* problems and *invigilator assignment* problems, *maximum* and *minimum* number of *events* (or *times*) for *teachers* are needed:

```
teacher {<TE>}|default maximum|minimum<N>
events|times[in period<PE>]
```

Teachers or *groups of teachers* may also have timetable *preferences*. So a *teacher* (or *group*) can exclude (or avoid with some *preference*) specific *time slots*:

```
teacher|group_teachers {<TE>|<GTE>}|default
specify|excludes slots {<S>}|default [preference <P>]
```

Usually *teachers* cannot have *double booking* and sometimes, depending on the meaning of *teacher groups*, these *groups* cannot have *double booking* too:

```
teacher|group_teacher {<TE>|<GTE>}|default cannot have
double_bookings [preference <P>]
```

If the problem includes assigning *teachers* to *events* (i.e. *staff allocation* or *invigilator assignment*) then, the following line should be included:

```
solve teachers assignment
solve invigilator assignment
```

The order in which different sub-problems will be solved does not concern this language. UniLang defines the complete problem and states the timetabling sub-problems to be solved. It is up to the solver, by analysing the complete problem description, to choose the order in which different sub-problems will be solved.

5.7 Time and Space Preferences

We have two different kinds of *time preferences*: regarding *time slots* and regarding *time periods* (mornings, afternoons, etc.). These preferences include pre-allocations (*specify*) and exclusion or avoidance (*excludes*), with *preferences*:

```
event {<E>}|default specify|excludes slots {<S>}
[preference <P>]
event {<E>}|default specify|excludes time_period{<TP>}
[preference<P>]
```

Preferences associated with space are just like *time slot preferences*. There are two types of those – *room preferences* and *room_type preferences*:

```
event {<E>}|default specify|excludes rooms {<R>}
[preference <P>]
event {<E>}|default specify|excludes room_types{<RT>}
[preference<P>]
```

5.8 Workload, Spreading, and Ordering Constraints

The workload and spreading constraints are concerned primarily with the utilization of *teachers* and *students* throughout the scheduling period. *Students* and *student groups* may have workload constraints stating that they cannot have more than a given number of *events* or a given number of *time slots* of work in a row (*consecutive*) or in a given *day*. Usually, these are *at most* constraints but the *exactly* and *at least* constraints can also be used in some, less frequent, problems. These constraints can be also applied to *teachers*:

```
students|student_group {<ST>|<GST>}|default have
exactly|atleast|atmost <N> [consecutive] events|times
in a day [preference <P>]
teacher {<TE>}|default have exactly|atleast|atmost <N>
[consecutive] events|times in a day [preference <P>]
```

Spreading constraints are typical in examination scheduling problems. Through them it is achieved that *students* (or *student groups*) have sufficient time intervals between *events*. Usually these are *at least* constraints:

```
students|student_group {ST|GST}|default have exactly|
atleast|atmost <N> times|days between events
[preference <P>]
```

Sometimes, spreading constraints state that *students* have *at most* a given number of *events* (or occupied *times*) in each number of *days* (or *times*):

```
students|student_group {ST|GST}|default have exactly|
atleast|atmost <N> events|times in each <N> times|days
[preference <P>]
```

The ordering constraints are concerned with the sequential order of some *events* in the scheduling period. An *event* can be scheduled *exactly*, *at least* or *at most*, a given number of *times* (or *days*) before another *event*. Another usual type of constraint states that there are *exactly*, *at least* or *at most* a given number of *times* or *days* of interval between two *events* (without any concern of which *event* comes first):

```
events {<E>} exactly|atleast|atmost <N> times|days
before|interval events {<E>} [preference <P>]
```

Another typical constraint states that a given number of *events* occurs or cannot occur *simultaneously* or *on the same day*:

```
events {<E>} are [not] simultaneously|on_the_same_day
[preference<P>]
```

5.9 Override and Missing Mechanisms

Two useful mechanisms in our language are the override and the missing mechanisms. The override mechanism allows the redefinition of some concept in a stronger way, overriding the previous definition. The missing mechanism enables some concepts that are not formally defined in our problem description to be deduced from the problem description. These mechanisms are fully employed in our translator from this specification language to a constraint logic program.

5.10 Evaluation Function

The evaluation function for a given timetabling problem is implicitly defined through the definition of the problem data and constraints. The hard constraints must be respected. The soft constraints (in which the keyword *preference* appears) should be respected to achieve a good-quality solution. For each of the soft constraints violated, the *preference* value (*<P>*) is added to the total penalty associated with that specific solution. The smaller this value is, the better the final solution is.

5.11 Representation of the Final Solution

The final solution of any timetabling problem always includes for each *event part*, a set of *time slots* (when the *event parts* happen), a set of *rooms* (where the *events* take place), a set of *teachers* (or *teacher groups*) that will lecture (supervise or participate) the *event* and a set of *students* (or *student groups*) that will attend the *event*. In UniLang this is specified using the following notation:

```
event_part <EP>|event_section <ES>|event <E> is in
slots {<S>}
event_part <EP>|event_section <ES>|event <E> is in
rooms {<R>}
```

```
event <E>|event_section <ES> is taught by {<TE|GTE>}
event <E>|event_section <ES> has students {<ST|GST>}
```

For each of the *events*, event sections (if they exist) or event parts (if they are specified), a set of *time slots* and a set of *rooms* may be specified. The solution is completed with a set of *teachers* (or *teacher groups*) and a set of *students* (or *student groups*) for each of the *events* (or *event sections* if they are specified). It is assumed that each *event section* has the same *teachers* and *students* for each of its *parts*. We have also developed a simple evaluator that takes as inputs a simple problem (specified using our language) and a solution for the problem and calculates (using this data) a numeric value denoting the solution quality.

6   Translating and Solving Timetabling Problems

To enable the resolution of any kind of timetabling problem, which is represented using the proposed language, we have implemented a simple translator that converts this representation to a Constraint Logic Program [27], [41] in ECLiPse language [21]. To solve the problem, the user shall then specify the labelling strategy. The solution is then automatically obtained. Our constraint logic programming approach uses, internally, finite-domain variables for the starting times of events, a finite-domain variable for single-room or single-teacher assignment problems and set variables [23] for rooms, invigilators/teachers or students in multi-assignment problems.

After the translation of the problem definition, the user may specify the ordering in which the different types of variables are labelled [36]. For example, in a complete examination timetabling problem, we may have finite-domain variables for exams starting times, set variables for rooms and set variables for invigilators. The user may specify that the rooms would be labelled first, then times and lastly invigilators. Another possibility is, for each event, to assign first, simultaneously, time and rooms and then, at the end, perform the easier invigilator assignment. All combinations of the different types of variables are available to the user. For more details about our methods for solving timetabling problems which are described using this language and using different constraint logic programming approaches, see [35], [36], [37].

7   Conclusions and Future Work

This is, to our knowledge, the first attempt to use the identification of the subclasses of the timetabling problem to guide a possible standardization of the data and constraints that define any complete timetabling problem. We do not ignore that there are lots of open issues in this "standard-like" definition and that some characteristics are still missing in our language to enable the complete representation of specific problems. Representation of soft constraints is not complete, in the sense that any user can have a specific individual quality measure. However, if it is specific and individual, then it should not be included in the standard. Some people may argue that the definition of the constraints should also appear explicitly (and not be simply implied by the terminology). The problem is that doing so, we would lose the simplicity of this verbose format and enter into a more complicated logic or functional format. We once more emphasize that, from the outset, one of our requirements has been that administration staff may use the proposed format to describe, themselves, their timetabling problems.

To enable the easy use of our specification language, we have already implemented a simple translator (including the override and missing mechanisms) from this definition language to a constraint logic programming language. The translator was tested against a complete examination together with some class–teacher timetabling problems (with different sub-problems to solve). Several real problems were represented in the language, successfully translated and, after selecting appropriate labelling strategies, they were successfully solved.

We foresee future work as being related to developing a library for complete educational timetabling problem descriptions. Based on UniLang, we are building such a library of timetabling problems, including the eight timetabling sub-problems presented. Moreover, complete timetabling problems, including several frequent combinations of the eight previous ones are also being addressed. We are analysing both real timetabling problems and randomly generated ones. For each problem, quality measures and the already known best solution are included. We believe that this future work will make possible an easy and fast comparison between the solutions of typical timetabling problems, achieved by different researchers, using different algorithms.

Acknowledgement. This work was partially supported by a research grant PRAXIS XXI, BD/5663/95 of the Portuguese Foundation for Science and Technology.

References

1. Abramson, D., Dang, H.: School Timetables: A Case Study Using Simulated Annealing, Applied SA. Lecture Notes in Economics and Mathematical Systems. Springer-Verlag, Berlin Heidelberg New York (1993) 104–124
2. Appleby J., Blake D., Newman E.: Techniques for Producing School Timetables on a Computer and Their Application to other Scheduling Problems. Comput. J. **3** (1960) 237–245
3. Bardadym, V. Computer-Aided Lessons Timetables Construction, A Survey. USIM – Management Systems and Computers, Vol. 8. (1991) 119–126
4. Barham, A., Westwood, J.: A Simple Heuristic to Facilitate Course Timetabling. J. Oper. Res. Soc. **29** (1978) 1055–1060
5. Burke, E.K., Ross, P. (eds.): The Practice and Theory of Automated Timetabling. Lecture Notes in Computer Science, Vol. 1153. Springer-Verlag, Berlin Heidelberg New York (1996)
6. Burke E.K., Elliman D.G., Weare R.F.: A Genetic Algorithm for University Timetabling. In: Proc. AISB Workshop on Evolutionary Computing (University of Leeds, UK, April 1994)
7. Burke E., Elliman D., Ford P., Weare R.: Examination Timetabling in British Universities – A Survey. 1st Int. Conf. Practice and Theory of Automated Timetabling PATAT'95 (Edinburgh, UK) (1995) 423–434
8. Burke, E., Carter M. (eds.): Practice and Theory of Automated Timetabling II, Lecture Notes in Computer Science, Vol. 1408. Springer-Verlag, Berlin Heidelberg New York (1998)

9. Burke, E., Jackson, K., Kingston, J., Weare, R.: Automated Timetabling: The State of the Artif. Comput. J. **40** (1996)
10. Burke, E., Kingston, J, Pepper,P. A Standard Data Format for Timetabling Instances. In: Burke, E., Carter M. (eds.): Practice and Theory of Automated Timetabling II, Lecture Notes in Computer Science, Vol. 1408. Springer-Verlag, Berlin Heidelberg New York (1998) 213–222
11. Carter, M., Laporte, G.: Recent Developments in Practical Examination Timetabling. In: Burke, E., Carter M. (eds.): Practice and Theory of Automated Timetabling II, Lecture Notes in Computer Science, Vol. 1408. Springer-Verlag, Berlin Heidelberg New York (1998) 3–21
12. Carter, M.: A Survey of Practical Applications of Examination Timetabling Algorithms. Oper. Res. **34** (1986) 193–202
13. Collingwood, E., Ross, P., Corne, D.: A Guide to GATT. University of Edinburgh (1996)
14. Cooper, T., Kingston, J.: A Program for Constructing High School Timetables. Proc. 1st Int. Conf. on the Practice and Theory of Automated Timetabling (Napier University, Edinburgh, UK, 1995)
15. Cooper, T., Kingston, J.: The Complexity of Timetable Construction Problems. In: PATAT'95: Proc. 1st Int. Conf. on the Practice and Theory of Automated Timttabling (Napier University, Edinburgh). Lecture Notes in Computer Science, Vol. 1153. Springer-Verlag, Berlin Heidelberg New York (1996)
16. Cooper,T., Kingston, J.: The Solution of Real Instances of the Timetabling Problems. Comput. J. **36** (1993) 645–653
17. Corne, D., Ross, P., Fang, H.: Fast Practical Evolutionary Timetabling, Proc. AISB Workshop on Evolutionary Computing. Springer-Verlag, Berlin Heidelberg New York (1994)
18. Cumming, A., Paechter, B.: Seminar: Standard Timetabling Data Format. Int. Conf. on the Practice and Theory of Automated Timetabling (Edinburgh, UK, 1995)
19. de Werra, D.: An Introduction to Timetabling. Eur. J. Oper. Res. **19** (1985) 151–162
20. de Werra, D.: Construction of School Timetables by Flow Methods. INFOR – Can. J. Oper. Res. and Inf. Process. **9** (1971) 12–22
21. Aggoun et al.: Eclipse User Manual, ECRC GmbH (1992). Int. Computers Ltd and IC-Parc (1998)
22. Fahrion, R., Dollanski G.: Construction of University Faculty Timetables Using Logic Programming Techniques. Discrete Appl. Math. **35** (1992) 221–236
23. Gervet, C.: Interval Propagation to Reason about Sets: Definition and Implementation of a Practical Language. Constraints **1** (1997) 191–246
24. Gotlieb, C.: The Construction of Class-Teacher Time-Tables. Proc. IFIP Congress (Munich, 1963) 73–77
25. Gudes E., Kuflik T., Meisels A.: On Resource Allocation by an Expert System. Eng. Applic. Artif. Intell. **3** (1990) 101–109
26. Gueret, C. et al.: Building University Timetables Using Constraint Logic Programming. Proc. 1st Int. Conf. on the Practice and Theory of Automated Timetabling (1995) 393–408
27. Hentenryck, P.: Constraint Satisfaction in Logic Programming, LP Series, MIT Press, Cambridge, MA (1989)
28. Hertz, A.: Finding a Feasible Course Schedule using Tabu Search. Discrete Appl. Math. **35** (1992) 225–270
29. Junginger, W.: Timetabling in Germany – a Survey. Interfaces **16** (1986) 66–74
30. Kang, L., White, G.: A Logic Approach to the Resolution of Constraints in Timetabling. Eur. J. Oper. Res. **61** (1992) 306–317
31. Kingston, J., Bardadym, V., Carter, M.: Bibliography on the Practice and Theory of Automated Timetabling. ftp://ftp.cs.usyd.edu.au/jeff/timetabling/timetabling.bib.gz (Sidney University, 1995)
32. Kovacic, M.: Timetable Construction with a Markovian Neural Network. Eur. J. Oper. Res. **69** (1993)

33. Lawrie, N.: An Integer Programming Model of a School Timetabling Problem. Comput. J. **12** (1969) 307–316

34. Potter, B. et al.: An Introduction to the Formal Specification Language Z. Prentice-Hall, Englewood Cliffs, NJ (1991)

35. Reis, L.P., Oliveira, E.: A Constraint Logic Programming Approach to Examination Scheduling. AICS'99, Artificial Intelligence and Cognitive Science Conference (Cork, Ireland, September 1999)

36. Reis, L.P., Oliveira, E.: Constraint Logic Programming using Set Variables for Solving Timetabling Problems, INAP'99, 12th Int. Conf. on the Applications of Prolog (Tokyo, Japan, September 1999)

37. Reis, L.P., Teixeira, P., Oliveira, E.: Examination Timetabling using Constraint Logic Programming, ECP'99, 5th Eur. Conf. on Planning (Durham, UK, September 1999)

38. Schaerf, A.: A Survey of Automated Timetabling. TR CS-R9567. CWI–Cent. Wiskunde en Informatica (1995)

39. Schmidt, G., Strohlein, T.: Timetable Construction – An Annotated Bibliography. Comput. J. **23** (1979) 307–316

40. Stamatopoulos, P. et al.: Nearly Optimum Timetable Construction Through CLP and Intelligent Search. Int. J. Artif. Intell. Tools **7** (1998) 415–442

41. Tsang, E.: Foundations of Constraint Satisfaction. Academic, London New York (1993)

42. Verbraeck, A.: A Decision Support System for Timetable Construction. Int. Conf. on Expert Planning Systems (Brighton, UK, 1990) 207–211

43. Welsh, D., Powell M.: An Upper Bound for the Chromatic Number of a Graph and its Applications to Timetabling Problems. Comput. J. **10** (1996) 85–86

44. Wren, A.: Scheduling, Timetabling and Rostering – A Special Relationship. In Burke, E.K., Ross P. (eds.): The Practice and Theory of Automated Timetabling. Lecture Notes in Computer Science, Vol. 1153. Springer-Verlag, Berlin Heidelberg New York (1996) 46–76

A Software Architecture for Timetable Construction

Jeffrey H. Kingston and Benjamin Yin-Sun Lynn

Basser Department of Computer Science,
The University of Sydney, 2006, Australia
jeff@cs.usyd.edu.au

Abstract. As timetable construction research moves beyond the limitations of single, isolated solution methods, software architectures are needed that allow the researcher to combine many different solution methods in arbitrary ways in the solution of a single problem. This paper describes a blackboard architecture, implemented as an object-oriented framework, which addresses this need. Also included is a first application of the framework, comparing two quite different solution methods on a tutor allocation problem.

1 Introduction

Several researchers have recently expressed the view that the way forward in automated timetable construction is to abandon the "magic bullet" approach, which searches for a single, universally successful method, and replace it by a much more flexible approach, in which several solution methods (possibly quite different in character from each other) are combined together, perhaps in quite sophisticated ways [3], [7].

The individual solution methods are well known, so the problem in realizing this approach is finding a suitable architecture for the software, one in which new combinations of methods can be created quickly, at the whim of the researcher or under program control. Fortunately, many of the ideas we need have already been developed by researchers into object-oriented design, under the name of "design patterns" [4].

In this paper we present a blackboard architecture for timetable construction, implemented as an object-oriented framework. Section 2 is an overview of the framework. Sections 3 and 4 analyse two issues that we see as particularly significant: how to achieve flexibility in the combination of solution methods, and how to incorporate constraint propagation efficiently. Section 5 presents a first application of the framework, comparing tabu search with a self-adjusting sequence of bipartite matchings on a tutor allocation problem. Finally, in Section 6 we map out the future course of this work.

There are many object-oriented frameworks for solving various combinatorial problems, but we know of none that attempt a degree of eclecticism in solution methods approaching ours. The job-shop scheduling framework of [1] shares several features with our design. A previous timetabling paper [5] used a blackboard architecture, but its agents of change were conceived as stakeholders (teachers, students, management).

2 Overview of the Software Architecture

In this section we present an overview of our architecture, particularly the four key classes, *BLACKBOARD*, *COMMAND*, *STRATEGY* and *OBSERVER*. These are abstract

E. Burke and W. Erben (Eds.): PATAT 2000, LNCS 2079, pp. 342–350, 2001.

classes; that is, they mainly define interfaces and omit implementations. A particular instantiation of the framework must contain concrete versions of the classes that inherit the abstract ones and supply the missing implementations.

The term "blackboard architecture" is well known to mean a software architecture in which data are held in a central place, able to be viewed and updated by all interested parties (metaphorically, written on a public blackboard). For us, objects of the *BLACK-BOARD* class fill this role: each represents an individual working solution, which may be incomplete or inconsistent, and which is visible in full detail to everyone. Solution methods that deal with a set of working solutions, such as A* search and genetic algorithms, will use a set of *BLACKBOARD* objects.

For convenience and efficiency, blackboard data may be held in a redundant form – allowing fast access, for example, to the set of all meetings scheduled for a particular time, the set of all times of a given meeting, and so on. Although anyone may change the blackboard, to maintain consistency within it we must prohibit direct assignment to its variables. Instead, the blackboard offers a small set of update routines that change the blackboard safely.

In our framework, each call to an update routine must be representable as data in the form of a *COMMAND* object. This "commands-as-objects" design pattern is recommended in the object-oriented design literature as greatly increasing the flexibility of software that utilizes it. We give two well known examples of this effect now.

First, if a solution method involves backtracking, the blackboard may be instructed to record the sequence of commands that took the blackboard from its initial state to the current state. As each command is applied, it records within itself enough information to allow it to undo itself later if required. Backtracking may then proceed by undoing each command in the recorded sequence in reverse order of application.

Second, commands are crucial to the implementation of forced moves discovered by forward checking. Execution of one command may trigger a cascade of others which are forced consequences of the first; these may be queued as commands and applied in turn. If backtracking is in use, forced moves may be undone when the triggering command is undone. It is next to impossible to combine forced moves with backtracking in any other way.

Although commands mainly represent point changes of state ("assign this resource to this slot" and so on), more elaborate composite commands are possible. We have implemented a command type which represents an arbitrary sequence of other commands. To apply it, apply each of them in turn; to undo it, undo each in turn in reverse order. Such composite commands are useful because there is a significant cost in adding a whole new command to the architecture, but no cost if it can be built as a sequence of existing commands. Swapping two assignments is a good example of such a composite, involving two deassignments followed by two assignments.

One must also be aware of the possibility of more exotic commands. A "kill" command can declare a blackboard to be dead, perhaps as the result of forward checking, thereby initiating backtracking in a tree search method, or culling in a method which is carrying forward a set of working solutions. A "do-not-assign" command ("domain reduction" in the terminology of constraint propagation) might declare a certain resource ineligible for assignment to a certain slot, achieving nothing positive itself but potentially triggering a cascade of forced consequences.

Turning now to solution methods, we observe that, although methods such as heuristic assignment, tabu search and bipartite matching have nothing in common, still, from the point of view of software architecture they are equivalent: they are all solution methods. By defining each as a child class of the abstract *STRATEGY* class (named for the object-oriented design pattern that we follow here) we can have arbitrarily many solution methods and treat them as data. We also treat the input of a file containing a solution or partial solution, and similarly the execution of a sequence of commands typed interactively by the user, as strategies.

The relationship between strategies and commands is interesting from the software design viewpoint. Commands are intended to be small, unqualified instructions to the blackboard to change its state in a specific way, whereas strategies are algorithms for finding solutions to whole or part instances. Nevertheless, both take blackboards from one state to another, and so it is at least arguable that each is a kind of the other.

The main challenge with strategies, as we said earlier, is to ensure that they can be combined flexibly. We defer discussion of this problem to Section 3.

Finally, objects of our *OBSERVER* class (again, named after the design pattern) are devoted to monitoring a particular blackboard, or a particular part of it. For example, one observer might monitor everything involving a given teacher, checking that he or she is not overloaded, that it is currently possible to find a suitable set of classes to assign to this teacher, suggesting meetings for this teacher that would be a good fit, and so on. An observer registers its interest with its blackboard, and when a change is made that affects that observer, the blackboard is responsible for sending a notification to the observer.

Observers in our framework collectively do what evaluation functions and forward checking do in other systems. By distributing the work among many objects, we acknowledge that many different types of evaluation are possible, that they differ greatly in their utility and cost, and that consequently there is a need for flexibility in their application. For example, tabu search would typically run with just a single observer, for calculating the current overall badness; tree search methods might carry out some observations continually, but others only occasionally.

The beauty of observers is that they can operate quite independently of strategies, and hence can be used behind the scenes with many strategies. A tree search strategy, for example, can be totally unaware of the observers that are checking its blackboard continually for consistency, possibly using elaborate forward checks. The tree search strategy applies a command, checks whether the blackboard is still alive, and if so recurses. It does not see the observer checks and consequent forced moves that occur between the command application and the "alive" check, and they are undone automatically as described earlier when the triggering command is undone.

Further discussion of observers may be found in Section 4, which examines techniques for their efficient implementation.

3   Combining Strategies

Our architecture is primarily motivated by a desire to combine strategies in highly flexible ways. Here is one example of a combined strategy:

Use tree search methods to assign times to the meetings of Year 12, Year 11, and so on down to Year 8. Then, for each successful assignment of those meetings, switch to exhaustive search to assign the times of the last year, Year 7. Then use a complex bin packing algorithm to assign teachers to all classes, followed by a simple heuristic assignment method for the easier problem of assigning rooms, and finish with a run of simulated annealing to reduce the overall badness of the solution, using a neighbourhood that preserves feasibility.

The literature also contains proposals for selecting strategies under program control [7].

Having strategies as objects goes a good way towards providing the necessary flexibility. One can have collections of strategies, pass one strategy as a parameter to another, and so on. We give three specific examples now of ways to combine strategies.

The simplest way to combine two strategies is sequentially: run one strategy to completion then continue with the other. Even this simple case has its complexities when there are multiple solution paths. A tree search may follow one strategy to a certain depth, at which point it ends and another strategy takes over, either continuing the tree structure or not as it desires. The alternative paths may be explored using backtracking, or generated in parallel by keeping multiple blackboards.

Another useful combination is to select one of a set of alternative strategies, depending on the state of the blackboard. For example, we might choose a heuristic assignment strategy when the number of assignments to be made is large, and a brute force search strategy when it is small.

Again, one strategy may be used as the single step of another. Cooper and Kingston [2] describe a strategy for assigning times to all meetings containing a given resource (these meetings' times must all be different, which is a good reason for assigning them all together). This whole strategy is used by them as the single step in a beam search. In our architecture it would be available as the single step for any strategy desiring to use it.

The most difficult problem, we believe, is that of constraining a strategy to a particular part of the timetable, as the bin packing strategy is constrained above to teacher assignments, for example. On another day, the researcher might want to apply bin packing to room assignments, or to just the teachers in small departments, and so on.

An initial idea is the fully general one of passing the strategy a parameter which is a list or iterator defining the set of slots that the strategy is expected to assign values to. However, many strategies would not know what to do with an arbitrary set of slots; they are specialized algorithms for assigning meetings to one resource, or times to related sets of meetings, and so on. It makes more sense to pass, at the time a strategy object is created, a strategy-specific description of its scope: a set of teachers, a time, a meeting, or whatever.

For strategies that are highly general, such as heuristic assignment or tabu search, the appropriate parameters may be an iterator which can be used to find all available assignments, or all elements of a neighbourhood, plus an observer from which the current badness of the working solution can be obtained. It could be argued that this design merely transfers the generality problem to the iterators, but one iterator can be used by many strategies, and in practice there are likely to be relatively few iterators (all time slots, all teacher slots, all meetings containing a given resource, and so on).

A second problem that arises when strategies are combined is what to do when a strategy finds that some slot it is supposed to fill already has a value. Local search methods would typically want to change this value; heuristic assignment and tree search would not.

Rather than leaving this decision to strategies that discover such slots, it is better to give the decision to the strategy that assigned the value to begin with, by associating a Boolean value with each slot recording whether its assignment is to be considered "fixed" or not. The assigning strategy is best placed to know whether there is any point in changing the assignment, and all strategies can then be written to follow a simple and uniform policy of leaving "fixed" slots untouched and freely reassigning others in their scope if they wish. One typically needs these "fixed" flags anyway to handle preassignments, and there is no loss of efficiency in using them more generally.

4 Efficiency and Effectiveness of Observers

Every inconsistency or forced move detected by an observer is probably worth its cost of detection, particularly when any kind of tree search is being used. We have already stated that an observer is to be notified only when it is affected by a change; but even so, we need to be concerned about frequent notifications to expensive observers that detect little.

Some observers are quite predictable in their effectiveness, and then it makes sense to call them in a fixed way. For example, any change involving a particular teacher will be important to that teacher's observer. Other observers can be quite variable in their performance. For example, a check that the supply of teachers of various capabilities is adequate to cover the overall demand, while obviously very effective initially, will be much less so subsequently (although not useless, since a decision to utilize a teacher in one capability will occasionally cause demand to exceed supply in another, overlapping one). Such observers could be run at fixed but infrequent intervals, such as after each major stage of a staged solution method. Another very interesting idea is to require observers to adjust the time they consume to their effectiveness. For example, an observer could be required to return immediately (without carrying out its observation) when notified, except on every kth notification. A successful detection of an inconsistency or forced move would set k to 1; otherwise k would be doubled.

The other potential efficiency problem arises when observations are re-built from scratch on every notification. It is often considerably more efficient to save the previous observation and incrementally change it. For example, [2] maintains several large bipartite matchings in this way. To save the cost of searching the blackboard to discover what has changed since the last observation, such "persistent" observers need access to a list of those commands that affect them that have been applied since the last observation. One way to provide this is to send the command which triggers each notification to the observer as part of the notification. If the observer is skipping this check, it will merely store this command, increment a counter, and return; if it is carrying out an observation, it will update its internal state to reflect all the commands it has stored since the last observation, then perform its observation. Naturally, every notification must have a matching de-notification when commands are being undone; but a careful implementation can rely on the blackboard's command history, and not require each observer to keep its own.

If an observer detects something interesting, such as an inconsistency, a forced assignment or a forced non-assignment (domain reduction), the question arises of who should be told, and how. Informing strategies or other observers directly, or even allowing such components access to the observer in order to make enquiries, ties components together, reduces flexibility, and hides potentially useful information from other components. So we prefer to require observers to post any interesting findings on the blackboard, either by changing a total badness value there, or by issuing commands. Whether this approach will cause the blackboard to become overloaded with marginally useful information remains to be seen. At present, the authors' blackboard stores only the total badness, the current assignments and the current domains of all slots.

5 Application to a Tutor Allocation Problem

In this section we describe a first application of our framework, in which two methods of solving a university tutor allocation problem (the July 1998 one from the authors' department) are compared. There were 58 tutors, requesting workloads of between four and 16 hours per week, and 15 courses, with various tutorial structures (Fig. 1). University management practices beyond our control are such that the times of all tutorials are fixed and very difficult to modify at the time the tutor allocation is performed.

Since our purpose is to show that our framework has actually been used on a real problem, not to rigorously investigate that problem, we omit many details from this description. Although, regrettably, our example does not show strategies being combined, it does show our framework in operation, applying two radically different approaches to one problem. We hope this gives a glimpse of the prospects opened up by this work.

Of 20 trials of tabu search, beginning with a random allocation, 19 found an exact solution within a few seconds, and one had to be terminated without finding a solution. Fig. 2 presents a trace of a typical successful run. It shows, as might have been anticipated [7], that the final step to perfection is quite difficult for tabu search: perfection took twice as long as near-perfection.

Our second method was a somewhat novel one. We took each course in sequence and built a bipartite graph, connecting tutors capable of teaching that course with the tutorials they were available to teach. When a tutor wished to take several tutorials, some edges had to be omitted to prevent solutions in which the tutor is assigned clashing tutorials. Then we found a matching for the course, made the indicated assignments, and proceeded to the next course. When some course failed to match, it was moved to the front of the entire sequence and the whole run restarted from the beginning. This idea, adapted from the move-to-front list adjusting heuristic [8], ensures that difficult courses are allocated first without requiring any a priori analysis. Of our 20 runs, all were successful (we randomize the order of insertion of edges, to randomize the matching found when there are several), taking between 2 and 13 s, averaging 5.6 s. Fig. 3 shows a typical run.

6 Conclusion

If we were asked why our architecture deserves adoption, we would answer with one word: flexibility. In place of a single, monolithic algorithm, our architecture offers an

Course	Tutorial structure	Tutorials
COMP 3102 User Interfaces	1	5
COMP 3101 Theory of Computation	1	2
COMP 3005 Database Systems	1	9
COMP 3004 Computer Graphics	1	5
COMP 3003 Computer Architecture	4	5
COMP 2904 Programming Practice (Advanced)	1	4
COMP 2004 Programming Practice	1	21
COMP 2903 Languages and Logic (Advanced)	1	3
COMP 2003 Languages and Logic	1	15
COMP 1902 Introductory Computer Science (Advanced)	3 + 2	4
COMP 1002 Introductory Computer Science	3 + 2	25
COMP 1001 Introductory Programming	3 + 2	4
COMP 1000 Information Technology Tools laboratories	2 + 2	25
COMP 1000 Information Technology Tools tutorials	1	26
Help Desk	1	15

Fig. 1. Tutorials in the case study problem. In the second column, 3 + 2 for example means one three-hour block plus a separate two-hour block. COMP 1000 laboratories and COMP 1002 are much harder to allocate than the other courses

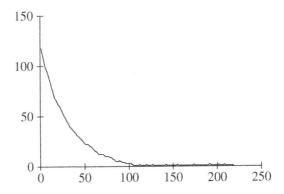

Fig. 2. A typical successful run of tabu search. The horizontal axis shows the number of moves made; the vertical shows the badness function, which is the number of timetable clashes plus the number of unassigned slots

eclectic collection of strategies, each usable in many ways: on limited regions of the problem, as one step in a larger strategy, and so on. Observers implement constraint propagation, but go beyond it to offer complex operations research models, detection of tight situations, and suggestions for useful moves; they work independently of strategies and may be switched on and off at will. The researcher may experiment with arbitrary combinations of all these options, and add new ideas incrementally without disturbing existing code.

It is fair to ask whether the generality and flexibility of this architecture requires too high a price in terms of execution time. High-level events such as switching from

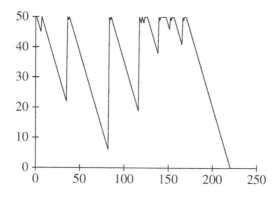

Fig. 3. A typical run of the matching algorithm. The horizontal axis shows the number of matches performed; the vertical shows the number of courses remaining to be matched (these include trivial "courses" such as lectures, which have been omitted elsewhere)

one strategy to another take negligible total time. What really matters is the "inner loop": deciding which command to execute next, and executing it. At this level the code is much the same as in any system capable of backtracking. The architecture encourages efficiency by holding out the prospect of wide re-use of code. For example, this has encouraged the authors to write fully incremental observers (Section 4) rather than the simpler but slower ones which rebuild their observations from scratch by searching the blackboard.

The current version of our architecture, as used in the application from Section 5, is written in Eiffel [6] and implements all of the major classes we have described. We have completed the handling of observers and forced moves, and tested this code on small puzzles including the eight queens problem. The most significant part of our design yet to be coded is the set of ideas that promote flexibility in combining strategies.

The next step is to develop a reasonably generally applicable concrete blackboard and command set for timetabling, based on the one we have now, and to transfer as many of the ideas of timetabling researchers as we can into strategies and observers. After that we will be in an excellent position to carry out experimental work comparing and combining these methods.

Further ahead lies the question of selecting strategies under program control. This is too large and too new an area to say much about now. But we are confident that our framework will support it well when the time comes.

References

1. Beck, J. C., Davenport, A. J., Davis, E. D., Fox, M. S.: The ODO Project: Toward a Unified Basis for Constraint-Directed Scheduling. J. Sched. **1** (1998) 89–125
2. Cooper, T. B. and Kingston, J. H.: The Solution of Real Instances of the Timetabling Problem. Comp. J. **36** (1993) 645–653
3. Dowsland, K. A.: Off-the-Peg or Made-to-Measure? Timetabling and Scheduling with SA and TS. In: Practice and Theory of Automated Timetabling II. Lecture Notes in Computer

Science, Vol. 1408. Springer-Verlag, Berlin Heidelberg New York (1998) 37–52

4. Gamma, E., Helm, R., Johnson, R., Vlissides, J.: Design Patterns: Elements of Reusable Object-Oriented Software. Addison-Wesley (1995)

5. Macdonald, R.: University Timetabling at the Blackboard. In: Proceedings 1st International Conference on the Practice and Theory of Automated Timetabling, Napier University, Edinburgh, UK (1995) 443–454

6. Meyer, B.: Object-Oriented Software Construction. 2nd edn. Prentice-Hall (1997)

7. Ross, P., Hart, E., Corne, D.: Some Observations about GA-based Exam Timetabling. In: Practice and Theory of Automated Timetabling II. Lecture Notes in Computer Science, Vol. 1408. Springer-Verlag, Berlin Heidelberg New York (1998) 115–129

8. Sleator, D. D. and Tarjan, R. E.: Amortized Efficiency of List Update and Paging Rules. CACM **28** (1985) 202–8

Other Timetabling Presentations

Other Timetabling Presentations

Other conference presentations are listed below together with the addresses of the authors.

Title: Constrained Satisfaction, Not So Constrained Satisfaction and the Timetabling Problem
Author: G.M. White
Address: School of Information Technology and Engineering, University of Ottawa, Ottawa, Canada K1N 6N5

Title: Generating University Timetables in an Interactive System: DIAMANT
Author: R. Gonzalez Rubio
Address: Université de Sherbrooke, Département de génie électrique et de génie informatique, Sherbrooke, Quebec, Canada J1K 2R1

Title: The Timetable Timeline Approach to Creating a Decentralised Collaborative University Timetabling Process
Authors: L.M. Parks and I.A. Newman
Address: Department of Computer Science, Loughborough University, Loughborough, UK

Title: Development of a Campus-Wide University Course Timetabling Application
Authors: A. Lim Leong Chye, Oon Wee Chong and Ang Juay Chin
Address: Department of Computer Science, National University of Singapore, 3 Science Drive 2, Singapore 117543

Title: Combined Automatic and Interactive Timetabling Using Constraint Logic Programming
Author: H.J. Goltz
Address: German National Research Center for Information Technology GMD-FIRST, Kekuléstrasse 7, D-12489 Berlin, Germany

Title: Stone Soup?
Authors: B. Paechter, T.C. Fogarty, A.Cumming and R.C. Rankin
Address: Napier University, Edinburgh, UK

Title: Post-publication Timetabling
Authors: A. Cumming, B. Paechter and R.C. Rankin
Address: Napier University, Edinburgh, UK

Title: Constraint-Based Timetabling with Student Schedules
Authors: H. Rudová and L. Matyska
Address: Faculty of Informatics, Masaryk University Botánicka 68a, Brno 602 00, Czech Republic

Title: Integrating Knowledge-Based Elaboration and Restructuring of Timetables at a Spanish Private University
Authors: P. Gervás and B. San Miguel
Address: Departmento de Inteligencia Artificial, Universidad Europea de Madrid, Villaviciosa de Odón, Madrid 28670, Spain

Title: Implementing a University Course and Examination Timetabling System in a Distributed Environment
Author: M. Dimopoulou and P. Miliotis
Address: Athens University of Economics and Business, 76 Patission Odos, Athens 10434, Greece

Title: The Temposcope: a Computer Instrument for the Idealist Timetabler
Authors: M. Beynon, A. Ward, S. Maad, A. Wong, S. Rasmequan and S. Russ
Address: Department of Computer Science, University of Warwick, Coventry, UK

Title: Examination Timetabling Using Set Variables
Authors: L.P. Reis and E. Oliveira
Addresses: LIACC – Artificial Intelligence and Computer Science Laboratory, University of Porto, Portugal

Title: Solving School Timetabling Problems by Microcanonical Optimization
Authors: M.J.F. Souza, N. Maculan and L.S. Ochi
Addresses: M.J.F. Souza: Universidade Federal de Ouro Preto, Departamento de Computação, 35400-000 Ouro Preto, Minas Gerais, Brazil
N. Maculan: Universidade Federal do Rio de Janeiro, Progr. de Engenharia de Sistemas e Computação, Rio de Janeiro, RJ, Brazil
L.S. Ochi: Universidade Federal Fluminense, Departamento de Ciência da Computação, Rua Passo da Pátria 156, 24210-240 Niterói, RJ, Brazil

Title: A Sequential Approach to Solve Hard School Timetabling Problems Using Column Generation
Authors: P. Eveborn and M. Rönnqvst
Address: Division of Optimization, Linkoping University, 581 83 Linkoping, Sweden

Title: Consecutive Graph Coloring for School Timetabling
Authors: K. Giaro, M. Kubale and D. Szyfelbein
Address: Technical University of Gdańsk, Foundations of Informatics Department Narutowicza 11/12, 80-952, Gdańsk, Poland

Title: Towards Constraint-Based Grammar School Timetabling
Authors: M. Marte
Address: Computer Science Department, University of Munich, Oettingenstrasse 67, 80538 Munchen, Germany

Title: Time Slot and Subject Group Assignment at Secondary Schools
Authors: R.J. Willemen, H.M.M. ten Eikelder
Address: Technische Universiteit Eindhoven, Department of Mathematics and Computing Science, PO Box 513, 5600 MB Eindhoven, The Netherlands

Title: Specifying Constraint Satisfaction Problems with HyperDataSheet
Author: M. Yoshikawa
Address: Distribution Industry Systems Development Division, NEC (NEC Solutions) 503402 Shiba, Minato-ku, Tokyo 108-8421, Japan

Title: Days-off Employee Scheduling Over a Three-Week Work Cycle
Author: H.K. Alfares
Address: System Engineering Department, King Fahd University of Petroleum and Minerals, PO Box 5067, Dhahran 31261, Saudi Arabia

Title: Personnel Timetabling Based on OR Models; Two Recent Case Studies
Author: J.A.M.Schreuder
Address: Department of Applied Mathematics, University of Twente, PO Box 217, 7500 AE Enschede, The Netherlands

Title: 'Floating' Personnel Demands in a Shift Based Timetable
Authors: P. De Causmaecker and G. Vanden Berghe
Address: KaHo St.-Lieven, Information Technology, Gebr. Desmetstraat 1, 9000 Gent, Belgium

Title: Solving Employee Timetabling Problems with Flexible Workload Using Tabu Search
Authors: M. Chirandini, A. Schaerf and F. Tiozzo
Addresses: M. Chirandini and A. Schaerf: Dipartimento di Ingegneria Elettrica, Gestionale e Meccanica, Università di Udine, via delle Science 208, Udine, Italy
F. Tiozzo: Cybertec Via Udine 11, I-34132, Trieste, Italy

Title: Efficient Generation of Rotating Workforce Schedules
Authors: N. Muslija, J. Gartner and W. Slany
Address: Technische Universität Wien, Institut fur Informationssysteme 184/2, Avteukybg fyr Datebbanken und Artificial Intelligence, Favoritenstrasse 9, A-1040 Wien, Austria

Title: Assigning Magistrates to Sessions of the Amsterdam Criminal Court
Authors: J.A.M. Schreuder
Address: University of Twente – TW, PO Box 217, 7500 AE Enschede, Netherlands

Title: Request-based Timetabling – Concept and Methods
Authors: M. Tanaka and S. Adachi
Addresses: M. Tanaka: Department of Information Systems and Management Science, Faculty of Science, Konan University, Okamoto 8-9-1, Higashinada-ku, Kobe 658-8501, Japan
S. Adachi: NEC Corporation, Tokyo, Japan

Title: Solving Constrained Staff Workload Scheduling Problems Using the Simulated Annealing Technique
Authors: M.L. Ng and H.B. Gooi
Address: School of Electrical and Electronic Engineering, Nanyang Technological University, Republic of Singapore

Title: Solving Single-Track Railway Scheduling Problems Using Constraint Programming
Authors: E. Oliveira and B.M. Smith
Address: School of Computer Studies, University of Leeds, Leeds, UK

Title: A Timetabling Process Model
Author: V.V. Bondarenko
Address: Taras Shevchenko Kyiv University, The Ukraine

Title: Delivering Timetables
Author: G.M. White
Address: School of Information Technology and Engineering, University of Ottawa, Ottawa, Canada K1N 6N5

Title: Using Problem Specific Knowledge to Improve the Efficiency of an Integer Programming Approach to two Scheduling Problems
Author: K.A. Dowsland
Address: Logistics and Scheduling Research Group, European Business Management School, University of Wales Swansea, Swansea SA2 8PP, UK

Title: Ontology for Timetabling
Authors: P. Causmaecker, P. Demeester, P. De Pauw-Waterschoot and G. Vanden Berghe
Address: KaHo St.-Lieven, Information Technology, Gebr. Desmetstraat 1, 9000 Gent, Belgium

Title: Introducing Optime: Examination Timetabling Software
Authors: B. McCollum and J. Newall
Address: Department of Computer Science, University of Nottingham, Nottingham, UK

Title: ConBaTT – Constraint-Based Timetabling
Authors: H.J. Goltz and D. Matzke
Address: GMD – German National Research Center for Information Technology, GMD-FIRST, Kekuléstrasse 7, D-12489 Berlin, Germany

Title: Syllabus Plus
Authors: Scientia Ltd
Address: St Johns Innovation Centre, Cowley Road, Cambridge, UK

Author Index

Lecture Notes in Computer Science

For information about Vols. 1–2034
please contact your bookseller or Springer-Verlag

Vol. 2076: F. Orejas, P.G. Spirakis, J. van Leeuwen (Eds.), Automata, Languages and Programming. Proceedings, 2001. XIV, 1083 pages. 2001.

Vol. 2077: V. Ambriola (Ed.), Software Process Technology. Proceedings, 2001. VIII, 247 pages. 2001.

Vol. 2078: R. Reed, J. Reed (Eds.), SDL 2001: Meeting UML. Proceedings, 2001. XI, 439 pages. 2001.

Vol. 2079: E. Burke, W. Erben (Eds.), Practice and Theory of Automated Timetabling III. Proceedings, 2001. XII, 359 pages. 2001.

Vol. 2080: D.W. Aha, I. Watson (Eds.), Case-Based Reasoning Research and Development. Proceedings, 2001. XII, 758 pages. 2001. (Subseries LNAI).

Vol. 2081: K. Aardal, B. Gerards (Eds.), Integer Programming and Combinatorial Optimization. Proceedings, 2001. XI, 423 pages. 2001.

Vol. 2082: M.F. Insana, R.M. Leahy (Eds.), Information Processing in Medical Imaging. Proceedings, 2001. XVI, 537 pages. 2001.

Vol. 2083: R. Goré, A. Leitsch, T. Nipkow (Eds.), Automated Reasoning. Proceedings, 2001. XV, 708 pages. 2001. (Subseries LNAI).

Vol. 2084: J. Mira, A. Prieto (Eds.), Connectionist Models of Neurons, Learning Processes, and Artificial Intelligence. Proceedings, 2001. Part I. XXVII, 836 pages. 2001.

Vol. 2085: J. Mira, A. Prieto (Eds.), Bio-Inspired Applications of Connectionism. Proceedings, 2001. Part II. XXVII, 848 pages. 2001.

Vol. 2086: M. Luck, V. Mařík, O. Štěpánková, R. Trappl (Eds.), Multi-Agent Systems and Applications. Proceedings, 2001. X, 437 pages. 2001. (Subseries LNAI).

Vol. 2087: G. Kern-Isberner, Conditionals in Nonmonotonic Reasoning and Belief Revision. X, 190 pages. 2001. (Subseries LNAI).

Vol. 2089: A. Amir, G.M. Landau (Eds.), Combinatorial Pattern Matching. Proceedings, 2001. VIII, 273 pages. 2001.

Vol. 2091: J. Bigun, F. Smeraldi (Eds.), Audio- and Video-Based Biometric Person Authentication. Proceedings, 2001. XIII, 374 pages. 2001.

Vol. 2092: L. Wolf, D. Hutchison, R. Steinmetz (Eds.), Quality of Service – IWQoS 2001. Proceedings, 2001. XII, 435 pages. 2001.

Vol. 2093: P. Lorenz (Ed.), Networking – ICN 2001. Proceedings, 2001. Part I. XXV, 843 pages. 2001.

Vol. 2094: P. Lorenz (Ed.), Networking – ICN 2001. Proceedings, 2001. Part II. XXV, 899 pages. 2001.

Vol. 2095: B. Schiele, G. Sagerer (Eds.), Computer Vision Systems. Proceedings, 2001. X, 313 pages. 2001.

Vol. 2096: J. Kittler, F. Roli (Eds.), Multiple Classifier Systems. Proceedings, 2001. XII, 456 pages. 2001.

Vol. 2097: B. Read (Ed.), Advances in Databases. Proceedings, 2001. X, 219 pages. 2001.

Vol. 2098: J. Akiyama, M. Kano, M. Urabe (Eds.), Discrete and Computational Geometry. Proceedings, 2000. XI, 381 pages. 2001.

Vol. 2099: P. de Groote, G. Morrill, C. Retoré (Eds.), Logical Aspects of Computational Linguistics. Proceedings, 2001. VIII, 311 pages. 2001. (Subseries LNAI).

Vol. 2100: R. Küsters, Non-Standard Inferences in Description Logocs. X, 250 pages. 2001. (Subseries LNAI).

Vol. 2101: S. Quaglini, P. Barahona, S. Andreassen (Eds.), Artificial Intelligence in Medicine. Proceedings, 2001. XIV, 469 pages. 2001. (Subseries LNAI).

Vol. 2102: G. Berry, H. Comon, A. Finkel (Eds.), Computer-Aided Verification. Proceedings, 2001. XIII, 520 pages. 2001.

Vol. 2103: M. Hannebauer, J. Wendler, E. Pagello (Eds.), Balancing Reactivity and Social Deliberation in Multi-Agent Systems. VIII, 237 pages. 2001. (Subseries LNAI).

Vol. 2104: R. Eigenmann, M.J. Voss (Eds.), OpenMP Shared Memory Parallel Programming. Proceedings, 2001. X, 185 pages. 2001.

Vol. 2105: W. Kim, T.-W. Ling, Y-J. Lee, S.-S. Park (Eds.), The Human Society and the Internet. Proceedings, 2001. XVI, 470 pages. 2001.

Vol. 2106: M. Kerckhove (Ed.), Scale-Space and Morphology in Computer Vision. Proceedings, 2001. XI, 435 pages. 2001.

Vol. 2109: M. Bauer, P.J. Gymtrasiewicz, J. Vassileva (Eds.), User Modelind 2001. Proceedings, 2001. XIII, 318 pages. 2001. (Subseries LNAI).

Vol. 2110: B. Hertzberger, A. Hoekstra, R. Williams (Eds.), High-Performance Computing and Networking. Proceedings, 2001. XVII, 733 pages. 2001.

Vol. 2111: D. Helmbold, B. Williamson (Eds.), Computational Learning Theory. Proceedings, 2001. IX, 631 pages. 2001. (Subseries LNAI).

Vol. 2116: V. Akman, P. Bouquet, R. Thomason, R.A. Young (Eds.), Modeling and Using Context. Proceedings, 2001. XII, 472 pages. 2001. (Subseries LNAI).

Vol. 2117: M. Beynon, C.L. Nehaniv, K. Dautenhahn (Eds.), Cognitive Technology: Instruments of Mind. Proceedings, 2001. XV, 522 pages. 2001. (Subseries LNAI).

Vol. 2118: X.S. Wang, G. Yu, H. Lu (Eds.), Advances in Web-Age Information Management. Proceedings, 2001. XV, 418 pages. 2001.

Vol. 2119: V. Varadharajan, Y. Mu (Eds.), Information Security and Privacy. Proceedings, 2001. XI, 522 pages. 2001.

Vol. 2120: H.S. Delugach, G. Stumme (Eds.), Conceptual Structures: Broadening the Base. Proceedings, 2001. X, 377 pages. 2001. (Subseries LNAI).

Vol. 2121: C.S. Jensen, M. Schneider, B. Seeger, V.J. Tsotras (Eds.), Advances in Spatial and Temporal Databases. Proceedings, 2001. XI, 543 pages. 2001.

Vol. 2123: P. Perner (Ed.), Machine Learning and Data Mining in Pattern Recognition. Proceedings, 2001. XI, 363 pages. 2001. (Subseries LNAI).

Vol. 2125: F. Dehne, J.-R. Sack, R. Tamassia (Eds.), Algorithms and Data Structures. Proceedings, 2001. XII, 484 pages. 2001.

Vol. 2126: P. Cousot (Ed.), Static Analysis. Proceedings, 2001. XI, 439 pages. 2001.

Vol. 2132: S.-T. Yuan, M. Yokoo (Eds.), Intelligent Agents. Specification. Modeling. and Application. Proceedings, 2001. X, 237 pages. 2001. (Subseries LNAI).